普通高等教育"十四五"规划教材

Theory of Combustion and Explosion

燃烧与爆炸学

Liu Jia　Zhang Yinghua　Huang Zhian
刘　佳　　张英华　　黄志安　　主编

扫码获得数字资源

Beijing
Metallurgical Industry Press
2023

Metallurgical Industry Press
39 Songzhuyuan North Alley, Dongcheng District, Beijing 100009, China

图书在版编目(CIP)数据

燃烧与爆炸学 = Theory of Combustion and Explosion：英文／刘佳，张英华，黄志安主编．—北京：冶金工业出版社，2023.3
普通高等教育“十四五”规划教材
ISBN 978-7-5024-9441-4

Ⅰ.①燃… Ⅱ.①刘… ②张… ③黄… Ⅲ.①燃烧学—高等学校—教材—英文 ②爆炸—理论—高等学校—教材—英文 Ⅳ.①O643.2

中国国家版本馆 CIP 数据核字(2023)第 045568 号

Theory of Combustion and Explosion

出版发行	冶金工业出版社	**电　　话**	(010)64027926
地　　址	北京市东城区嵩祝院北巷 39 号	**邮　　编**	100009
网　　址	www.mip1953.com	**电子信箱**	service@mip1953.com

责任编辑 郭冬艳　美术编辑 吕欣童　版式设计 郑小利
责任校对 郑 娟　责任印制 禹 蕊
北京建宏印刷有限公司印刷
2023 年 3 月第 1 版，2023 年 3 月第 1 次印刷
787mm×1092mm　1/16；18.75 印张；457 千字；290 页
定价 69.00 元

投稿电话　(010)64027932　投稿信箱　tougao@cnmip.com.cn
营销中心电话　(010)64044283
冶金工业出版社天猫旗舰店　yjgycbs.tmall.com
(本书如有印装质量问题，本社营销中心负责退换)

Preface

Controllable combustion and explosion will bring high efficiency to production and convenience to life; however, uncontrolled combustion and explosion may cause accidents, such as various fires, dust explosions, gas explosions, etc. Therefore, the study of combustion and explosion theory and process plays a vital role in reducing and avoiding fire and explosion accidents in production and life, as well as in environmental protection, safe production and effective fuel use. To strengthen readers' understanding of concepts, exercises and questions are provided at the end of each chapter in combination with the contents described in each chapter. This book, striving to be concise and straightforward, focuses on introducing fundamental theories and basic knowledge, elaborating the theoretical basis of combustion and explosion and their occurrence process in detail, and trying to enlighten readers by combining the research achievements of the editor for many years.

Jia Liu, Yinghua Zhang and Zhian Huang are the chief editors of this book, responsible for the unified collation and revision of the whole book. Yukun Gao and Huanjuan Zhao are the deputy associate editors. The specific division of work for the revision of this book is as follows: Chapter 1 revised by Yinghua Zhang and Lexi Wang; Chapter 2 revised by Jia Liu and Xue Wang; Chapter 3 revised by Zhian Huang and Xiaoyu Xia; Chapter 4 revised by Yukun Gao and Qian Feng; Chapter 5 revised by Huanjuan Zhao and Xingyu Zhang; and Chapter 6 revised by Yinghua Zhang and Lexi Wang. Zhengqing Zhou, Zhiming Bai, Zhibo Zhang, Yuan Tian, Yichao Yin and several others also participated in the revision work.

The publication of *Combustion and Explosives* was been funded by the textbook construction fund of the University of Science and Technology Beijing which provided a strong support and assistance in the compilation and publishing of (the funded books). The revision was also supported and helped by relevant leaders and teachers of the School of Civil and Resource Engineering of the University to whom I would like to express my sincere thanks.

Due to the limited (ability) of editors, there is inevitably some errors with the book, and we sincerely hope that readers can criticize and correct them.

Liu Jia

2023. 1

Contents

Chapter 1 Chemical Basis of Combustion and Explosion

扫码获得
数字资源

1.1 Nature and conditions of combustion and explosion

1.1.1 Combustion

Combustion refers to the exothermic reaction between combustibles and oxidants, usually accompanied by flame, luminescence and smoke. The temperature in the combustion zone is very high, which makes the incandescent solid particles and the electron of some unstable (or excited) intermediate substances have energy level transitions, thus emitting light of various wavelengths; The luminous gas phase combustion zone is the flame, and its existence is the most obvious sign in the combustion process; Due to incomplete combustion and other reasons, the product will be mixed with some small particles, thus forming smoke.

In essence, combustion is a redox reaction, but its basic characteristics such as exothermic, luminous, smoke and flame show that it is different from the general redox reaction. In the redox reaction, the substance losing electrons is oxidized and the substance obtaining electrons is reduced. For example, hydrogen burns in chlorine. The chlorine atom gains an electron and is reduced, while the hydrogen atom loses an electron and is oxidized. In this reaction, although there is no oxygen involved in the reaction, what happens is a fierce redox reaction, accompanied by light and heat. This reaction is also combustion.

When an electric lamp is illuminated, it emits light and heat, but there is no chemical reaction, which cannot be called combustion. Although the reaction of copper with dilute nitric acid has electron gain and loss, it does not produce light and heat, nor can it be called combustion. To sum up, the combustion process has five characteristics: (1) New substances are produced (that is, combustion is a chemical reaction); (2) Exothermic; (3) Flame; (4) Luminescence; (5) Smoke.

1.1.2 Combustion theory

1.1.2.1 Activation energy theory

The primary condition for chemical reaction between molecules of matter is mutual collision. However, in order to generate oxidation reaction between the two gas molecule of combustible and comburant, it is not enough to only rely on the collision of two molecules, and general repulsive force will be generated between the colliding molecules. Under normal conditions, these molecules do not have enough energy to carry out oxidation reaction. Only when a certain number of molecules obtain enough energy, can they cause significant vibration of molecular components during collision, and weaken the bonding between atoms or atomic groups in the molecules. The rearrangement of various parts of the molecule is possible, that is, the molecules that may cause chemical reactions are called activated molecules. The energy of an activated molecule is higher than that of an ordinary molecule, and the excess energy can activate the molecule and allow it to participate in the reaction. The energy necessary to turn ordinary molecules into active molecules is called activation energy.

The ordinate in Fig. 1-1 represents the molecular energy of the studied system, and the abscissa represents the reaction process. If system state Ⅰ changes to state Ⅱ. Since the energy of state Ⅰ is greater than that of state Ⅱ, the process is exothermic. The reaction heat effect is equal to Q_V, which is equal to the energy level difference between state Ⅰ and state Ⅱ. The size of state K is equivalent to the energy necessary to make the reaction take place. Therefore, the difference between the energy levels of state K and state Ⅰ is equal to the activation energy of the positive reaction (ΔE_1), the difference between the energy levels of state K and state Ⅱ is equal to the activation energy of the reverse reaction (ΔE_2), and the difference between ΔE_2 and ΔE_1 ($\Delta E_2 - \Delta E_1$) is equal to the thermal effect of the reaction.

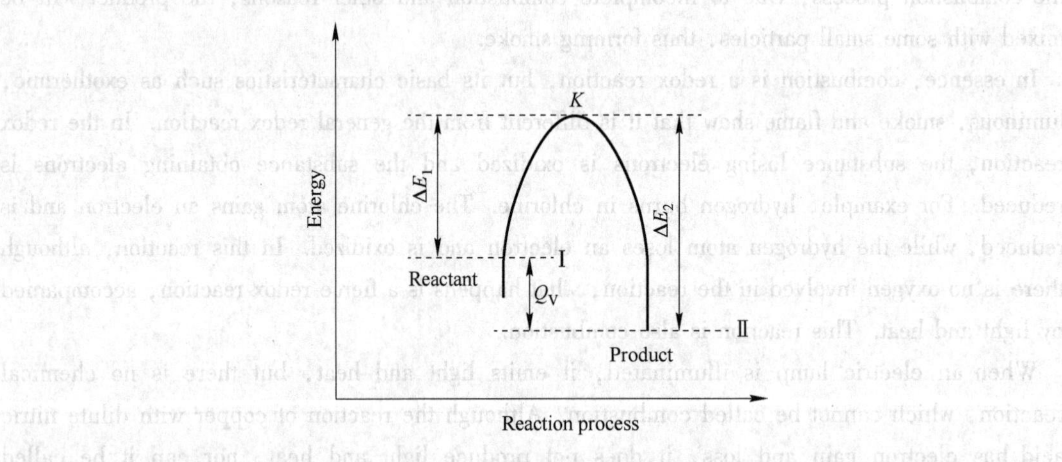

Fig. 1-1 Schematic diagram of activation energy

The activation energy theory points out the possibility and conditions of oxidation reaction between the two gas molecule of combustible and comburant.

1.1.2.2 Peroxide theory

According to peroxide theory, molecules can be activated under the action of various energies (thermal energy, radiation energy, electric energy, chemical reaction energy). For example, in the combustion reaction, firstly, the oxygen molecule (O=O) is activated under the action of heat energy, and one of the double bonds of the activated oxygen molecule is broken to form peroxy—O—O—, which can be added to the oxidized molecule to form peroxide:

$$A + O_2 = AO_2$$

Peroxy group (—O—O—) is included in the peroxide, and the oxygen atom in this group is more unstable than that in the free molecule. Therefore, peroxide is a strong oxidant, which can not only oxidize substance A forming peroxide, but also oxidize other substances B that are difficult to oxidize with molecular oxygen:

$$AO_2 + A = 2AO$$

$$AO_2 + B = AO + BO$$

For example, the combustion reaction of hydrogen and oxygen is usually directly expressed as:

$$2H_2 + O_2 = 2H_2O$$

According to the peroxide theory, hydrogen and oxygen first generate hydrogen peroxide, and then hydrogen peroxide reacts with hydrogen to generate H_2O. The reaction formula is as follows:

$$H_2 + O_2 = H_2O_2$$

$$H_2O_2 + H_2 = 2H_2O$$

Organic peroxides can usually be regarded as derivatives of hydrogen peroxide H—O—O—H, in which one or two hydrogen atoms are replaced by hydrocarbon groups to become H—O—O—R or R—O—O—R. Therefore, peroxides are the initial products when combustible substances are oxidized. They are unstable compounds that can decompose under heating, impact, friction and other conditions to produce free radicals and atoms, thus promoting the oxidation of new combustible substances.

The peroxide theory explains, to some extent, why substances are likely to be oxidized in the gaseous state. It assumes that oxygen molecules only undergo single bond destruction, which is easier than double bond destruction. Because it only needs 29.3 ~ 33kJ of energy to destroy the single bond of 1mol of oxygen. However, if we consider that the C—H bond must also be broken, and oxygen molecules must be added to hydrocarbons to form peroxides, the oxidation process is still very difficult. Therefore, Bach put forward another statement, that is, the combustible substance that is easy to oxidize has enough "free energy" to destroy the single bond in oxygen, so it is not the combustible substance itself but its free radicals that are oxidized. This view is the basis of the modern chain reaction theory about oxidation.

1.1.2.3 Chain reaction theory

According to the above principle, an activated molecule (Group) can only react with one molecule. However, the reason why one photon can be introduced into the reaction of hydrogen chloride to generate 100000 hydrogen chloride molecules is the result of chain reaction. According to the chain reaction theory, the interaction between gaseous molecules is not the direct interaction between two molecules to get the final product, but the interaction between activated molecular free radicals and another molecule. As a result, new groups are produced, and the new groups quickly participate in the reaction. In this way, a series of chain reactions are formed. This is how chlorine reacts with hydrogen:

$$Cl_2 + h\nu \text{ (Light quantum)} \longrightarrow Cl\cdot + Cl\cdot \qquad \text{Chain initiation}$$

$$Cl\cdot + H_2 \longrightarrow HCl + H\cdot$$

$$H\cdot + Cl_2 \longrightarrow HCl + Cl\cdot \quad \text{Chain transfer}$$
$$Cl\cdot + H_2 \longrightarrow HCl + H\cdot$$
$$H\cdot + Cl_2 \longrightarrow HCl + Cl\cdot$$
$$Cl\cdot + Cl\cdot \longrightarrow Cl_2 \quad \text{Chain break}$$
$$H\cdot + H\cdot \longrightarrow H_2$$

The above reaction formula shows that the initial free radical (or active center, action center, etc.) is generated under the action of some kind of energy, and the energy for generating free radical can be thermal decomposition or light, oxidation, reduction, catalysis, radiation, etc. Because free radicals have more activation energy than the average kinetic energy of ordinary molecules, they have very strong activity ability and are unstable under general conditions. They are easy to react with other material molecules to form new free radicals or combine into stable molecules by themselves. Therefore, when some kind of energy is used to make the reactants produce a small amount of active centers - free radicals, these initial free radicals can cause a chain reaction, so that the combustion can continue until all the reactants have reacted. In the chain reaction, if the action center disappears, the chain reaction will be interrupted, and the reaction will be weakened until the combustion stops.

Generally speaking, the chain reaction mechanism can be roughly divided into three stages: (1) Chain initiation, that is, the formation of free radicals, makes the chain reaction begin; (2) Chain transfer, free radicals are used as other compounds involved in the reaction to produce new free radicals; (3) Chain termination, the consumption of free radicals, terminates the chain reaction. There are many reasons for the consumption of free radicals, such as the collision of free radicals to generate molecules, the side reaction with impurities in the mixture, the collision with non activated similar molecules or inert molecules to disperse energy, and the collision with the wall of the impactor to be adsorbed.

1.1.3 Conditions of combustion

Combustion is very common, but it must have certain conditions. As a special oxidation-reduction reaction, the combustion reaction must involve oxidant and reductant, in addition to the energy to initiate combustion.

(1) Combustible (reductant). Whether gas, liquid or solid, metal or nonmetal, inorganic or organic, all substances that can react with oxygen or other oxidants in the air are called combustibles, such as hydrogen, acetylene, alcohol, gasoline, wood, paper, etc.

(2) Comburant (oxidizer). All substances that can cause and support combustion in combination with combustibles are called comburant, such as air, oxygen, chlorine, potassium chlorate, sodium peroxide, etc. Air is the most common comburant. In the future, unless otherwise specified, the combustion of combustibles refers to the combustion in air.

(3) Ignition source. All ignition energy sources that can cause material combustion are collectively referred to as ignition sources, such as open fire, high-temperature surface, friction and impact, spontaneous combustion, chemical reaction heat, electric spark, photothermal

ray, etc.

These three conditions are often referred to as the three elements of combustion. However, even if the three elements are combined and interact with each other, combustion does not necessarily occur. Other conditions must be met for combustion, such as certain quantity and concentration of combustibles and combustion supporting substances, certain temperature and sufficient heat of ignition source, etc. When combustion can occur, the three elements can be represented as closed triangles, which are usually called ignition triangles, as shown in Fig. 1-2 (a).

The classical ignition triangle is generally enough to explain the principle that combustion can occur and continue. However, according to the chain reaction theory of combustion, there are continuous free radicals as "intermediates" in many combustion. Therefore, the ignition triangle should be expanded to include an additional dimension indicating that free radicals participate in the combustion reaction, so as to form an ignition tetrahedron, as shown in Fig. 1-2 (b). The triangle adds the last two basic elements to become the famous explosive pentagon, as shown in Fig. 1-2(c). If a dust gas cloud is ignited in a confined space or semi-confined container, area or building, it will burn rapidly and may explode.

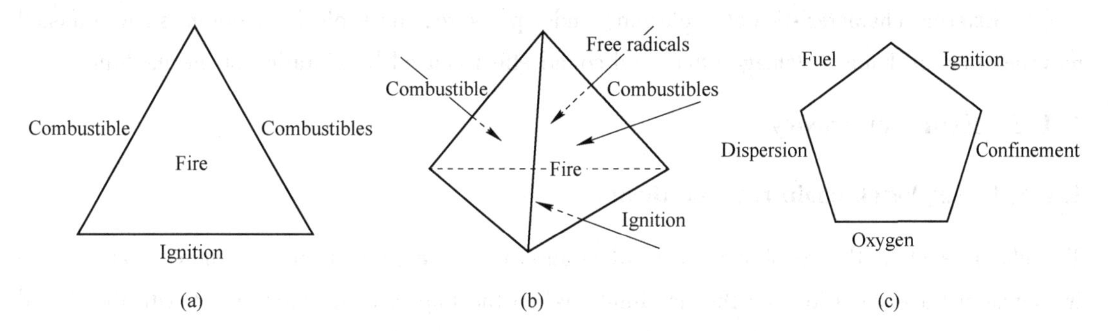

Fig. 1-2 Fire triangle, fire tetrahedron and explosive pentagon

1. 1. 4 Explosion and its characteristics

1. 1. 4. 1 Explosion

Explosion refers to the sudden change of substances from one state to another through physical or chemical changes, and the release of huge energy, while generating light, heat or mechanical work. When a substance "suddenly changes" from one state to another, its physical state or chemical composition changes sharply, so that its own energy (potential energy) is released at a rapid rate, and the surrounding objects are subjected to violent impact and destruction.

Lightning and volcanic eruption are natural explosion phenomena; Blasting in engineering construction is an explosion controlled by man and beneficial to mankind; unexpected explosions in people's production and life are accidental explosions, such as underground gas explosion, boiler and pressure vessel explosion, grain dust explosion, etc.

The several explosion phenomena mentioned above have a common feature, that is, the sudden increase of the surrounding pressure at the explosion site will disturb the surrounding medium, destroy the adjacent substances, and be accompanied by large or small sound effects.

1.1.4.2 Characteristics of explosion

As mentioned above, explosion is a sharp physical and chemical change of substances. In the process of change, it is accompanied by the rapid release of the energy contained in the material, which becomes the compression energy or motion energy of the material itself, the change product or the surrounding medium. The pressure of the material system increases sharply during explosion.

Generally, the explosion has the following characteristics:

(1) The internal characteristics of the explosion: a large amount of energy of the system explosion is suddenly released or transformed in a limited volume, and accumulates in a limited volume in a very short time, resulting in high temperature and high pressure, forming a sharp pressure jump and subsequent complex movement to the adjacent media.

(2) External characteristics of explosion: under pressure, the explosive medium shows unusual movement or mechanical damage effect and sound effect caused by vibration of the medium.

1.1.5 Explosion theory

1.1.5.1 Explosion chain reaction theory

The chain reaction theory of explosion illustrates a very important phenomenon, that is, the influence of trace impurities on the fire limit. When the impurity molecules meet with the linked intermediate activated molecules, the chain can be interrupted. Naturally, all the reactions that may occur as the chain continues to develop will not occur in the future. When impurities and molecules meet, they may produce new chains, so they can cause a new chain reaction. The chain reaction theory links the research field of explosion with the physical theory of molecular structure and its interaction.

The chain reaction theory can be used to explain the combustion limit, that is, the maximum and minimum limits of the combustible content in the mixture (the substances outside the limit cannot explode) and the phenomenon of diluting the combustible mixture with inert mixture so that it cannot catch fire. This phenomenon can be explained by chain breaking and gas cooling, because most of the heat is used to unnecessarily heat the inert gas that does not participate in the reaction during explosion expansion.

The chain reaction theory also shows that the explosion does not occur immediately when the critical condition for ignition is reached, but after a certain period of time necessary for the development of the chain. Therefore, any explosion has a time delay, which depends on the development process of the chain and external conditions, and can range from a few hundredths of a second to several hours.

1.1.5.2 Explosion wave theory

Explosion wave theory can be used to explain the explosion of combustible gas, steam air mixture.

The main content of this theory is: when the external impact acts on the mixture with explosion risk, if the impact force is enough to make the substance decompose rapidly, various phenomena of accelerating explosion will occur in turn. In the substances with explosion risk, all the energy that can cause explosion is changed into heat energy to cause impact. The impact is related to the acceleration of the gas molecules generated in the reaction. The impact of gas can heat and decompose a layer of explosives, which will turn into gas and impact on a new layer in turn. It can be seen that the explosion radiates outward from the impact and alternates mechanical, thermal and chemical interactions. This is the origin of the term "explosion wave".

1.1.5.3 Explosive electron nature hypothesis

The electron theory explains the instability of explosive materials by the weak bonding between atoms. In ordinary chemical reactions, electrons outside can also jump from one atom to another. Then, it can be assumed that in some particularly sensitive explosive compounds, the binding of valence electrons is weaker. Therefore, in the detonator, even under a small impact, molecular changes will occur. At the same time, it will not only release energy in the form of heat, but also release free electrons with kinetic energy.

1.1.5.4 Hydrodynamic explosion theory

According to the explosion theory of fluid dynamics, the explosion is caused by the propagation of shock wave in explosives. There may be two different situations for shock wave propagation in explosives. One is similar to that in inert media, that is, it does not cause chemical changes in explosives. If there is no continuous effect of external factors in this process, it is impossible to maintain constant speed propagation. This is because when the shock wave front passes through, the medium is irreversibly compressed, and the entropy increases, resulting in the irreversible loss of energy. Therefore, it must be attenuated in the propagation. On the other hand, the violent compression of the shock wave causes a rapid chemical reaction of the explosive, and the energy released by the reaction supports the propagation of the shock wave, which can maintain a constant speed without attenuation. This kind of shock wave following the chemical reaction, or the shock wave accompanied by the chemical reaction, is called detonation wave. Detonation is the process of detonation wave propagation in the explosive.

1.1.5.5 Hydrodynamic theory of gaseous detonation

This theory assumes an ideal detonation process, and the explosive gas obeys the ideal gas law before and after the explosion wave passes through, and assumes that the isentropic index of the gas is independent of temperature and composition. Under this condition, according to the law of conservation of energy and the law of ideal gas, a relationship between the initial parameters of explosives and explosion parameters is established. This relationship is used to express the change of internal energy caused by the change of medium state parameters (such as pressure and volume) before and after the explosion wave passes through.

1.1.6 Classification of explosions

There are many classification methods for explosion accidents in industrial production, which can be classified according to the changes before and after the material explosion, the type of explosion accident process, and the phase of explosion reaction.

1.1.6.1 Classification according to changes of substances before and after explosion

(1) Physical explosion. Physical explosion refers to the explosion caused by sudden change of state or pressure of substances. It is obviously different from chemical explosion in that the properties and chemical compositions of substances before and after physical explosion do not change. Such as boiler explosion, pressure vessel explosion, automobile tire explosion, etc.

(2) Chemical explosion. Chemical explosion refers to the explosion caused by exothermic chemical reaction of substances with extremely fast reaction speed and high temperature and high pressure. Before and after the chemical explosion, the composition and properties of the substance changed fundamentally.

Chemical explosion can be divided into three categories according to the chemical changes of substances during explosion:

1) Simple decomposition explosion. Explosives that cause simple decomposition do not necessarily undergo combustion reaction during explosion. The energy required for explosion is provided by the decomposition heat released during the decomposition of the explosives themselves. For example, the explosive reaction of acetylene, silver, mercury and other substances belongs to this category. This kind of substance is extremely unstable and can cause explosion when it is vibrated. It is a relatively dangerous explosive substance. Some gases generate a lot of heat due to decomposition, which may also cause decomposition explosion under certain conditions. It is easier to explode under pressure. Such as decomposition and explosion of ethylene and acetylene stored under high pressure.

2) Complex decomposition explosion. Under the action of strong external excitation energy (such as detonation wave), explosive substances can produce high-speed exothermic reaction, and form strongly compressed gas as the high-temperature and high-pressure gas source causing explosion. The explosion of such substances is accompanied by combustion. The oxygen required for combustion is generated by its own decomposition. After the explosion, the nearby combustible substances are often ignited, causing a fire. For example, many kinds of explosives and some organic peroxide explosions fall into this category. Compared with simple decomposition explosives, these substances are less sensitive to external stimuli and have slightly lower risk.

3) Explosion of explosive mixture. The explosion of all combustible gases, vapors and explosive

mixtures formed by dust and air belongs to this category. The explosion of such substances requires certain conditions (sufficient explosive substance content, oxygen content, ignition energy, etc.) at the same time, and its risk is lower than the above two categories. However, as it widely exists in many fields of industrial production, it also causes more explosion accidents and great harm. The classification of mixture explosion is shown in Table 1-1.

Table 1-1 Classification of mixture explosion

Class and level	Maximum experimental safety gap	Minimum ignition current ratio	Ignition temperature (℃) and group					
			T_1	T_2	T_3	T_4	T_5	T_6
	MESG(mm)	*MICR*	$T>450$	$450\geqslant T>300$	$300\geqslant T>200$	$200\geqslant T>135$	$135\geqslant T>100$	$100\geqslant T>85$
I	*MESG*=1.14	*MICR*=1.0	Methane					
I A	0.9<*MESG* <1.14	0.8<*MICR* <1.0	Ethane, propane, acetone, styrene, vinyl chloride, aminobenzene, toluene, benzene, ammonia, methanol, carbon monoxide, ethyl acetate, acetic acid, acrylonitrile	Butane, ethanol, propylene, butanol, butyl acetate, acetic acid, amyl acetate, acetic anhydride	Pentane, hexane, heptane, decane, octane, gasoline, hydrogen sulfide, cyclohexane	Ether, acetalde hyde		Nitrous acid, ethyl ester
I B	0.5<*MESG* ≤0.9	0.45<*MICR* ≤0.8	Diethyl ether, civil gas, cyclopropane	Ethylene oxide, propylene oxide, butadiene, ethylene	Isoprene			
I C	*MESG*≤0.5	*MICR*≤0.45	Water gas, hydrogen, coke oven gas	Acetylene			Carbon disulfide	Ethyl nitrate

Note: The relationship between the maximum experimental safety gap and the minimum ignition current ratio in classification is only approximately equal.

1.1.6.2 Classification by accident explosion process type

Explosion accidents always have certain causes and processes. By analyzing the origin and development of explosion accidents with system safety engineering and taking specific, accurate and applicable explosion protection measures, explosion accidents can be avoided or mitigated. According to the type of accident explosion, it can be divided into 6 types. Namely: ignition destructive explosion, leakage ignition explosion, spontaneous combustion ignition explosion, reaction out of control explosion, heat transfer steam explosion, balance destructive steam explosion.

(1) Ignition type explosion: hazardous substances inside containers, pipelines, tower grooves, etc. (hereinafter referred to as containers) are energized by ignition sources, causing chemical reactions such as ignition, combustion and decomposition, causing a sharp rise in pressure and causing explosion damage to containers.

(2) Leakage explosion: dangerous substances inside the container leak to the outside due to damage such as valve opening or container cracks, and catch fire when they contact with the ignition source, causing explosive fire.

(3) Spontaneous combustion explosion: due to the accumulation of chemical reaction heat, the temperature rises, the reaction speed accelerates, and the temperature rises even more. When the ignition temperature of this substance is reached, spontaneous combustion will occur and cause explosion.

(4) Uncontrolled reaction explosion: the accumulation of chemical reaction heat will increase the temperature and speed up the reaction, so that the vapor pressure or decomposition gas pressure of the substance will rise sharply, causing a destructive explosion of the container.

(5) Heat transfer explosion: when the superheated liquid contacts with other high-temperature substances, rapid heat transfer occurs, and the liquid is heated, making it temporarily in an overheated state, resulting in a vapor explosion accompanied by rapid gasification.

(6) Equilibrium destructive vapor explosion: this is an explosion due to the evaporation of superheated liquid. That is, when the vapor pressure of the liquid in the closed container is balanced under high pressure, if the container is damaged, the vapor will be ejected, and the balance will be lost due to the sharp decline of the internal pressure, so that the liquid will be temporarily in an unstable overheating state. Due to the rapid gasification, the residual liquid breaks through the container wall, and the impact pressure causes the container to be damaged again, resulting in steam explosion.

If the above six types of explosions are further summarized, they can also be divided into those requiring ignition sources and those not requiring ignition sources. The explosion requiring ignition source includes: ignition destructive explosion, leakage ignition explosion; The explosions that do not need ignition sources include: spontaneous ignition explosion of chemical reaction heat accumulation, reaction runaway explosion, heat transfer steam explosion of superheated liquid evaporation and balance breaking steam explosion.

1.1.6.3 Classification by explosive reaction phase

Gas phase explosion, liquid phase explosion and solid phase explosion are classified according to the phase causing explosion reaction.

(1) Gas phase explosion includes explosion of mixture of combustible gas and combustion supporting gas, thermal decomposition explosion of substances, explosion caused by fog drops of

combustible liquid (fog explosion), etc. The decomposition explosion does not need combustion supporting gas.

(2) Liquid phase explosion includes polymerization explosion, evaporation explosion and explosion caused by mixing of different liquids.

(3) Solid phase explosion includes explosion of explosive solid substances, explosion caused by mixing and melting of solid substances, and cable explosion caused by current overload. In addition, there is dust explosion, which is an explosion caused by violent combustion of combustible dust scattered in the air. Such as the explosion caused by aluminum powder and flour flying in the air.

In addition, there is dust explosion, which is an explosion caused by violent combustion of combustible dust scattered in the air. Such as the explosion caused by aluminum powder and flour flying in the air.

1.1.7 Conditions for explosion

The explosion conditions are complex, and the explosion process of different explosive substances has its unique characteristics.

1.1.7.1 Conditions for physical explosion

Physical explosion is a very rapid process in which physical energy is released out of control. In this process, the substances in the system release the energy contained in it at a very fast speed and transform it into mechanical work, light, heat and other energy forms. Considering the root cause of physical explosion, the occurrence conditions of explosion can be summarized as follows: the explosion system contains high-pressure gas or the rapid expansion of high-temperature and high-pressure gas or steam generated at the moment of explosion, and there is a sharp pressure jump between the explosion system and its surrounding media. Boiler explosion, pressure vessel explosion, large amount of rapid gasification of water, etc. all belong to this kind of explosion.

1.1.7.2 Conditions for chemical explosion

The chemical explosion process has two characteristics: the reaction process is exothermic, the reaction process is very fast and can spread automatically. These two characteristics are necessary for chemical reaction to become explosive reaction, and they are interrelated and indispensable conditions. The importance and significance of each condition are briefly discussed below.

A Exothermic property of reaction process

This is the most important basic condition for whether the chemical reaction can become the explosion reaction, and it is also the energy source of the explosion process. Without this condition, the explosion process cannot occur, and of course, the reaction cannot continue on its own. Therefore, it is impossible for the explosion process to propagate automatically. For example:

$$ZnC_2O_4 \longrightarrow 2CO_2 + Zn, \quad \Delta_r H_m^\ominus = -20.5kJ$$
$$PbC_2O_4 \longrightarrow 2CO_2 + Pb, \quad \Delta_r H_m^\ominus = -69.9kJ$$

The decomposition of zinc oxalate and lead oxalate is an endothermic reaction. They need external heat to carry out the reaction. Therefore, they cannot do work to the outside world, so they cannot explode. Another example is the decomposition reaction of ammonium nitrate:

$$NH_4NO_3 \longrightarrow NH_3 + HNO_3, \quad \Delta_r H_m^\ominus = -170.7kJ \text{ (Low temperature heating)}$$
$$NH_4NO_3 \longrightarrow N_2 + 2H_2O + 0.5O_2, \quad \Delta_r H_m^\ominus = +126.4kJ \text{ (Detonate with a detonator)}$$

The low-temperature heating reaction formula of ammonium nitrate is a slow decomposition reaction that occurs in the farmland when it is used as a fertilizer. The reaction process is endothermic and cannot explode at all. When ammonium nitrate is detonated by detonator, exothermic decomposition reaction occurs. It is a kind of explosive commonly used in mine blasting.

The heat released during the explosion reaction process is called explosion heat (or explosion heat). It is the constant volume thermal effect of reaction, the symbol of explosive damage ability, and also an important dangerous characteristic of explosives. The detonation heat of common explosives is about 3700~7500kJ/kg; For mixed explosives, the heat of explosion is the heat of combustion. The combustion heat of organic combustibles is about 48000kJ/kg.

B High speed of reaction process

The mixed explosive substance is a system in which the oxidizer and reductant are fully mixed in advance. The oxidizer and reductant of many explosives coexist in one molecule. Therefore, they can produce a rapid layer by layer transfer chemical reaction, so that the explosion process can be carried out at a very fast speed, which is the most important difference between explosion reaction and general chemical reaction. General chemical reactions can also be exothermic, and many chemical reactions give off much more heat than the explosion of explosive substances. However, the fundamental reason why they fail to form an explosion is that their reaction speed is slow. For example, the combustion heat of 1kg wood is 16700kJ, and it takes 10min for it to burn completely; The explosion heat of 1kg TNT explosive is only 4200kJ, and its explosion reaction takes only tens of microseconds; The time difference between the two is tens of millions of times.

Due to the extremely fast reaction speed of explosive substances, it can be approximated that the energy released by the explosive reaction is too late to escape, and all of it accumulates in the volume occupied by the explosive substances before the explosion, resulting in the energy density that can not be achieved by the general chemical reaction. It is for this reason that the explosive material explosion has huge power and strong destructive effect.

For example, the combustion heat of 1kg coal block and 1kg gas is 29000kJ. It takes about 10min for a 1kg coal block to burn completely, which is a combustion process. However, after mixing 1kg gas and air, it only takes 0.2s to burn out, which belongs to the explosion process. Similarly, the mixture of gas and air can be reacted in only 0.7ms under the condition of explosive detonation. According to the relationship that the power is inversely proportional to the work time, their power can be calculated as follows: the power generated when the coal block with 1kg calorific value of 29000kJ is burned is 48kW; 1kg the power generated when the mixture of gas and air with heating capacity of 29000kJ deflagration is 1.4×10^5kW; 1kg the power emitted when the mixture of gas and air with heating capacity of 29000kJ explodes is 4.1×10^7kW. This example clearly shows that the high speed of the explosion process and the corresponding high speed of releasing reaction heat are the main characteristics of the explosion process.

In addition, most explosive reaction processes must form gaseous products. Gases are more less dense than solid and liquid matter at normal atmospheric pressure. It has compressibility, and has much larger volume expansion coefficient than solid and liquid, which is an excellent working medium. The explosive substance generates a large number of gas products at the moment of explosion. Because the explosive reaction speed is very fast, they have no time to diffuse and expand, and are compressed in the volume occupied by the explosive substance. The explosion releases a lot of heat as it forms gaseous products. These heat are added to the generated gas products before they can escape. In this way, the gas with high temperature and high pressure is formed in the original volume occupied by the explosive substance. As a working medium, this kind of gas expands and does work in an instant. Due to its huge power, it will cause great damage to surrounding objects, equipment and houses. For example, 1L explosive can produce about 1000L of gas products at the moment of explosion. They are strongly compressed in the original volume. Due to the high temperature of 3000 ~ 5000℃, high-temperature and high-pressure gas sources are formed. When they expand instantaneously, they have huge power and strong destructive power. It can be seen that the generation of gas products during explosion is an important condition for explosion.

Under normal conditions, gaseous products produced in explosion process are also one of the important conditions for explosion. This conclusion can also explain that the generation of gas products is a necessary condition for explosion through the fact that some strong exothermic reactions that do not generate gas products do not have explosion effect. For example, thermite reaction:

$$2Al + Fe_2O_3 = Al_2O_3 + 2Fe, \quad \Delta_r H_m^{\ominus} = +841kJ$$

The thermal effect of this reaction is large enough to heat the product to a high temperature of 3000℃, and the reaction speed is also quite fast, but it does not have the ability to explode because it does not form gas products.

1.2 Theory of combustion reaction rate

The combustion reaction rate equation is used in the analysis of ignition conditions, the estimation of fire development speed, the study of combustion process and the analysis of fire extinguishing conditions. This equation can be obtained from the theory of chemical kinetics.

1.2.1 Basic concept of reaction rate

The speed of chemical reaction can be measured by the number of moles consumed by reactants or produced by products in unit volume per unit time, and it is called the reaction rate (consumption rate for reactants and production rate for products), which is expressed as:

$$\omega = \frac{dn}{V \cdot dt} = \frac{dc}{dt} \tag{1-1}$$

Where ω——the reaction rate, mol/($m^3 \cdot s$);

V——volume, m^3;

dn, dc——the changes in the number of moles of the substance and the molar concentration, mol and mol/m^3;

dt——time of change, s.

Although the reaction rate is different by the change of reactant concentration and the change of product concentration, there is a single value stoichiometric relationship between them, which is determined by the chemical reaction formula. If any reaction $aA+bB \rightarrow eE+fF$ is known, the reaction rate can be written as:

$$\left.\begin{aligned} \omega_A &= -\frac{dc_A}{dt} \quad \omega_B = -\frac{dc_B}{dt} \\ \omega_E &= -\frac{dc_E}{dt} \quad \omega_F = -\frac{dc_F}{dt} \end{aligned}\right\} \tag{1-2}$$

The relationship between the above four reaction rates is as follows:

$$\frac{\omega_A}{a} = \frac{\omega_B}{b} = \frac{\omega_E}{e} = \frac{\omega_F}{f} = \omega \tag{1-3}$$

Where, ω Represents the chemical reaction rate of the reaction system, and its value is unique, which is called the system reaction rate. Obviously, the reaction rate of each substance in the reaction system is:

$$\left.\begin{aligned} \omega_A &= \omega \cdot a \quad \omega_B = \omega \cdot b \\ \omega_E &= \omega \cdot e \quad \omega_F = \omega \cdot f \end{aligned}\right\} \tag{1-4}$$

1.2.2 Law of mass action

The stoichiometric equation expresses the quantitative relationship between reactants and products before and after the reaction. However, this expression only describes the overall situation of the reaction and does not explain the actual process of the reaction, that is, it does not give the intermediate process experienced in the reaction process. For example, the reaction of hydrogen and oxidation to produce water can be expressed as $2H_2+O_2 \rightarrow 2H_2O$, but in fact, H_2 and O_2 need several steps to be converted into H_2O.

The reaction in which reactant molecules are converted into product molecules in one step during collision is called elementary reaction. The transformation of a chemical reaction from reactant molecules to final product molecules often requires several elementary reactions. The experimental results show that for single-phase chemical elementary reactions, under isothermal conditions, the rate of any instantaneous chemical reaction is directly proportional to the product of a certain power of the concentration of each reactant at that instant. In the elementary reaction, the power of the concentration of each reactant is equal to the stoichiometric coefficient of the reactant.

The law of the relationship between the rate of chemical reaction and the concentration of reactants is called the law of mass action. The simple explanation is that the chemical reaction is generated after the collision between the reactant molecules. Therefore, the more the number of molecules per unit volume, that is, the greater the reactant concentration, the more the collision times between the reactant molecules, and the faster the reaction process. Therefore, the chemical reaction speed has a positive relationship with the reactant concentration.

For the reaction formula aA+bB→eE+fF, according to the law of mass action, the equation of chemical reaction rate is (in fact, it refers to the forward reaction rate):

$$V = KC_A^a C_B^b \tag{1-5}$$

Where K ——ratio constant, or reaction rate constant, whose value is equal to the reaction rate when the reactant is a unit concentration;

a, b——the reaction order of the chemical reaction.

It must be emphasized that the law of mass action is only applicable to elementary reactions, because only elementary reactions can represent the real way of reaction. For non elementary reactions, the mass action law can be applied one by one only when they are decomposed into several elementary reactions.

1.2.3 Arrhenius law

A large number of experiments have proved that the reaction temperature has a great influence on the chemical reaction speed, and this influence is also very complex, but the most common situation is that the reaction speed increases with the increase of temperature. The Van't Hoff approximation rule holds that for the general reaction, if the initial concentration is equal, the reaction speed will increase about 2~4 times when the temperature increases by 10℃.

The effect of temperature on the reaction rate is mainly reflected in the reaction rate constant K. Svante August Arrhenius (1859~1927) proposed the following relationship between reaction rate constant K and reaction temperature T:

$$K = K_0 \exp\left(-\frac{E}{RT}\right) \tag{1-6}$$

Where K ——arrhenius reaction rate constant, $m^3/(s \cdot mol)$;

E——reactant activation energy, kJ/mol;

R——universal gas constant, value 8.314×10^{-3} kJ/(mol · K);

T——temperature, K;

K_0——frequency factor, $m^3/(s \cdot mol)$.

In the above formula, relative to $\exp\left(-\frac{E}{RT}\right)$, the effect of temperature on K_0 is negligible.

The relationship expressed in formula (1-6) is usually called Arrhenius' law, which is not only applicable to elementary reactions, but also to complex reactions with definite reaction order and velocity constant.

Take logarithms from both sides of equation (1-6) to obtain:

$$\ln K = -\frac{E}{RT} + \ln K_0 \tag{1-7}$$

Perhaps

$$\lg K = -\frac{E}{2.303RT} + \lg K_0 \tag{1-7'}$$

From the above formula: $\ln K$ or $\lg K$ can obtain a straight line by plotting $1/T$, E can be obtained from its slope, and K_0 can be obtained from its intercept.

According to the law of mass action and Arrhenius law, the velocity equation of elementary reaction can be obtained, that is

$$V = K_0 C_A^a C_B^b \exp\left(-\frac{E}{RT}\right) \tag{1-8}$$

1.2.4 Combustion reaction rate equation

It is assumed that in the combustion reaction, the concentration of combustible is C_F and the reaction coefficient is x; The comburant (mainly air) is C_{ox} and the reaction coefficient is y; The frequency factor is K_{0s}; The activation energy is E_s; The reaction temperature is T_s. In this way, the combustion reaction rate equation can be written according to formula (1-8):

$$V_s = K_{0s} C_F^x C_{ox}^y \exp\left(-\frac{E_s}{RT_s}\right) \tag{1-9}$$

When dealing with some combustion problems, it is often assumed that the concentration of reactants is constant, so the concentration ratio of various substances is also constant. The concentration of one substance can be expressed by the concentration of another substance. For example, in equation (1-9), let $C_{ox} = m \cdot C_F$, m be a constant, and the reaction order be n, that is, $n=x+y$, so equation (1-9) can be expressed as:

$$V_s = K_{ns} C_F^n \exp\left(-\frac{E_s}{RT_s}\right) \tag{1-10}$$

Where, $K_{ns} = m^y \cdot K_{os}$.

For most hydrocarbon combustion reactions, the reaction order is approximately equal to 2, and $x=y=1$, so the combustion reaction rate equation can be written as:

$$V_s = K_{0s} C_F C_{ox} \exp\left(-\frac{E_s}{RT_s}\right) \tag{1-11}$$

Assuming that the concentration of reactants is constant (the reaction order is approximately equal to 2), according to equation (1-10), the combustion reaction rate equation can be written as:

$$V_s = K_{ns} C_F^2 \exp\left(-\frac{E_s}{RT_s}\right) \tag{1-12}$$

In practical work, it is convenient to use relative mass concentration to express the concentration of substances. Thus, equation (1-11) can be expressed in the following common form:

$$V_s = K'_{0s}\rho_\infty^2 f_F f_{ox} \exp\left(-\frac{E_s}{RT_s}\right) \tag{1-13}$$

Equation (1-12) can be expressed as follows:

$$V_s = K'_{ns}\rho_\infty^2 f_F^2 \exp\left(-\frac{E_s}{RT_s}\right) \tag{1-14}$$

The derivation process of equation (1-13) is:

It is assumed that the molar mass of combustibles and combustion supporting substances are M_F, M_{ox} respectively, and the mass concentration is ρ_F, ρ_{ox}; total mass concentration of combustion reaction process ρ_∞; The relative mass concentrations of combustibles and combustion supporting substances are f_F, f_{ox} respectively. According to the relationship between mass concentration and molar concentration, there are:

$$C_F = \frac{\rho_F}{M_F}, C_{ox} = \frac{\rho_{ox}}{M_{ox}}$$

and

$$f_F = \frac{\rho_F}{\rho_\infty}, f_{ox} = \frac{\rho_{ox}}{\rho_\infty}$$

so

$$C_F = \frac{f_F\rho_\infty}{M_F}, C_{ox} = \frac{f_F\rho_\infty}{M_{ox}}$$

By substituting these two equations into equation (1-11), we get:

$$V_s = K_{0s} \cdot \frac{1}{M_F} \cdot \frac{1}{M_{ox}} \rho_\infty^2 f_F f_{ox} \exp\left(-\frac{E_s}{RT_s}\right)$$

Let $K'_{0s} = K_{0s}\frac{1}{M_F} \cdot \frac{1}{M_{ox}}$, the above formula becomes the form of equation (1-13).

In particular, it should be pointed out that since the combustion reactions are not elementary reactions but complex reactions, they do not strictly obey the mass action law and Arrhenius' law. Therefore, in the above formula, $K_{0s}(K'_{0s})$ and E_s are no longer of direct physical significance, and they are only the apparent data obtained from the test. The K'_{0s} and E_s of some common combustible substances are listed in Table 1-2.

Table 1-2 K'_{0s} and E_s values of common combustible substances

Substance name	K'_{0s}(mol · s^{-1})	E_s(kJ/mol)
propane + air	2×10^{14}(387K)	129.58
methane + air	2×10^{14}(558K)	121.22
butane + oxygen	5.4×10^{13}(400K)	87.78
isooctane + air	5.4×10^{13}(400K)	16.72
n-octane + air	5.4×10^{13}(400K)	16.72
n-hexane + air	5.4×10^{13}(400K)	23.32

Continuted Table 1-2

Substance name	K'_{0s}(mol · s^{-1})	E_s(kJ/mol)
benzene + air	5. 4×10^{13}(400K)	172. 22
ethylene + air	5. 4×10^{13}(400K)	172. 22
ammonia + oxygen	5. 4×10^{13}(400K)	206. 91
ammonia + Air	1. 6×10^{12}(313K)	41. 80
hydrogen + oxygen	1. 6×10^{12}(313K)	75. 24
hydrogen + fluorine	1. 6×10^{12}(313K)	209. 00

The above combustion reaction rate equation is an approximate formula derived from gaseous substances, from which some useful conclusions can be drawn. For example, in the fire scene, the lower the concentration of combustible and oxygen, the slower the combustion reaction speed; The lower the temperature at the fire site, the slower the combustion reaction speed, which is the basis of the cooling fire extinguishing method; The higher the activation energy (the energy required to destroy the internal chemical bond of the reactant molecule) of the combustible reaction, the slower the combustion reaction speed, and so on.

Compared with gaseous combustibles, the combustion reaction process of liquid and solid combustibles is more complex, because there are evaporation, melting, cracking and other phenomena. Therefore, the mass action law and Arrhenius law used to describe the combustion reaction of these two substances are far from the actual situation. The combustion reaction rate of liquid and solid combustibles can not be expressed by the above equation, but by other expressions.

1. 3 Calculation of combustion air volume

As we all know, the air contains nearly 21% oxygen by volume (23. 2% by weight), and general combustibles can burn in case of ignition source. When the amount of air or oxygen is insufficient, the combustibles cannot burn or the ongoing combustion will gradually extinguish. As the basic parameter of combustion reaction, air demand represents the air mass or volume required for the combustion of a certain amount of combustibles. Its calculation is carried out under the condition that the combustibles are completely burned.

1. 3. 1 Theoretical air volume

Theoretical air quantity refers to the minimum air quantity required for complete combustion of a unit quantity of fuel, which is usually also called theoretical air demand (commonly expressed as L_0). At this time, the combustibles in the fuel completely react with the oxygen in the air to obtain complete oxidation products.

1.3.1.1 Theoretical air demand for solid and liquid combustibles

In general, for solid and liquid combustibles, it is customary to express their composition by mass percentage, and their composition is:

$$w(\mathrm{C})\% + w(\mathrm{H})\% + w(\mathrm{O})\% + w(\mathrm{N})\% + w(\mathrm{S})\% + w(\mathrm{A})\% + w(\mathrm{W})\% = 100\% \tag{1-15}$$

Where, $w(\mathrm{C})$, $w(\mathrm{H})$, $w(\mathrm{O})$, $w(\mathrm{N})$, $w(\mathrm{S})$, $w(\mathrm{A})$ and $w(\mathrm{W})$ are the mass percentages of carbon, hydrogen, oxygen, nitrogen, sulfur, ash and water of combustible, where C, H and S are combustible components;

N, A and W is a non combustible ingredient; O is the combustion supporting component.

To calculate the theoretical air volume, the oxygen volume required for the complete combustion of combustible elements (carbon, hydrogen, sulfur, etc.) in the fuel should be calculated first. Therefore, the stoichiometric equation of complete combustion of these elements should be used.

According to the chemical reaction formula of complete combustion, the quantitative relationship of carbon combustion is:

$$\mathrm{C} + \mathrm{O_2} = \mathrm{CO_2}$$

Molecular weight 12 32 44

That is, the amount of O_2 required for complete combustion of 1kg C is 8/3kg. Similarly, the amount of O_2 required for the complete combustion of 1kg H_2 is 8kg, and the amount of O_2 required for the complete combustion of 1kg S is 1kg.

The amount of oxygen required for complete combustion of unit mass C, H and S is:

$$G_{0,\mathrm{O_2}} = \frac{8}{3} w(\mathrm{C}) \times 10^{-2} (\mathrm{kg/kg})$$

$$G_{0,\mathrm{O_2}} = 8 w(\mathrm{H}) \times 10^{-2} (\mathrm{kg/kg})$$

$$G_{0,\mathrm{O_2}} = w(\mathrm{S}) \times 10^{-2} (\mathrm{kg/kg})$$

To sum up, the amount of oxygen required for complete combustion of combustibles per unit mass is:

$$G_{0,\mathrm{O_2}} = \left(\frac{8}{3} w(\mathrm{C}) + 8 w(\mathrm{H}) + w(\mathrm{S}) - w(\mathrm{O}) \right) \times 10^{-2} (\mathrm{kg/kg}) \tag{1-16}$$

It is assumed that the gas involved in the calculation is an ideal gas, that is, the volume of 1kmol gas in the standard state (273K, 0.1013MPa) is 22.4$\mathrm{m^3}$. Then the volume of oxygen required for complete combustion of a unit weight of fuel is:

$$V_{0,\mathrm{O_2}} = \frac{G_{0,\mathrm{O_2}}}{32} \times 22.4 = 0.7 \times \left(\frac{8}{3} \varphi(\mathrm{C}) + 8 \varphi(\mathrm{H}) + \varphi(\mathrm{S}) - \varphi(\mathrm{O}) \right) \times 10^{-2} (\mathrm{m^3/kg}) \tag{1-17}$$

Therefore, the volume of air required for complete combustion of combustibles per unit mass is:

$$V_{0,\mathrm{air}} = \frac{V_{0,\mathrm{O_2}}}{0.21} = \frac{0.7}{0.21} \times \left(\frac{8}{3} \varphi(\mathrm{C}) + 8 \varphi(\mathrm{H}) + \varphi(\mathrm{S}) - \varphi(\mathrm{O}) \right) \times 10^{-2} (\mathrm{m^3/kg}) \tag{1-18}$$

【**Example 1-1**】 Calculate the theoretical air volume required for the complete combustion of 4kg wood. The mass percentage composition of known wood is: C 43%, H 7%, O 41%, N 2%, W 6%, A 1%.

【**Solution**】 According to the above formula, the theoretical oxygen volume required to burn 1kg of this wood is:

$$V_{0,O_2} = 0.7 \times \left(\frac{8}{3} \times 43 + 8 \times 7 - 41\right) \times 10^{-2} = 0.91(m^3)$$

Therefore, the theoretical air volume required to burn 4kg of this wood is:

$$V_{0.\,air} = \frac{0.91}{0.21} \times 4 = 17.33(m^3)$$

Answer: the theoretical amount of air required for the complete combustion of 4kg wood is $17.33m^3$.

1.3.1.2 Theoretical air volume of gaseous combustibles

For gaseous combustibles, it is customary to express their composition by volume percentage, and their composition is:

$$\varphi(CO)\% + \varphi(H_2)\% + \varphi\sum(C_nH_m)\% + \varphi(H_2S)\% +$$
$$\varphi(CO_2)\% + \varphi(O_2)\% + \varphi(N_2)\% + \varphi(H_2O)\% = 100\% \quad (1\text{-}19)$$

Where $\varphi(CO)$, $\varphi(H_2)$, $\varphi(C_nH_m)$, $\varphi(H_2S)$, $\varphi(CO_2)$, $\varphi(O_2)$, $\varphi(N_2)$, $\varphi(H_2O)$ respectively represent the volume percentage of corresponding components in gaseous combustibles. C_nH_m represents the general formula of hydrocarbon, which may be CH_4, C_2H_2, C_2H_4 and other combustible gases.

According to the reaction equation of complete combustion of combustibles, it is as follows:

$$\left.\begin{aligned} &H_2 + \frac{1}{2}O_2 = H_2O \\ &H_2S + \frac{3}{2}O_2 = H_2O + SO_2 \\ &C_nH_m + \left(n + \frac{m}{4}\right)O_2 = nCO_2 + \frac{m}{2}H_2O \end{aligned}\right\} \quad (1\text{-}20)$$

From the above reaction equation, it can be concluded that 1/2mol O_2 is required for complete combustion of 1mol of CO, and $1/2m^3$ O_2 is required for combustion of $1m^3$ of CO according to the ideal gas equation of state. Similarly, complete combustion of $1m^3$ H_2, H_2S and C_nH_m requires $1/2m^3$, $3/2m^3$ and $(n+m/4)m^3$ of O_2 respectively. Therefore, the volume of oxygen required for complete combustion of each 1 m^3 of combustible is:

$$V_{0,O_2} = \left[\frac{1}{2}\varphi(CO) + \frac{1}{2}\varphi(H_2) + \frac{3}{2}\varphi(H_2S) + \sum\left(n + \frac{m}{4}\right)\varphi(C_nH_m) - \varphi(O_2)\right] \times 10^{-2}(m^3) \quad (1\text{-}21)$$

The theoretical air volume required for complete combustion of combustibles per $1m^3$ is:

$$V_{0,air}=\frac{V_{0,O_2}}{0.21}=4.76\times\left[\frac{1}{2}\varphi(CO)+\frac{1}{2}\varphi(H_2)+\frac{3}{2}\varphi(H_2S)+\sum\left(n+\frac{m}{4}\right)\varphi(C_nH_m)-\varphi(O_2)\right]\times10^{-2}(m^3) \quad (1\text{-}21')$$

【Example 1-2】 Calculate the theoretical air volume required for $1m^3$ coke oven gas combustion. The volume percentage array of known coke oven gas is: CO 6.8%, H_2 57%, CH_4 22.5%, C_2H_4 3.7%, CO_2 2.3%, N_2 4.7%, H_2O 3%.

【Solution】 from the general hydrocarbon formula:

$$\sum\left(n+\frac{m}{4}\right)\varphi(C_nH_m)=\left(1+\frac{4}{4}\right)\times22.5+\left(2+\frac{4}{4}\right)\times3.7=56.1$$

The theoretical air volume required for complete combustion of $1m^3$ of this gas is:

$$V_{0,air}=4.76\times\left(\frac{1}{2}\times6.8+\frac{1}{2}\times57+56.1\right)\times10^{-2}=4.19(m^3)$$

That is, the theoretical air volume required for $1m^3$ coke oven gas combustion is $4.19m^3$.

1.3.2 Actual air volume and excess air coefficient

In the actual combustion process, the amount of air supplied ($V_{a,air}$) is often not equal to the theoretical amount of air required for combustion ($V_{0,air}$). The ratio of the theoretical air quantity actually supplied to the theoretical air quantity required for combustion is called the excess air coefficient,

$$V_{a,air}=\alpha\cdot V_{0,air} \quad (1\text{-}22)$$

For 1kg of fuel, the excess air coefficient is usually expressed as:

$$\alpha=\frac{L}{L_0} \quad (1\text{-}22')$$

Where L_0, L respectively represent the theoretical air quantity required for 1kg fuel combustion and the actual air supplied for 1kg fuel combustion.

Therefore, the relationship between actual air demand and theoretical air demand is:

$$L=\alpha\cdot L_0 \quad (1\text{-}23)$$

The value of α is generally between 1 and 2, and the empirical value when the substances in all States are completely burned is: gaseous combustible $\alpha=1.02\sim1.2$; Liquid combustible $\alpha=1.1\sim1.3$; Solid combustible $\alpha=1.3\sim1.7$. See Table 1-3 below for the amount of air required for the combustion of common combustibles.

When $\alpha=1$, indicates that the actual air supply is equal to the theoretical air supply. Theoretically, all combustible substances in the fuel can be oxidized, and the ratio of fuel and oxidant conforms to the equivalent relationship of chemical reaction equation. The ratio of fuel to air at this time is called the chemical equivalent ratio.

When $\alpha>1$, which means that the actual air supply is more than the theoretical air supply. In the actual combustion device, this air supply mode is adopted in most cases, because it can not only save fuel, but also have other beneficial effects.

Whether $\alpha=1$ or $\alpha>1$, the combustion of fuel is complete combustion, and the main difference lies in the composition ratio of products formed after combustion. When $\alpha>1$, after the reaction between fuel and oxidant is completed, some oxidants that did not participate in the reaction remain in the product, which should be paid attention to when analyzing combustion products.

When $\alpha<1$ indicates that the actual air supply is less than the theoretical air supply. This kind of combustion process cannot be complete. There are still combustible substances left in the combustion products, but the oxygen is consumed, which is bound to cause fuel waste. However, in some cases, such as ignition, more fuel is often supplied to make ignition successful. Generally, it should be avoided $\alpha<1$. Air volume required for combustion of common combustible substances is shown in as Table 1-3.

Table 1-3 Air volume required for combustion of common combustible substances

Substance name	Air demand		Substance name	Air demand	
	m^3/m^3	kg/m^3		m^3/m^3	kg/m^3
Acetylene	11. 9	15. 4	Acetone	7. 53	9. 45
Hydrogen	2. 38	3. 00	Benzene	10. 25	13. 20
Carbon monoxide	2. 38	3. 00	Toluene	10. 30	13. 30
Methane	9. 52	21. 30	Petroleum	10. 80	14. 00
Propane	23. 8	30. 60	Gasoline	11. 10	14. 35
Butane	30. 94	40. 00	Kerosene	11. 50	14. 87
Water gas	2. 20	2. 84	Wood	4. 60	5. 84
Coke oven gas	3. 68	4. 76	Dry peat	5. 80	7. 50
Ethylene	14. 28	18. 46	Sulfur	3. 33	4. 30
Propylene	21. 42	27. 70	Phosphorus	4. 30	5. 56
Butene	28. 56	36. 93	Potassium	0. 70	0. 90
Hydrogen sulfide	7. 14	9. 23	Naphthalene	10. 00	12. 93

To sum up, excess air coefficient α it is a parameter indicating the fuel air ratio in the combustible mixture composed of liquid or gaseous fuel and air, and its value has a great influence on the combustion process, α too large or too small is not conducive to combustion.

1. 3. 3 Fuel air ratio and excess air coefficient

In the actual combustion process, the parameters representing the composition ratio of fuel and air in the combustible mixture, except α. In addition, there are fuel air ratio f and excess fuel coefficient β.

1.3.3.1 Fuel air ratio f

The fuel air ratio is the ratio of the actual fuel quantity supplied to the air quantity during the combustion process:

$$f = \frac{\text{Fuel quantity}}{\text{Air quality}} \tag{1-24}$$

It indicates the actual kilogram of fuel per kilogram of air. This parameter is often used for the combustible mixture formed by liquid fuel, which is customarily called "oil gas ratio". According to the definition of fuel air ratio, its excess air coefficient can be obtained α. The relationship is:

$$f = \frac{1}{\alpha L_0} \tag{1-25}$$

For a certain fuel, L_0 is a definite value, so f and α inversely proportional. When $\alpha = 1$, oil-gas ratio $f = 1/L_0$. For general hydrocarbon liquid fuels, such as gasoline, diesel, heavy oil and kerosene, the theoretical air volume L_0 is about 13~14kg. So, when $\alpha = 1$, the corresponding oil-gas ratio is $\frac{1}{14} \sim \frac{1}{13}$.

1.3.3.2 Excess fuel factor β

This definition refers to the ratio of actual fuel supply to theoretical fuel supply. The theoretical fuel quantity refers to the maximum fuel quantity consumed to make 1kg air burn completely, which is the reciprocal of the theoretical air quantity, namely:

$$\text{Theoretical fuel quantity} = \frac{1}{L_0} \tag{1-26}$$

It can be seen that the reciprocal of the actual air volume is $1/(\alpha L_0)$ is the actual fuel quantity, that is, the actual fuel quantity supplied when the combustion consumes 1kg air. Therefore, the excess fuel factor β is:

$$\beta = \frac{1/\alpha L_0}{1/L_0} = \frac{1}{\alpha} \tag{1-27}$$

Obviously, the excess fuel factor β and excess air coefficient α reciprocal. Some gas thermodynamic property data are based on excess fuel coefficient β listed as variables.

1.4 Combustion products and their calculation

The formation of new substances is one of the basic characteristics of combustion reaction. Combustion products are new products of combustion reaction, which have great harmful effects. The calculation of combustion products mainly includes product quantity calculation, product percentage composition calculation and product density calculation. The composition and quantity of combustion products are not only related to the completeness of combustion, but also to the excess air coefficient α. Therefore, it should be divided into complete combustion and incomplete combustion according to specific conditions.

1.4.1 Composition and toxicity of combustion products

1.4.1.1 Composition of combustion products

The gas, liquid and solid substances generated by combustion are called combustion products, which can be divided into complete combustion products and incomplete combustion products. The so-called complete combustion means that in the combustibles, C becomes CO_2(gas), H becomes H_2O (liquid), S becomes SO_2(gas), and N becomes N_2(gas); CO, NH_3, alcohols, ketones, aldehydes and ethers are incomplete combustion products.

Combustion products mainly exist in gaseous form, and their composition mainly depends on the composition of combustibles and combustion conditions. Most combustibles belong to organic compounds, which are mainly composed of carbon, hydrogen, oxygen, nitrogen, sulfur, phosphorus and other elements. Under the condition of sufficient air, the combustion products are mainly complete combustion products, and the amount of incomplete combustion products is very small; If the air is insufficient or the temperature is low, the amount of incomplete combustion products increases relatively.

Under normal conditions, nitrogen does not participate in combustion reaction, but precipitates in free state (N_2). However, under specific conditions, nitrogen can also be oxidized to NO or combined with some intermediate products to form CN and HCN, etc. Table 1-4 lists the common combustibles and their combustion products in building fires.

Table 1-4 Common combustibles and their combustion products in building fires

Combustible	Combustion products
All carbonaceous combustibles	CO_2, CO
Polyurethane, nitrocellulose, etc	NO, NO_2
Sulfur and sulfur containing (rubber) combustibles	SO_2, S_2O_3
Rayon, rubber, carbon disulfide, etc	H_2S
Phosphorous substances	P_2O_5, PH_3
PVC, fluoroplastics, etc	HF, HCl, Cl_2
Nylon, melamine, ammonia plastic, etc	NH_3, HCN
Polystyrene	Benzene
Wool, rayon, etc	Carboxylic acids (formic acid, acetic acid, caproic acid)
Wood, phenolic resin, polyester	Aldehydes and ketones
Thermal decomposition of polymer materials	Hydrocarbons(CH_4, C_2H_2, C_2H_4, etc)

Among the products of combustion, there is a special kind of substance, which is smoke. It is the product of combustion or pyrolysis that is suspended in the atmosphere and can be seen by people. The main components of smoke are some very small carbon black particles, whose diameter is generally between $10^{-7} \sim 10^{-5}$ m. Large diameter particles are easy to fall from the smoke and become smoke or carbon black.

The formation process of carbon particles is very complicated. For example, during the combustion of hydrocarbon combustibles, a series of intermediate products will be produced due to thermal cracking, and the intermediate products will be further cracked into smaller "fragments". These small "fragments" will undergo dehydrogenation, polymerization, cyclization and other reactions, and finally form graphitized carbon particles, forming smoke.

The formation of carbon particles is affected by oxygen supply, the molecular structure of combustibles and the ratio of hydrocarbon in molecules. The oxygen supply is sufficient. The carbon in the combustibles mainly reacts with oxygen to generate CO_2 or CO, and the generation of carbon particles is less, or even no carbon particles; Aromatic organic compounds belong to ring structure, and their carbon generating capacity is higher than that of straight chain aliphatic organic compounds; The higher the ratio of hydrocarbon to hydrocarbon in the combustible molecules, the stronger the carbon generating ability.

1.4.1.2 Toxic effects of combustion products

On the fire site, the existence of combustion products has a great toxic effect, which is mainly reflected in the following aspects:

A Hypoxia and asphyxia

At the scene of a fire, the combustion of combustibles consumes oxygen in the air, which makes the oxygen content in the air much lower than the value required for people's normal physiology, thus causing harm to human body. Table 1-5 lists the harm to human body caused by the decrease of oxygen concentration.

Table 1-5 Harm of oxygen concentration decrease to human body

Oxygen concentration(%)	Hazards to human body
16~12	Breathing and pulse quicken, causing headache
14~9	Decreased judgment, general collapse, cyanosis
10~6	Unconsciousness, causing spasm and death within 6~8 min
6	5 min lethal concentration

Carbon dioxide is the main product of the combustion of many combustibles. Too high CO_2 content in the air will stimulate the respiratory system, accelerate breathing, and produce asphyxia. Table 1-6 lists the effects of different concentrations of CO_2 on human body.

Table 1-6 Effects of different concentrations of CO_2 on human body

CO_2 concentration(%)	Hazards to human body
1~2	Discomfort
3	Respiratory center is stimulated, breathing is accelerated, pulse is accelerated, and blood pressure rises
4	Headache, dizziness, tinnitus, palpitation
5	Dyspnea, poisoning symptoms within 30 min
6	Shortness of breath, in a difficult state
7~10	A few minutes of unconsciousness, purple spots, death

B Toxic, irritant and corrosive effects

Combustion products contain a variety of toxic and irritant gases. In places such as fire rooms, the content of these gases is very easy to exceed the minimum concentration allowed by people′s physiological normal, causing poisoning or irritant hazards. In addition, some products or their aqueous solutions have strong corrosive effects, which will cause human tissue necrosis or chemical burns. The toxic effects of several typical products are introduced below.

(1) Carbon monoxide (CO). This is a very toxic gas, and the poisoning death caused by CO accounts for a large proportion in the fire. This is because it can replace oxygen from the oxygen heme in the blood and combine with heme to form hydroxyl compounds, which makes the blood lose its oxygen delivery function. Table 1-7 lists the effects of different concentrations of CO on human body.

Table 1-7 Effects of different concentrations of CO on human body

CO concentration(%)	Impact on human body
0.04	Mild headache in 2~3h
0.08	Headache and vomiting before 1~2h, headache after 2.5~3h
0.16	Headache, dizziness, vomiting, spasm within 45min, and blindness within 2h
0.32	Headache, dizziness, vomiting and spasm within 20min, and death within 10~15min
0.64	Headache, dizziness, vomiting and spasm within 1~2min, and death within 10~15min
1.28	Death within 1~3min

(2) Sulfur dioxide (SO_2). This is a product released when sulfur-containing combustibles (such as rubber) are burned. SO_2 is toxic. It is a kind of gas with great harm in air pollution. It can irritate people′s eyes and respiratory tract, cause cough and even cause death. At the same time,

SO_2 is easy to form an acidic corrosive solution. Table 1-8 lists the effects of SO_2 with different concentrations on human body.

Table 1-8 Effects of SO_2 with different concentrations on human body

SO_2 content		Impact on human body
%	mg/L	
0.0005	0.0146	No danger after long-term action
0.001~0.002	0.029~0.058	Irritation of trachea, cough
0.005~0.01	0.146~0.293	1h no direct danger
0.05	1.46	Life threatening for a short time

(3) Hydrogen chloride (HCl). HCl is a gas with strong toxicity and irritation. Because it can absorb water in the air and become acid mist, it has strong corrosivity, and will strongly stimulate people's eyes at high concentrations, causing respiratory tract inflammation and pulmonary edema. Table 1-9 lists the effects of different concentrations of HCl on human body.

(4) Hydrogen cyanide (HCN). This is a highly toxic gas, which is mainly the combustion product of protein substances such as polypropylene eye, nylon, silk and hair. HCN can be mixed with water in any proportion to form highly toxic hydrocyanic acid. Table 1-10 lists the effects of different concentrations of HCN on human body.

(5) Oxides of nitrogen. Nitrogen oxides mainly include no and NO_2, which are combustion products of nitrocellulose and other nitrogen-containing organic compounds. Nitric acid and explosive products of nitrates also contain no, NO_2, etc. They are toxic and irritant gases, which can stimulate the respiratory system, cause pulmonary edema and even death. Table 1-11 lists the effects of nitrogen-containing compounds on human body.

In addition, H_2S, P_2O_5, PH_3, Cl_2, HF, NH_3 and other gas products, benzene, hydroxy acid, aldehyde, ketone and other liquid products as well as soot particles also have certain toxicity, irritation and corrosiveness.

Table 1-9 Effects of different concentrations of HCl on human body

HCl concentration	Impact on human body
$(0.5\sim1)\times10^{-4}$	Slightly irritating
5×10^{-4}	Irritant to nasal cavity with unpleasant sensation
10×10^{-4}	Strong irritation to nasal cavity, unable to endure for more than 30min
35×10^{-4}	Short term irritation to throat
50×10^{-4}	Critical concentration tolerated for a short time
1000×10^{-4}	Life threatening

Table 1-10 Effects of HCN of different concentrations on human body

HCN concentration	Impact on human body
$(18\sim36)\times10^{-4}$	Mild poisoning symptoms occurred after several hours
$(45\sim54)\times10^{-4}$	Withstand 0. 5~1h without major injury
$(110\sim125)\times10^{-4}$	Life threatening or fatal within 0. 5~1. 1h
135×10^{-4}	Death within 30min
181×10^{-4}	Death within 10min
270×10^{-4}	Immediate death

Table 1-11 Effects of nitrogen oxides on human body

Nitrogen oxide content		Impact on human body
%	mg/L	
0. 004	0. 019	No obvious reaction after long-term action
0. 006	0. 29	Short term tracheal irritation
0. 01	0. 48	If it irritates the trachea and coughs for a short time, it is life-threatening if it continues to work
0. 025	1. 20	Quick death in a short time

C Thermal damage of high temperature gas

People's tolerance to high temperature environment is limited. Relevant data show that it can be tolerated for a short time at 65℃; Unrecoverable damage will occur in a short time at 120℃; The damage time is shorter when the temperature is further increased. In the fire room, the high-temperature gas can reach several Baidu; In underground buildings, the temperature is above 1000℃. Therefore, the heat damage of high temperature gas to human is very serious.

1. 4. 2 Calculation of product quantity in complete combustion

When the fuel is completely burned, the composition and volume of the flue gas can be obtained from the reaction equation and according to the elemental composition or composition of the fuel. The products involved in the calculation mainly include CO_2, H_2O, SO_2, N_2 and steam. The amount of flue gas generated is also calculated according to the unit amount of fuel. If the combustion is incomplete, there is still O_2 in the residual products, and the volume of flue gas composed of the above substances is:

$$V_P = V_{CO_2} + V_{SO_2} + V_{N_2} + V_{H_2O} + V_{O_2} \tag{1-28}$$

When $\alpha = 1$, there is no more O_2 in the flue gas. This flue gas volume is called the theoretical flue gas volume, which is represented by $V_{0,P}$. Therefore,

$$V_{0,P} = V_{0,CO_2} + V_{0,SO_2} + V_{0,H_2O} + V_{0,N_2} \tag{1-29}$$

Where, V_{0,N_2} and V_{0,H_2O} respectively represent the theoretical nitrogen and theoretical water vapor volume in the flue gas after complete combustion when the theoretical air volume (dry air) is supplied.

1.4.2.1 Calculation of flue gas volume from combustion of solid and liquid fuels

A Volume calculation of carbon dioxide and sulfur dioxide

The known composition of combustibles is $\varphi(C)\% + \varphi(H)\% + \varphi(O)\% + \varphi(N)\% + \varphi(S)\% + \varphi(A)\% + \varphi(W)\% = 100\%$. According to the chemical reaction formula of complete combustion, the quantity relationship of carbon combustion is as follows:

$$C + O_2 = CO_2$$

molecular weight 12 32 44

According to the above formula, when 1kg of carbon is completely burned, $\frac{11}{3}$kg of CO_2 can be generated, which is expressed as the volume in the standard state is $\frac{11}{3} \times \frac{22.4}{44} = \frac{22.4}{12}(m^3)$, so the volume of CO_2 generated when 1kg of combustible is completely burned is:

$$V_{0,CO_2} = \frac{22.4}{12} \times \frac{\varphi(C)}{100} (m^3/kg)$$

Similarly, the volume of SO_2 generated when 1kg combustible is completely burned is:

$$V_{0,SO_2} = \frac{22.4}{32} \times \frac{\varphi(S)}{100} (m^3/kg)$$

B Theoretical nitrogen volume

Theoretical nitrogen includes nitrogen generated by the nitrogen component contained in the fuel and nitrogen substituted by combustion supporting air,

$$V_{0,N_2} = \frac{22.4}{28} \times \frac{\varphi(N)}{100} + 0.79V_{0,air} (m^3/kg)$$

Where 0.79 is the volume fraction of nitrogen in dry air.

C Theoretical volume of water vapor

This part consists of the following two parts:

(1) Water vapor from complete combustion of hydrogen in fuel:

$$V_{0,H_2O} = \frac{22.4}{2} \times \frac{\varphi(H)}{100} (m^3/kg)$$

(2) Water vapor generated after vaporization of water contained in fuel:

$$V_{0,H_2O} = \frac{22.4}{18} \times \frac{\varphi(W)}{100} (m^3/kg)$$

By adding the above two parts, the theoretical amount of water vapor in the flue gas is:

$$V_{0,H_2O} = \left(\frac{22.4}{2} \times \frac{\varphi(H)}{100} + \frac{22.4}{18} \times \frac{\varphi(W)}{100}\right) (m^3/kg)$$

So far, the theoretical flue gas volume is:

$$V_{0,P}=V_{0,CO_2}+V_{0,SO_2}+V_{0,H_2O}+V_{0,N_2}$$
$$=\left(\frac{\varphi(C)}{12}+\frac{\varphi(S)}{32}+\frac{\varphi(H)}{2}+\frac{\varphi(W)}{18}+\frac{\varphi(N)}{28}\right)\times$$
$$\frac{22.4}{100}+\frac{79}{100}\times V_{0,air}(m^3/kg)$$

Generally, the flue gas generated after fuel combustion includes water vapor, which is called "wet flue gas". The flue gas after deducting moisture is called "dry flue gas". Therefore, the theoretical dry flue gas volume $V_{0,yq}$ can be written as:

$$V_{0,yq}=V_{0,CO_2}+V_{0,SO_2}+V_{0,N_2}$$

When $\alpha>1$, the actual amount of air supplied in the combustion process is more than the theoretical amount of air, and the fuel combustion is complete at this time. In addition to the theoretical amount of flue gas, the amount of excess air and the amount of water vapor brought in with the excess air shall be added to the amount of flue gas generated. The amount of water vapor is usually calculated according to the saturated moisture content d (unit: g/kg dry air) at the air temperature.

The excess air volume is:

$$\Delta V_{air}=(\alpha-1)V_{0,air} \tag{1-30}$$

The amount of water vapor carried in is:

$$V'_{0,H_2O}=\frac{\frac{d}{1000}\times\frac{22.4}{18}}{\frac{1}{1.293}}\times V_{\alpha,air}=0.00161dV_{\alpha,air}(m^3/kg) \tag{1-31}$$

(Note: Under standard state, when $T_0=273K$, $p_0=101.3kPa$, the density of dry air with normal composition $\rho_0=1.293kg/m^3$)

Similarly, the water vapor volume can also be deducted from the actual flue gas, and the obtained flue gas volume is called the actual dry flue gas volume, namely:

$$V_{\alpha,yq}=V_{0,yq}+\Delta V_{air}=V_{0,CO_2}+V_{0,SO_2}+V_{\alpha,N_2}+\Delta V_{O_2} \tag{1-32}$$

Where V_{α,N_2} is the actual nitrogen volume in the flue gas: $V_{\alpha,N_2}=V_{0,N_2}+\Delta V_{N_2}=V_{0,N_2}+0.79(\alpha-1)V_{0,air}$. ΔV_{O_2} is the volume of free oxygen in flue gas: $\Delta V_{O_2}=0.21(\alpha-1)V_{0,air}$

So, when $\alpha>1$, the actual dry flue gas volume is:

$$V_{\alpha,yq}=V_{0,yq}+\Delta V_{air}=\left(\frac{\varphi(C)}{12}+\frac{\varphi(S)}{32}+\frac{\varphi(N)}{28}\right)\times$$
$$\frac{22.4}{100}+\frac{79}{100}\times V_{0,air}+(\alpha-1)V_{0,air}(m^3/kg) \tag{1-33}$$

This means that the actual dry flue gas volume is equal to the sum of the theoretical dry flue gas volume and the excess air volume.

1.4.2.2 Calculation of flue gas quantity in gas fuel combustion

For gaseous combustibles, the composition can be expressed as:

$$\varphi(CO)\% + \varphi(H)_2\% + \sum \varphi(C_nH_m)\% + \varphi(H_2S)\% +$$
$$\varphi(CO_2)\% + \varphi(O)_2\% + \varphi(N)_2\% + \varphi(H_2O)\% = 100\% \tag{1-34}$$

According to the chemical reaction equation (1-20) of complete combustion, the volumes of CO_2, SO_2, H_2O and N_2 generated by combustion of combustible per $1m^3$ are respectively:

$$V_{0,CO_2} = \left(\varphi(CO) + \varphi(CO_2) + \sum n\varphi(C_nH_m)\right) \times 10^{-2}(m^3)$$

$$V_{0,SO_2} = H_2S \times 10^{-2}(m^3)$$

$$V_{0,H_2O} = \left(\varphi(H_2) + \varphi(H_2O) + \varphi(H_2S) + \sum \frac{m}{2}\varphi(C_nH_m)\right) \times 10^{-2}(m^3)$$

$$V_{0,N_2} = \varphi(N_2) \times 10^{-2} + 0.79 \times V_{0,air}(m^3)$$

Therefore, the total volume of combustion products is:

$$V_{0,P} = V_{0,CO_2} + V_{0,SO_2} + V_{0,H_2O} + V_{N_2}$$
$$= \Big[\varphi(CO) + \varphi(CO_2) + \varphi(H_2) + 2\varphi(H_2S) + \varphi(H_2O) +$$
$$\varphi(N_2) + \sum\left(n + \frac{m}{2}\right)\varphi(C_nII_m) \times 10^{-2} + 0.79V_{0,air}(m^3)\Big] \tag{1-35}$$

When $\alpha>1$, as for the calculation of solid and liquid fuels, in addition to the theoretical air volume, the excess air volume and the amount of water vapor brought in by this part of air should be added. The analysis method and calculation formula are the same as above.

In addition, the calculation method of theoretical dry flue gas volume and actual dry flue gas volume after combustion of gaseous fuel is the same as that of solid and liquid fuel. The analysis method and calculation formula are the same as above.

1.4.3 Calculation of flue gas volume in incomplete combustion

Having enough oxygen doesn't always make a fuel burn, but based on the complete mixing of fuel and oxygen. Therefore, incomplete combustion may occur at any air coefficient.

(1) When $\alpha>1$, incomplete combustion will also occur due to imperfect combustion equipment, poor mixing of fuel and air and other factors. Therefore, after incomplete combustion, there will still be combustibles and some oxygen in the combustion products.

The combustible substances in incomplete combustion flue gas mainly include CO, H_2 and CH_4. The reaction equation of each mole of these combustible substances burning in air is:

$$CO + 0.5O_2 + 1.88N_2 = CO_2 + 1.88N_2$$
$$H_2 + 0.5O_2 + 1.88N_2 = H_2O + 1.88N_2$$
$$CH_4 + 2O_2 + 7.52N_2 = CO_2 + 2H_2O + 7.52N_2$$

From the above reaction formula, it can be seen that in $\alpha > 1$, the volume of incomplete combustion flue gas is increased by $(0.5V_{CO} + 0.5V_{H_2})$,

$$V_{\alpha,P}^{b} = V_{\alpha,P} + (0.5V_{CO} + 0.5V_{H_2}) \tag{1-36}$$

Where $V_{\alpha,P}^{b}$——incomplete combustion flue gas volume;

$V_{\alpha,P}$——amount of flue gas completely burned.

When calculating the amount of dry flue gas during incomplete combustion, the reduction of moisture generation shall also be considered. According to the above relationship:

$$V^{b}_{\alpha,yq} = V_{\alpha,yq} + (0.5V_{CO} + 1.5V_{H_2} + 2V_{CH_4}) \tag{1-37}$$

Therefore, in the presence of excess air, if incomplete combustion occurs, the volume of flue gas will be larger than that in the case of complete combustion. The more serious the incomplete combustion is, the larger the volume of flue gas will increase.

(2) Stay $\alpha<1$, there are two main cases of incomplete combustion:

1) The fuel and air are evenly mixed, all O_2 is consumed, and CO, H_2, CH_4 and other components are left in the flue gas.

It can be seen from the above reaction equation that the volume of flue gas generated by every $1m^3$ fuel is reduced by $1.88V_{CO} + 1.88V_{H_2} + 9.52V_{CH_4}$.

Which is:

$$V^{b}_{\alpha,P} = V_{0,P} - (1.88V_{CO} + 1.88V_{H_2} + 9.52V_{CH_4}) \tag{1-38}$$

Dry flue gas volume:

$$V^{b}_{\alpha,yq} = V_{0,yq} - (1.88V_{CO} + 0.88V_{H_2} + 7.52V_{CH_4}) \tag{1-39}$$

Therefore, when $\alpha<1$. When all the oxygen in the air is consumed, the amount of flue gas is reduced. The more serious the incomplete combustion is, the more severe the reduction of flue gas is.

2) The oxygen supply is insufficient, and there is incomplete combustion caused by poor mixing of fuel and air, that is, there is free oxygen in the flue gas. Let the volume of this part of oxygen be V_{O_2}, and the equivalent air volume be $V_{O_2}/0.21 = 4.76V_{O_2}$. When free oxygen is not zero, the amount of flue gas generated is:

$$V^{b}_{\alpha,P} = V_{0,P} - (1.88V_{CO} + 1.88V_{H_2} + 9.52V_{CH_4}) + 4.76V_{O_2} \tag{1-40}$$

Therefore, the change of actual flue gas generation depends on ($1.88V_{CO} + 0.88V_{H_2} + 7.52V_{CH_4}$) the difference between $4.76V_{O_2}$. If it is positive, then $V_{0,P}>V^{b}_{\alpha,P}$ otherwise $V_{0,P}<V^{b}_{\alpha,P}$. However, in most cases, the amount of residual oxygen is very small, so the amount of flue gas during incomplete combustion is reduced.

1.5 Calculation of combustion heat

1.5.1 Heat capacity

Heat capacity refers to the amount of heat required for a certain amount of material to rise by one degree without phase change and chemical reaction. If the amount of the substance is unit mole, the heat capacity at this time is called molar heat capacity, which is called heat capacity for short, and the unit is J/(mol · K). If the amount of the substance is 1kg, the heat capacity at this time is called the specific heat, and the unit is J/(kg · K).

1.5.1.1 Constant pressure heat capacity, constant volume heat capacity

Since heat is a path variable and is related to the path, the heat required for the same amount of substances to rise by one degree in the constant pressure process and constant volume process is different. Therefore, the size of constant pressure heat capacity and constant volume heat capacity are different, which are introduced respectively.

A Constant pressure heat capacity

Under the condition of constant pressure, the heat required for a certain amount of material to rise by one degree is called the constant pressure heat capacity, which is expressed by c_p. Assuming that the heat required for n mol substance to rise from T_1 to T_2 under constant pressure is Q_P, it is called constant pressure heat. Then:

$$Q_P = n \cdot \int_{T_1}^{T_2} c_p \mathrm{d}T \tag{1-41}$$

The heat required for each degree of increase of a substance at different temperatures is different, so the heat capacity is a function of temperature. The function form is as follows:

$$\left.\begin{aligned} c_P &= a + bT \\ c_P &= a + bT + cT^2 \\ c_P &= a + bT + cT^2 + dT^3 \\ c_P &= a + bT + c'/T^2 \end{aligned}\right\} \tag{1-42}$$

Where a, b, c, c', dare characteristic constants determined by experiments. The Relationship between constant pressure heat capacity and temperature of some gases is shown as Table 1-12.

Table 1-12 Relationship between constant pressure heat capacity and temperature of some gases
($c_p = a+bT+cT^2$ kJ/(kmol · K))

Gas	a	b	c	Temperature range(K)
Oxygen	28.17	6.297×10^3	-0.7494×10^6	273~3800
Nitrogen	27.32	6.226×10^3	-0.9502×10^6	273~3800
Water vapor	29.16	14.49×10^3	-2.022×10^6	273~3800
Sulfur dioxide	25.76	57.91×10^3	-38.09×10^6	273~1800
Carbon monoxide	26.537	7.6831×10^3	-1.172×10^6	300~1500
Carbon dioxide	26.75	42.258×10^3	-14.25×10^6	300~1500
Hydrogen	26.88	4.347×10^3	-0.3265×10^6	273~3800
Ammonia	27.550	25.627×10^3	-9.9006×10^6	273~1500
Methane	14.15	75.496×10^3	-17.99×10^6	298~1500

B Constant volume heat capacity

Under constant volume conditions, the amount of heat required to raise the temperature of a certain amount of matter by one degree is called the constant volume heat capacity, which is expressed by c_V.

Under the condition of constant pressure, when the temperature of a substance rises, its volume will expand. As a result, the substance will do work on the environment, and its internal energy will increase accordingly. Therefore, when a certain amount of substance rises one degree at the same temperature, the constant pressure process needs to absorb more heat than the constant volume process, that is, c_P is greater than c_V.

For an ideal gas: $c_p - c_V = R$. For solids and liquids, $c_p = c_V$ because the volume does not expand much when the temperature rises. The ratio of the heat capacity at constant pressure to that at constant volume of a gas is called the heat capacity ratio, which is expressed by γ, that is, $\gamma = c_p / c_V$. The heat capacity ratio γ is different for different substances. The heat capacity ratio of air is 1.4.

1.5.1.2 Average heat capacity

A Average heat capacity at constant pressure

Under the condition of constant pressure, when a certain amount of substance rises from temperature T_1 to T_2, the heat required for every 1℃ rise is called the average heat capacity at constant pressure. $\bar{c}_p$ express, in kJ/(m · K). See Table 1-13 for the average heat capacity at constant pressure of various gases.

Although it is accurate to calculate the constant pressure heat Q_P by using the specific functional relationship between heat capacity and temperature, the calculation process is more complex. The average heat capacity at constant pressure is often used in practical calculation. It is related to the heat capacity at constant pressure by

$$\bar{c}_p = \frac{\int_{T_1}^{T_2} c_p \mathrm{d}T}{T_2 - T_1} \tag{1-43}$$

So,

$$Q_P = n \cdot \bar{c}_p (T_2 - T_1) \quad \text{or} \quad Q_P = V \cdot \bar{c}_p (T_2 - T_1) \tag{1-44}$$

Where n——molar mass of the substance;

V——molar volume of the substance.

Table 1-13 Average heat capacity at constant pressure of various gases (kJ/(m^3 · K))

Temperature (K)	Air	CO_2	H_2O	N_2	O_2	SO_2	CO	H_2	CH_4	C_2H_6
273	1.297	1.600	1.494	1.299	1.306	1.733	1.299	1.277	1.548	2.207
373	1.300	1.700	1.505	1.300	1.318	1.813	1.302	1.291	1.640	2.492
473	1.307	1.787	1.522	1.304	1.335	1.888	1.307	1.297	1.757	2.771
573	1.317	1.863	1.542	1.311	1.356	1.955	1.317	1.299	1.884	3.040
673	1.329	1.930	1.565	1.321	1.377	2.018	1.329	1.302	2.013	3.304

Continuted Table 1-13

Temperature (K)	Air	CO_2	H_2O	N_2	O_2	SO_2	CO	H_2	CH_4	C_2H_6
773	1.343	1.989	1.590	1.332	1.393	2.068	1.343	1.305	2.138	2.548
873	1.357	2.041	1.615	1.345	1.417	2.114	1.357	1.308	2.258	3.773
973	1.371	2.088	1.641	1.359	1.434	2.152	1.372	1.312	3.374	3.981
1073	1.384	2.131	1.668	1.372	1.465	2.181	1.386	1.317	2.491	4.176
1173	1.398	2.169	1.696	1.385	1.478	2.215	1.400	1.323	2.599	4.356
1273	1.410	2.204	1.723	1.397	1.478	2.236	1.413	1.329	2.696	4.524
1373	1.421	2.235	1.750	1.407	1.489	2.261	1.425	1.336	2.783	4.218
1473	1.433	2.264	1.777	1.420	1.501	2.278	1.436	1.343	2.859	4.819
1573	1.443	2.394	1.803	1.431	1.511		1.447	1.351		
1673	1.453	2.314	1.828	1.441	1.520		1.457	1.359		
1773	1.462	2.333	1.853	1.450	1.529		1.466	1.367		
1873	1.471	2.355	1.876	1.459	1.528		1.475	1.375		
1973	1.479	2.374	1.900	1.467	1.546		1.483	1.383		
2073	1.487	2.392	1.821	1.475	1.554		1.490	1.392		
2173	1.494	2.407	1.942	1.482	1.562		1.497	1.400		
2273	1.501	2.422	1.963	1.489	1.569		1.504	1.408		
2373	1.507	2.436	1.982	1.496	1.576		1.510	1.415		
2473	1.514	2.448	2.001	1.502	1.583		1.516	1.432		
2573	1.519	2.460	2.019	1.507	1.590		1.521	1.430		
2673	1.525	2.471	2.036	1.513	1.596		1.527	1.437		
2773	1.530	2.481	2.053	1.518	1.603		1.532	1.445		

B Average heat capacity at constant volume

Under the condition of constant volume, when a certain amount of substance rises from t_1 to t_2, the heat required for an average increase of 1℃ is called the average heat capacity at constant volume. $\bar{c}_V$ in J/(mol · K). Table 1-14 shows the calculation formula of the average heat capacity of some gases when the temperature rises from 0℃ to t ℃.

Table 1-14 Calculation formula of average heat capacity of some gases from 0℃ to *t* ℃

Name of the gas	$\bar{c}_V$(kJ/(kmol · K))
Mono-atomic gases (Ar, He, metal vapors and others)	20.84
Diatomic gases (N_2, O_2, H_2, CO, NO)	$20.8 + 0.00288t$
Triatomic gases (CO_2, SO_2)	$37.66 + 0.00243t$
H_2O, H_2S	$16.74 + 0.00900t$
Tetraatomic gases (NH_3 and others)	$41.84 + 0.00188t$
Pentaatomic gases (CH_4 and others)	$50.21 + 0.00188t$

1.5.2 Combustion heat

1.5.2.1 Heat generation

In a chemical reaction, the heat of reaction when a compound is formed by the reaction of a stable simple substance is called the heat of formation of the compound. The isobaric heat of reaction for the formation of 1 mol of a substance from a stable simple substance in a standard state is called the standard heat of formation of the substance. $\Delta H^{\ominus}_{f,298}$ express. Standard heats of formation for some substances are given in Table 1-15. It is clear that the heat of formation of stable simple substances is zero.

Table 1-15 Standard heat of formation of substances (0.1013MPa, 25℃)

Name	Molecular Formula	Status	Heat of formation (kJ/mol)	Name	Molecular Formula	Status	Heat of formation (kJ/mol)
Carbon monoxide	CO	Gas	−110.54	Propane	C_3H_8	Gas	−103.85
Carbon dioxide	CO_2	Gas	−393.51	N-butane	C_4H_{10}	Gas	−124.73
Methane	CH_4	Gas	−74.85	Isobutane	C_4H_{10}	Gas	−131.59
Acetylene	C_2H_2	Gas	226.90	N-pentane	C_5H_{12}	Gas	−146.44
Ethylene	C_2H_4	Gas	52.55	Hexane	C_6H_{14}	Gas	−167.19
Benzene	C_6H_6	Gas	82.93	N-heptane	C_7H_{16}	Gas	−187.82
Benzene	C_6H_6	Liquid	48.04	Propylene	C_3H_6	Gas	20.42
Octane	C_8H_{18}	Gas	−208.45	Formaldehyde	CH_2O	Gas	−113.80
N−octane	C_8H_{18}	Liquid	−249.95	Acetaldehyde	C_2H_4O	Gas	−166.36

Continuted Table 1-15

Name	Molecular Formula	Status	Heat of formation (kJ/mol)	Name	Molecular Formula	Status	Heat of formation (kJ/mol)
N-octane	C_8H_{18}	Gas	-208.45	Methanol	CH_3OH	Liquid	-238.57
Calcium oxide	CaO	Crystal	-635.13	Ethanol	C_2H_6O	Liquid	-277.65
Calcium carbonate	$CaCO_3$	Crystal	-1211.27	Formic acid	CH_2O_2	Liquid	-409.2
Oxygen	O_2	Gas	0	Acetic acid	$C_2H_4O_2$	Liquid	-487.02
Nitrogen	N_2	Gas	0	Oxalic acid	CH_2O_4	Solid	-826.76
Carbon (Graphite)	C	Crystal	0	Carbon tetrachloride	CCl_4	Liquid	-139.33
Carbon (diamond)	C	Crystal	1.88	Aminoacetic acid	C_2HsO_2N	Solid	-528.56
Water	H_2O	Gas	-241.84	Ammonia	NH_3	Gas	-41.02
Water	H_2O	Liquid	-285.85	Hydrogen bromide	HBr	Gas	35.98
Ethane	C_2H_6	Gas	-84.68	Hydrogen iodide	HI	Gas	25.10

1.5.2.2 Heat of reaction

In the process of chemical reaction, the chemical composition of the system changes before and after the reaction, accompanied by the change of energy distribution in the system, which shows the difference between the total energy contained in the products and the total energy contained in the reactants after the reaction. This energy difference is dissipated to or absorbed from the environment in the form of heat, which is the heat of reaction. Its value is equal to the difference between the sum of the product enthalpies and the sum of the reactant enthalpies. The heat of reaction in the standard state is called the standard heat of reaction. $\Delta H^{\ominus}_{R,298}$ express.

1.5.2.3 Heat of combustion

Combustion reaction is a kind of chemical reaction in which combustibles and combustion-supporting substances react to produce stable products, and the reaction heat of this reaction is called combustion heat. The most common combustion-supporting substance is oxygen. In the standard state, the heat of reaction at constant pressure when 1mol of a substance is completely burned is called the standard heat of combustion of the substance. $\Delta H^{\ominus}_{c,298}$ express. Table 1-16 gives the standard heats of combustion for some substances.

Table 1-16 Heat of combustion of some fuels (0.1013MPa, 25℃, Products N_2, H_2O (1) and CO_2)

Name	Molecular Formula	Status	Heat of combustion (kJ/mol)	Name	Molecular Formula	Status	Heat of combustion (kJ/mol)
Carbon (Graphite)	C	Solid	−392.88	Methanol	CH_3OH	Liquid	−712.95
Hydrogen	H_2	Gas	−285.77	Benzene	C_6H_6	Liquid	−3273.14
Carbon monoxide	CO	Gas	−282.84	Cycloheptane	C_7H_{14}	Liquid	−4549.26
Methane	CH_4	Gas	−881.99	Cyclopentane	C_5H_{10}	Liquid	−3278.59
Ethane	C_2H_6	Gase	−1541.39	Acetic acid	$C_2H_4O_2$	Liquid	−876.13
Propane	C_3H_8	Gase	−2201.61	Benzoic acid	$C_7H_6O_2$	Solid	−3226.7
Butane	C_4H_{10}	Liquid	−2870.64	Ethyl acetate	$C_4H_8O_2$	Liquid	−2246.39
Pentane	C_5H_{12}	Liquid	−3486.95	Naphthalene	$C_{10}H_8$	Solid	−5155.94
Heptane	C_7H_{16}	Liquid	−4811.18	Sucrose	$C_{12}H_{22}O_{11}$	Solid	−5646.73
Octane	C_8H_{18}	Liquid	−5450.50	Cidone	$C_{10}H_{16}O$	Solid	−5903.62
Dodecane	$C_{12}H_{26}$	Liquid	−8132.43	Toluene	C_7H_8	Liquid	−3908.69
Hexadecane	$C_{16}H_{34}$	Solid	−1070.69	Monomethyl benzene	C_8Hg	Liquid	−4567.67
Ethylene	C_2H_4	Gas	−1411.26	Urethane	$C_5H_7NO_2$	Solid	−1661.88
Ethanole	C_2H_5OH	Liquid	−1370.94	Styrene	C_8H_8	Liquid	−4381.09

1.5.2.4 Calculation of heat of combustion

When the constant pressure or constant volume is maintained in the whole chemical reaction process and the system does not do any non-volume work, the heat of chemical reaction only depends on the beginning and final state of the reaction, and has nothing to do with the specific way of the process. This law is called Hess's law, which is a very important law in thermochemistry. According to Hess's law, the isobaric heat of reaction of any reaction is equal to the sum of the heats of formation of the products minus the sum of the heats of formation of the reactants, that is,

$$Q_P = \Delta H = \left(\sum \varphi_i \Delta H^{\ominus}_{f,298,i}\right)_{Product} - \left(\sum \varphi_i \Delta H^{\ominus}_{f,298,i}\right)_{Reactant} \quad (1\text{-}45)$$

The standard heat of combustion of a substance can be obtained from the above formula, where V_i is the coefficient of component i in the reaction formula.

The heat of combustion of a gaseous mixture can be roughly calculated using the following equation,

$$\Delta H^{\ominus}_{c,m} = \sum V_i \Delta H^{\ominus}_{c,m,i} \quad (1\text{-}46)$$

Where V_i——volume percent of component i in the mixture;

$\Delta H^{\ominus}_{c,m,i}$——heat of combustion of component i, kJ/mol.

[Example 1-3] Determined the standard heat of combustion of ethanol at 25 degree.

The combustion reaction equation of ethanol is:

$$C_2H_5OH(1) + 3O_2(g) \longrightarrow CO_2(g) + 3H_2O(1)$$

From Table 1-15:

$$\Delta H^{\ominus}_{f,298,CO_2(g)} = -393.1 kJ/mol \qquad \Delta H^{\ominus}_{f,298,H_2O(1)} = -285.85 kJ/mol$$

$$\Delta H^{\ominus}_{f,298,C_2H_5OH(1)} = -277.65 kJ/mol \quad \Delta H^{\ominus}_{f,298,O_2(g)} = 0$$

Using the above formula:

$$Q_P = \Delta H^{\ominus}_{298} = [2 \times (-395.51) + 3 \times (-285.85)] - [1 \times (-277.65) + 0]$$
$$= -1366.8(kJ/mol)$$

According to the definition of standard combustion heat, the standard combustion heat of ethanol is 1366.8 kJ/mol.

[Example 1-4] Determine the standard heat of combustion of coke oven gas. The composition of the coke oven gas in percentage by volume is as follows: $\varphi(CO)6.8\%$, $\varphi(H_2)57\%$, $\varphi(CH_4)$ 22.5%, $\varphi(C_2H_4)3.7\%$, $\varphi(CO_2)$ 2.3%, $\varphi(N_2)$ 4.7% and $\varphi(H_2O)3\%$.

[Explanation] According to Table 1-16, the standard combustion heat of each combustible component in the gas is:

$$\Delta H^{\ominus}_{c,298,CO(g)} = -282.84 kJ/mol$$
$$\Delta H^{\ominus}_{c,298,H_2(g)} = -285.77 kJ/mol$$
$$\Delta H^{\ominus}_{c,298,CH_4(g)} = -881.99 kJ/mol$$
$$\Delta H^{\ominus}_{c,298,C_2H_4(g)} = -1411.26 kJ/mol$$

From the above formula, the standard combustion heat of the gas is:

$$Q_P = \Delta H^{\ominus}_{298} = -282.84 \times 0.068 - 285.77 \times 0.57 - 881.99 \times 0.225 - 1411.26 \times 0.037$$
$$= -432.79(kJ/mol)$$

The standard combustion heat of coke oven gas is 432.79kJ/mol.

1.5.2.5 Calculation of calorific value

Calorific value is another representation of the heat of combustion and is commonly used in practice. The so-called calorific value refers to the heat generated by the complete combustion of combustibles per unit mass or volume, which is usually expressed by Q. Mass calorific value Q_m (kJ/kg) for liquid and solid combustibles; for gaseous combustibles , expressed as the volumetric calorific value Q_v(kJ/m^3).

The heat released by the combustion of some substances can be expressed by the heat of combustion or the calorific value. The conversion relationship between the two is as follows:

For liquid and solid combustibles:

$$Q_m = \frac{1000 \cdot \Delta H_C}{M}(kJ/kg) \tag{1-47}$$

For gaseous combustibles:

$$Q_V = \frac{1000 \cdot \Delta H_C}{22.4} (\text{kJ/m}^3) \quad (1\text{-}48)$$

Where M——molar mass of liquid or solid combustible, g/mol.

It is worth noting that if the combustible contains moisture and hydrogen, the calorific value can be high or low. The high calorific value Q_H is the calorific value of the water generated by the combustion of water and hydrogen in the combustibles in liquid state, while the low calorified value Q_L is the calorified value of the water generated by the combustion of water and hydrogen in the combustibles in gaseous state. See Table 1-17 for the commonly used low calorific values in the study of fire combustion.

Table 1-17 Lower calorific values of some gases

Combustible gas	Molecular Formula	Lower heating value (kJ/m³)	Combustible gas	Molecular Formula	Lower heating value (kJ/m³)
Hydrogen	H_2	10780	N-butane	n-C_4H_{10}	123495
Carbon monoxide	CO	12628	Isobutane	j-C_4H_{10}	122706
Methane	CH_4	35861	Pentene	C_5H_{10}	148652
Acetylene	C_2H_2	56418	N-pentane	C_5H_{12}	156538
Ethylene	C_2H_4	50408	Benzene	C_6H_6	155576
Ethane	C_2H_6	64317	Propane	C_3H_8	93128
Propylene	C_3H_6	87559	Hydrogen Sulfide	H_2S	23354
Butylene	C_4H_8	117548			

At present, for combustibles whose molecular structure is very complex and molar mass is difficult to determine, such as oil, coal, wood and so on, the heat released by their combustion is generally only expressed by calorific value, and usually calculated by empirical formula. The most commonly used is Mendeleev's formula:

$$Q_H = 4.18 \times [81 \times w(C) + 300 \times w(H) - 26 \times (w(O) - w(S))] \times 10^2 (\text{kJ/kg}) \quad (1\text{-}49)$$

$$Q_L = Q_H - 6 \times (9w(H) + w(W)) \times 4.18 \times 10^2 (\text{kJ/kg}) \quad (1\text{-}50)$$

In the formula, $w(C)$, $w(H)$, $w(S)$ and $w(W)$ are respectively the mass percentage contents of carbon, hydrogen, sulfur and water in the combustibles; $w(O)$ is the total mass percent of oxygen and nitrogen in the combustible.

【Example 1-5】 Find the high and low calorific values of 4kg wood combustion: $w(C)43\%$, $w(H)7\%$, $w(O)41\%$, $w(W)6\%$, $w(A)1\%$.

【Solution】 From the formula (1-49), the high and low calorific values of 1kg wood combustion are:

$$\begin{aligned} Q_H &= 4.18 \times [81 \times w(C) + 300 \times w(H) - 26 \times (w(O) - w(S))] \\ &= 4.18 \times [81 \times 43 + 300 \times 7 - 26 \times (41 - 0)] = 18881 (\text{kJ/kg}) \end{aligned}$$

$Q_L = Q_H - 6 \times (9H + W) \times 4.18 = 18881 - 6 \times (9 \times 7 + 6) \times 4.18 = 17150(kJ/kg)$

Then the high calorific value of 4 kg wood combustion is: $18881 \times 4 = 75524(kJ)$

Lower calorific value is: $17150 \times 4 = 68602(kJ)$

1.6 Calculation of combustion temperature

1.6.1 Classification of combustion temperature

The temperature of the flue gas produced by the combustion of combustibles is called the combustion temperature of combustibles. In the actual building fire, the high temperature gas in the burning room can reach hundreds of degrees, and in the underground building, the temperature is as high as more than 1000 degrees. Therefore, it is of great practical significance to study the smoke temperature in fire.

Combustion temperature is classified according to different conditions, which can be divided into theoretical combustion temperature, calorimeter combustion temperature, theoretical heating temperature and actual combustion temperature.

If the combustion is complete under adiabatic conditions and the work exchange between the system and the outside world is not considered, the temperature obtained at this time is called the theoretical combustion temperature of the combustible. On the basis of the theoretical combustion temperature, if the high temperature dissociation of combustion products is not considered, the temperature obtained at this time is called the calorimeter combustion temperature. If the combustion is carried out under the condition of complete combustion with $\alpha = 1$, and the initial temperature of combustible and air is 0℃, the temperature obtained at this time is called the theoretical heating temperature. The temperature measured in an actual fire is called the actual combustion temperature.

Theoretically, the combustion temperature is the highest under the condition of $\alpha = 1$ and complete combustion; When $\alpha<1$, the combustion is incomplete due to excessive fuel, so that the chemical energy of the fuel can not be fully released, thus reducing the combustion temperature; When $\alpha>1$, the amount of air supplied is excessive, and the amount of heat released by the fuel is basically a definite value, so the combustion temperature should also be reduced.

1.6.2 Calculation of combustion temperature

According to the theory of heat balance, combined with the formula: $Q_P = n \cdot \int_{T_1}^{T_2} c_p dT$, the formula for calculating the theoretical combustion temperature can be obtained as follows:

$$Q_l = \sum n_i \cdot \int_{298}^{T} c_{pi} dT \quad (1\text{-}51)$$

Where, Q_l is the lower heating value of the combustible substance; n_i is the number of kilomoles of the i-th product; and c_{pi} is the heat capacity at constant pressure of the i-th product.

The result of the above method is more accurate, but the result of the above integral is a cubic equation, so it is more troublesome to get a specific solution. For this purpose, the average heat capacity at constant pressure is used $\bar{c}_{pi}$. The formula for solving the combustion temperature is obtained as follows:

$$Q_1 = \sum V_i \cdot \bar{c}_{pi} \cdot (T - 298) \tag{1-52}$$

Or

$$Q_1 = \sum V_i \cdot \bar{c}_{pi} \cdot (t - 25) \tag{1-52'}$$

Where, V_i is the volume of the i-th product.

Since the average heat capacity of the product at constant pressure $\bar{c}_{pi}$ depends on the temperature, while the theoretical combustion temperature t is unknown, so $\bar{c}_{pi}$ also an undetermined quantity. In the specific calculation, a theoretical combustion temperature t_1 is usually assumed first, and the corresponding value is found from the "average heat capacity at constant pressure" table. $\bar{c}_{pi}$ substitute the above formula to find the corresponding Q_{l_1}; then assume a second theoretical combustion temperature T_2, find the corresponding $\bar{c}_{pi}$ and Q_{l_2}. Finally, the theoretical combustion temperature t is obtained by interpolation,

$$t = t_1 + \frac{t_2 - t_1}{Q_{l_2} - Q_{l_1}} \cdot (Q_1 - Q_{l_1}) \tag{1-53}$$

Generally, for the convenience of calculation, it is assumed that the initial temperature of combustibles and air before combustion is 0℃, then the above formula becomes

$$Q_1 = \sum V_i \cdot \bar{c}_{pi} \cdot t \tag{1-54}$$

Table 1-18 gives the combustion temperatures for some substances.

Table 1-18 Combustion temperature of certain substances

Name of substance	Combustion temperature (℃)	Name of substance	Combustion temperature (℃)	Name of substance	Combustion temperature (℃)
Methane	1800	Acetone	1000	Sodium	1400
Ethane	1895	Ether	2861	Paraffin	1427
Propane	1977	Crude oil	1100	Carbon monoxide	1680
Butane	1982	Gasoline	1200	Sulphur	1820
Pentane	1977	Kerosene	700~1030	Carbon disulfide	2195
Hexane	1965	Heavy oil	1000	Liquefied gas	2110
Benzene	2032	Bituminous-coal	1647	Natural gas	2020
Toluene	2071	Hydrogen	2130	Petroleum gas	2120
Acetylene	2127	Gas	1600~1850	Phosphorus	900
Methanol	1100	Wood	1000~1177	Ammonia	700
Ethanol	1180	Magnesium	3000	—	—

Assuming t_1 = 1900℃, the average heat capacities at constant pressure of CO_2, H_2O and N_2 are respectively as follows from Table 1-13:

$\bar{c}_{p,CO_2} = 2.407$kJ/(m^3 · K) $\bar{c}_{p,H_2O} = 1.942$kJ/(m^3 · K) $\bar{c}_{p,N_2} = 1.482$kJ/(m^3 · K)

Substitute the above data into the formula (1-54) to obtain:

$$Q_{1_1} = 1900 \times (0.803 \times 2.407 + 0.856 \times 1.942 + 3.433 \times 1.482) = 16497\ (\text{kJ})$$

Since $Q_1 > Q_n$, $t > t_1$. If t_2 = 2000℃, the corresponding average heat capacities at constant pressure in Table 1-13 are:

$\bar{c}_{p,CO_2} = 2.422$kJ/(m^3 · K) $\bar{c}_{p,H_2O} = 1.963$kJ/(m^3 · K) $\bar{c}_{p,N_2} = 1.489$kJ/(m^3 · K)

Substitute the above data into the formula (1-54):

$$Q_{1_2} = 2000 \times (0.803 \times 2.422 + 0.856 \times 1.963 + 3.433 \times 1.489) = 17474\ (\text{kJ})$$

Since $Q_{11} < Q_1 < Q_{12}$, $t_1 < t < t_2$, the theoretical combustion temperature of wood is obtained by using the formula (1-53):

$$t = 1900 + \frac{2000 - 1900}{17474 - 16497} \times (16933 - 16497) = 1945(℃)$$

That is, the theoretical combustion temperature of wood is 1945℃.

1.7 Chemical process of fire

Fire is one of the most frequent and devastating disasters in all kinds of disasters. The direct economic loss caused by fire is second only to drought and flood, and its frequency ranks first in all kinds of disasters. Especially in recent years, due to high-rise buildings, large-scale and complex uses, the factors of fire occurrence are also increasing, and the scale of fire is also expanding day by day. It is important to study the chemical process in fire for making reasonable fire protection measures and building structural buildings with strong fire protection ability.

1.7.1 Fire spreading speed

Combustion is a complex chemical reaction. A large number of experiments have proved that the reaction temperature has a great influence on the chemical reaction rate, and the common situation is that the reaction rate increases with the increase of temperature. By Arrhenius law:

$$K = K_0 \exp\left(-\frac{E}{RT}\right) \tag{1-55}$$

It can be seen that after the room is on fire, the indoor temperature increases, and the burning speed is constantly accelerating. When the temperature reaches about 250℃, the window glass expands and deforms, but limited by the window frame, the glass breaks by itself, and the flame escapes from the window and spreads outward. First, the heat radiation of the flame passes through the window and scorches the opposite building. Second, the flames burn directly to the eaves of the fire on the upper floor, so that they spread upward layer by layer, causing the whole building to catch fire.

When a fire breaks out in a building, the general indoor furniture is burned down. The furniture is generally wood, and the combustibles in wood are mainly carbon, so the burning of furniture can be considered to be basically the burning of carbon. The combustion of carbon is a heterogeneous reaction characterized by the fact that the material actually begins in a condensed phase on the surface of the solid and liquid and ends in a gas-phase flame, under the action of fire source, it is first evaporated into vapor, then the vapor is oxidized and decomposed, and then burned in the gas phase. Such as wood, when heated, it first decomposes, releasing gaseous and liquid products, and then the vapors of gaseous and liquid products are oxidized and burned. The combustion reaction rate equation mentioned above is an approximate formula derived from gaseous substances. In the actual fire scene, because most of the combustion is solid or liquid, if this formula is used to calculate, it is far from the actual situation.

The out-of-phase burning rate is different from the in-phase burning rate. The out-of-phase combustion rate refers to the amount of substance initiating the combustion reaction on a unit surface per unit time ($kg/(m^2 \cdot s)$), while the in-phase combustion rate refers to the amount of substance initiating the combustion reaction in a volume per unit time ($kg/(m^3 \cdot s)$). The overall speed of out-of-phase combustion is the speed at which the entire out-of-phase process progresses and depends on the speed of the slowest stage of the out-of-phase process. As with the in-phase reaction, the rate of out-of-phase combustion can be expressed in terms of either carbon or oxygen consumption.

During the reaction, the consumption of oxygen obeys the mass action law. Both the experimental results and theoretical analysis show that the heterogeneous chemical reactions on the solid surface are 0~1 order reactions, that is:

$$V_m = kC^n \tag{1-56}$$

Due to the complexity of carbon combustion, it is often impossible to consider all the influencing factors at the same time in the study, so different degrees of simplification are often used. It is generally assumed that the chemical reaction on the carbon surface is a first-order reaction, and the amount of oxygen consumed by the reaction during combustion should be:

$$V_m^{O_2} = kC_b \tag{1-57}$$

Where C_b——oxygen concentration on the carbon surface.

1.7.2 Fire products

When a house is on fire, the general combustion substance is carbon, and the general combustion equation of carbon is:

$$C + O_2 \longrightarrow CO_2 \tag{1-58}$$

$$2C + O_2 \longrightarrow 2CO \tag{1-59}$$

The reaction mechanism is as follows:

(1) Primary reaction: $x(CO_2)/x(CO) = 1$ at temperature<1200~1300℃

$$4C + 3O_2 \longrightarrow 2CO_2 + 2CO \tag{1-60}$$

Or $x(CO_2)/x(CO) = 1/2$ when the temperature is higher than 1500~1600℃

$$3C + 2O_2 \longrightarrow CO_2 + 2CO \tag{1-61}$$

(2) Secondary reaction:

Heterogeneous reaction

$$C + CO_2 \longrightarrow 2CO \tag{1-62}$$

In-phase reaction

$$CO + O_2 \longrightarrow 2CO_2 \tag{1-63}$$

When water vapor is present, the following reactions also occur:

$$C + H_2O \longrightarrow CO + H_2 \tag{1-64}$$

In the combustion process, the above reactions cross at the same time, which speed is high and which speed is weak can be ignored, depending on the situation.

When the actual house is on fire, because the combustion occurs in a limited space, and the combustion reaction time is fast, the air is not enough, the general combustibles are not completely burned, and the products are mainly CO or carbon particles, so many casualties in the fire are caused by poisoning or asphyxiation, and generally a large number of carbon black is accumulated on the walls of the house where the fire occurs. It is also caused by incomplete combustion.

1.7.3 Fire temperature

In the previous section, we have obtained the calculation formula of theoretical combustion temperature, but when the actual house is on fire, there are many kinds of indoor combustibles, generally furniture, clothing, curtains, etc., so in the actual fire, the calculation of combustion temperature becomes very complex, simply using the temperature calculation formula mentioned above, only the theoretical combustion temperature of a single substance can be calculated. The actual indoor temperature should also take into account various factors in order to obtain the contribution rate of each substance to the temperature.

The calculation of actual temperature is of great significance for the reliability assessment of buildings after fire. Relevant scholars have calculated the strength characteristics of reinforced concrete after fire by simulating the temperature field. Readers can search for relevant information to read.

Exercises

1. Explain the following basic concepts:

   (1) Combustion; (2) Fire; (3) Smoke; (4) Heat capacity; (5) Heat of formation; (6) Standard heat of combustion; (7) Calorific value; (8) Lower calorific value.

2. What is the nature of combustion? What are its characteristics? Illustrate these features.
3. How to correctly understand the conditions of combustion ? What fire prevention and extinguishing methods can be proposed based on the combustion conditions?
4. Try to find outtheoretical amount of air required for complete combustion of 1 kg benzene (C_6H_6) at p = 1 atm and T = 273K.

5. The composition of the wood is $w(C)$ 48%, $w(H)$ 5%, $w(O)$ 40%, $w(N)$ 2%, $w(W)$ 5%. Try to find the actual required air volume, actual smoke volume and smoke density when burning 5kg of this kind of wood under the conditions of 1.5atm and 30℃ (The air consumption coefficient is 1.5).
6. The composition of the gas is known to be: $\varphi(C_2H_4)$ 4.8%, $\varphi(H_2)$ 37.2%, $\varphi(CH_4)$ 26.7%, $\varphi(C_3H_6)$ 1.3%, $\varphi(CO)$ 4.6%, $\varphi(CO_2)$ 10.7%, $\varphi(N_2)$ 12.7%, $\varphi(O_2)$ 2.0%, assuming that $p = 1$atm, $T = 273$K, air is in a dry state, and $1m^3$ gas is burned:
   (1) What is the theoretical air volume in m^3?
   (2) What are the various combustion products volume in m^3?
   (3) What is the total combustion product volume in m^3?
7. The composition of wood is known to be: $w(C)$ 43%, $w(H)$ 7%, $w(O)$ 41%, $w(N)$ 2%, $w(W)$ 7%. Calculate the calorific value of 5kg wood at 25℃.
8. What is the heat Q_P required for 1000 kg methane from 260℃ to 538℃ at normal pressure?

Chapter 2 Physical Basis of Combustion and Explosion

扫码获得
数字资源

Other processes in combustion and explosion arethe transferring of heat and mass, which constitutes the physical basis of combustion and explosion. The engineering fields of thermal science include thermodynamics and heat transfer. The role of heat transfer is to supplement thermodynamic analysis with laws that predict the rates of energy transfer, since the latter deals only with systems in equilibrium. These additional laws are based on three basic modes of heat transfer, namely conduction, convection, and radiation.

In terms of heat conduction, the flat plate heat conduction experiment obtained by French physicist Biot in 1804 was the earliest expression of the law of heat conduction. Later, Fourier from France used mathematical methods to express it more accurately as what was later called the differential forms of Fourier's law.

In the field of thermal convection, when Newton estimated the temperature of red-hot iron rods in 1701, he brought forward what was later recognised as Newton's law of cooling. However, it did not reveal the mechanism of convective heat transfer. German physicist Prandtl's boundary-layer theory in 1904 and Nusselt's dimensional analysis in 1915 laid the foundation for the correct understanding and quantitative analysis of convective heat transfer in both theoretical and experimental fields.

In terms of thermal radiation, in 1860, Kirchhoff demonstrated that the emissivity (blackness) of a Black Body was the highest at the same temperature by simulating an absolute Black Body with an artificial cavity, and pointed out that the emissivity of an object was equal to the absorption rate of the object at the same temperature, which was called Kirchhoff's law by later generations. In 1878, Stefan discovered the fact that the radiation rate is proportional to the absolute temperature to the power of four. In 1884, it was theoretically proved by Boltzmann, which was called Stefan-Boltzmann law, commonly known as the fourth power law. In 1900, when Planck studied the cavity Black Body radiation, he discovered Planck's law of thermal radiation. This law not only described the relationship between Black Body radiation, temperature and frequency, but also demonstrated Wien displacement law of Black Body energy distribution.

2.1 Heat conduction

According to the heat transfer theory, if there is a temperature gradient inside objects, heat will be transferred from high temperature areas to a low temperature areas, which is heat conduction. Heat conduction is a heat-transferring phenomenon mainly associated with solids. Although it also occurs in liquids, its occurence is often masked by heat convection. Solids conduct thermal energy in two forms: free electron migration and lattice vibrations. For a good electrical conductor, there are a large number of free electrons moving between lattice structures, which can transfer heat energy from high temperature region to low temperature region. Because of the lack of free electrons, heat can only be transmitted by mechanical vibrations of the material's lattice, which usually transmit less energy than free electrons.

2.1.1 Fourier's law of heat conduction

The heat conduction obeys the Fourier law, that is, in the non-uniform temperature field, the heat flux density at a certain place formed by heat conduction is proportional to the temperature gradient at the sametime and place. In one-dimensional temperature fields, its mathematical expression is:

$$q_x = -k\frac{\mathrm{d}T}{\mathrm{d}x} \tag{2-1}$$

Where q_x——heat flux, the amount of heat transferred per unit area per unit time, also known as heat flux density, in $\mathrm{J/(m^2 \cdot s)}$;

$\frac{\mathrm{d}T}{\mathrm{d}x}$——temperature gradient in the x-axis direction, in K/m, with a negative sign indicating that the direction of heat flow is opposite to the direction of temperature increase;

k——thermal conductivity, in $\mathrm{W/(m \cdot K)}$.

The thermal conductivity k represents the ability of a substance to conduct heat, that is, the heat flux per unit temperature gradient. Different thermal conductive materials have different thermal conductivity, and the thermal conductivity of the same material will also change with the change of materials' structure, density, humidity, temperature and other factors.

2.1.2 Differential equation of heat conduction

An object whose temperature at each point changes with time while it is being heated or cooled; in periodic heating and cooling, the temperature change at each point is also periodic. Both of the above processes belong to the astable heat conduction process. If we can make one side of the object heated and the other side cooled, and make the temperature of each space point in the object unchanged with time, then we make the steady-state heat conduction process.

The task of heat conduction theory is to study the temperature change law of each point in the object at each time, and to establish and solve the temperature field in the object with mathematical expression, which is $T = f(x, y, z, t)$.

The differential equation of heat conduction is the equation that describes the relationship between the temperature of each point in an object and its spatial and time coordinates. The mathematical solution of the temperature field of a body can be obtained by integrating the differential equation of heat conduction with the boundary conditions and time conditions (usually the initial conditions) of the body.

Assuming that:

(1) The object under study is an isotropic continuum.

(2) Thermal conductivity, specific heat capacity and density are all known constants.

(3) There is an internal thermal source in the object with following features: its intensity is q_v (W/m^3) which indicates the amount of heat released per unit volume of heat conductor per unit time, and this internal heat source is evenly distributed.

When there is an exothermic heat source or an endothermic heat source in the infinitesimal body of an object, the heat release of the infinitesimal body in dt time is:

$$dQ' = q' \cdot dx \cdot dy \cdot dz \cdot dt \tag{2-2}$$

Where, q' is the heat release per unit volume per unit time, $J/(m^3 \cdot s)$ (at the time if heat is released q' positive, when heat is absorbed q' is negative). The differential equation for heat conduction in each case is as follows:

(1) The unsteady heat conduction differential equation of the three-dimensional temperature field for a homogeneous and isotropic object with an internal heat source is:

$$\frac{\partial T}{\partial t} = \alpha\left(\frac{\partial^2 T}{\partial x^2} + \frac{\partial^2 T}{\partial y^2} + \frac{\partial^2 T}{\partial z^2}\right) + q'/(c\rho) \tag{2-3}$$

Where, $\alpha = k/\rho c$ it is called the coefficient of thermal diffusivity (a physical parameter).

(2) The unsteady heat conduction differential equation of three-dimensional temperature field without internal heat source is:

$$\frac{\partial T}{\partial t} = \alpha\left(\frac{\partial^2 T}{\partial x^2} + \frac{\partial^2 T}{\partial y^2} + \frac{\partial^2 T}{\partial z^2}\right) \tag{2-4}$$

(3) The steady-state heat conduction differential equation of three-dimensional temperature field with internal heat source is

$$\alpha\left(\frac{\partial^2 T}{\partial x^2} + \frac{\partial^2 T}{\partial y^2} + \frac{\partial^2 T}{\partial z^2}\right) + q/(c\rho) = 0 \tag{2-5}$$

(4) The steady-state heat conduction differential equation of three-dimensional temperature field without internal heat source is:

$$\alpha\left(\frac{\partial^2 T}{\partial x^2} + \frac{\partial^2 T}{\partial y^2} + \frac{\partial^2 T}{\partial z^2}\right) = 0 \tag{2-6}$$

(5) The unsteady heat conduction differential equation of one-dimensional temperature field without internal heat source is:

$$\frac{\partial T}{\partial t} = \alpha \frac{\partial^2 T}{\partial x^2} \tag{2-7}$$

(6) The steady-state heat conduction differential equation of one-dimensional temperature field without internal heat source is:

$$\frac{\partial^2 T}{\partial x^2} = 0 \tag{2-8}$$

This is the simplest differential equation for heat conduction.

The coefficient of thermal diffusivity, α, characterizes the ability of an object to be heated or cooled with uniform temperatures at all points within it. When an object with a large α is heated, the temperature everywhere tends to be uniform more quickly.

In order to solve the heat conduction differential equation, for the unsteady heat conduction problem, the definite solution condition has two aspects: one is to give the initial condition of the temperature distribution at the initial time; the other is to give the boundary condition of the temperature or heat transfer on the boundary of the heat-conducting object. The heat conduction differential equation and its boundary conditions constitute a complete mathematical description of a specific heat conduction problem. However, for the steady-state heat conduction problem, the definite solution condition has no initial condition, only the boundary condition.

Common boundary conditions are divided into three categories: a given temperature value on the boundary, called the first type of boundary conditions; a given heat flux value on the boundary, called the second boundary condition; the given heat transfer coefficient between the object on the boundary and the surrounding fluid, and the temperature of the surrounding fluid, which are called the third boundary condition.

Finally, the scope of application of the differential equation of heat conduction should be explained. For the unsteady heat conduction problems in general engineering technology, generally the heat flux density is not very high while the temperature is not very low, and the duration of the process is quite long, so the Fourier law and the unsteady heat conduction equation are fully applicable.

The solution of unsteady heat conduction problem is far more complicated than that of steady heat conduction problem, so in general engineering, if only an approximate estimation of heat conduction is needed, the average temperature value can be used, and we can solve them the same as the steady-state heat conduction problem. Normally, if it is not specified as an unsteady condition, it is generally referred to as a steady-state heat transfer problem.

2.1.3 Unsteady heat conduction

Fire is a transient process. Not only the analysis process of fire, such as ignition and spread, but also the overall process of fire, such as the response of buildings to developing and fully developed fires, must be described by unsteady heat transfer equations. The basic equation of unsteady heat conduction is established by considering the flow of heat through a tiny volume element dxdydz and applying the principle of heat balance to it. For a one-dimensional problem, the differential form is:

$$\frac{\partial^2 T}{\partial x^2} = \frac{1}{\alpha}\frac{\partial T}{\partial t} + \frac{q'}{k} \tag{2-9}$$

Where, α is the coefficient of thermal conductivity, $q'$is the heat release rate per unit volume. For most problems $q' = 0$ for solid combustibles, q' can be positive or negative, with "positive" indicating exothermicity and "negative" indicating water evaporation and pyrolysis.

Equation (2-9) can be directly used to solve the heat conduction problem of an infinite flat plate and an infinite-sized solid and to infer the theoretical solution of the one-dimensional unsteady heat conduction problem. Some geometrically symmetric problems can be transformed into one-dimensional problems by simple coordinate transformation (such as polar coordinates or cylindrical coordinates). As shown in Fig. 2-1 below, consider an infinite flat plate with a thickness of $2L$ and an internal initial temperature of T_0, with both sides exposed to $T=T_\infty$ in the environment of, let $\theta = T - T_\infty$ then equaction (2-9) can be reduced to (without internal heat source):

$$\frac{\partial^2\theta}{\partial x^2} = \frac{1}{\alpha}\frac{\partial\theta}{\partial t} \tag{2-10}$$

Initial conditions: $\theta = \theta_0 = T - T_\infty (t = 0)$

Boundary conditions: $\left(\frac{\partial\theta}{\partial x}\right) = 0(x = 0)$

$$\frac{\partial\theta}{\partial x} = h\theta/k \quad (x = \pm L, \text{ convective boundary condition})$$

The solution is:

$$\frac{\theta}{\theta_0} = 2\sum_{n=1}^{\infty}\frac{\sin\lambda_n L}{\lambda_n L + (\sin\lambda_n L)\cdot(\cos\lambda_n L)}\exp(-\lambda_n^2\alpha t)\cos\lambda_n x \tag{2-11}$$

Where, λ_n is the solution for K in the equation $\cot(kL) = kL/B_i$; B_i is the Biot number (hL/K). It can be seen from equation (2-11) that θ/θ_0 is a function of three dimensionless numbers: the Biot number B_i, the Fourier number F_0, and x/L.

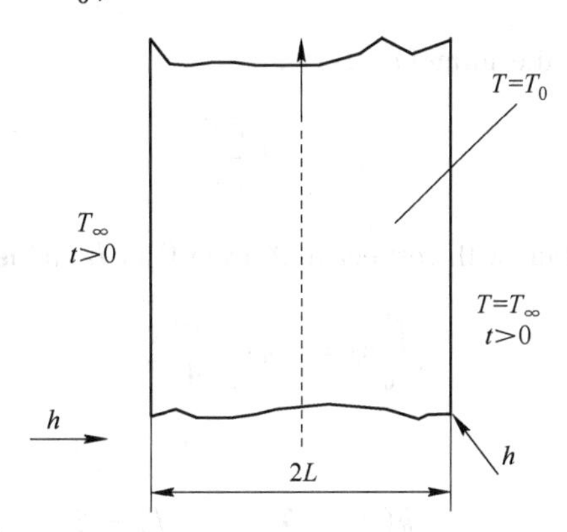

Fig. 2-1 Transient heat conduction of an infinite flat plate

B_i represents the ratio of the conductive thermal resistance inside the object to the convective thermal resistance at the surface of the object; F_0 represents the ratio of the approximate penetration depth of the temperature wave to the characteristic size of the object at a given time t; x/L represents the distance from the central line. For convenience, the calculation results of the theoretical solution of equation (2-9) are usually expressed as graphs, and the variation curves of θ/θ_0 with F_0 and B_i under different x/L conditions are drawn respectively. This results in a series of calculation diagrams, which can be found in the relevant books on heat transfer.

If B_i is very small, that is, for a very thin plate with a very large thermal conductivity coefficient, the internal thermal conductivity resistance is so small that it can be ignored compared with the surface convective thermal resistance, which means that the internal temperature of the solid approximately tends to be the same. Generally, when $B_i < 0.1$, the internal temperature gradient of the solid can be ignored, so the lumped heat capacity method can be used to approximately analyze the heat conduction process. From the energy balance relation in dt time:

$$q = Ah(T_\infty - T)\mathrm{d}t = V\rho c\mathrm{d}t$$

It can be concluded that:

$$\frac{T_\infty - T}{T_\infty - T_0} = \exp\left(-\frac{2ht}{\rho cL}\right) \tag{2-12}$$

Where, $L = 2V/A$, is the thickness of the plate (convective heat transfer on both sides of the plate); A is the convective surface area; V is the corresponding volume. This method can also be used to analyze the ignition problem and propagation process of thin fuel layers (such as paper and fiber) heated by radiation on one sideor cooled by convection on both sides.

【Example 2-1】 A flat plate with a thickness of 50 mm is maintained at $T_1 = 573$K and $T_2 = 373$K respectively on the two sides. Calculate the heat transferred through the unit cross-sectional area under the following conditions: (1) If the material is copper, K=374W/(m · K); (2) If the material is steel, K=36.3 W/(m · K); (3) If the material is chromic brick, K=2.32W/(m · K).

【Solution】 According to the formula (2-1),

$$q_x = -k\frac{\mathrm{d}T}{\mathrm{d}x}$$

Integrating the above equation with respect to X from 0 to L yields:

$$q_x\int_0^L \mathrm{d}x = -k\int_{T_1}^{T_2}\frac{\mathrm{d}t}{\mathrm{d}x}\mathrm{d}x$$

Collated:

$$q_x = \frac{-k(T_2 - T_1)}{L} = k\frac{T_1 - T_2}{L}$$

The above formula is the heat calculation formula for one-dimensional steady-state heat conduction when the thermal conductivity is constant. Substitute known values into the formula above:

$$q_{\text{copper}} = 374 \times \frac{300 - 100}{50 \times 10^{-3}} = 1.495 \times 10^6(\text{W/m}^2)$$

$$q_{\text{steel}} = 36.3 \times \frac{300 - 100}{50 \times 10^{-3}} = 1.452 \times 10^5(\text{W/m}^2)$$

$$q_{\text{chromic brick}} = 2.32 \times \frac{300 - 100}{50 \times 10^{-3}} = 9.28 \times 10^6(\text{W/m}^2)$$

2.2 Thermal convection

Convective heat transfer refers to the exchange of heat between a fluid and the surrounding solid or fluid in the process of flow. The convective mode in which a fluid consecutively flows over a solid wall under the action of an external force is called forced convection, and the flow of a hot gas near a hot solid driven by buoyancy is called natural convection.

2.2.1 Boundary layer

When the fluid flows through the solid wall, a boundary layer must be formed in the thin layer near the wall, and the velocity in the boundary layer has a large gradient in the direction perpendicular to the wall. This phenomenon is due to the viscous action of the fluid. The fluid layers in the laminar boundary layer do not mix with each other, the streamlines are roughly parallel to the wall, and the molecular thermal motion transfers momentum between adjacent layers. When the temperature of the moving fluid is different from that of the wall, a thermal boundary layer is formed within the flow boundary layer in the case with high Re. The thickness of the thermal boundary layer and the thickness of the velocity boundary layer are of the same order of magnitude, and there is a certain relationship. The temperature of the fluid close to the wall is the same as the wall temperature, which conforms to the wall adhesive condition. The temperature of the fluid at the outer boundary of the boundary layer is the temperature of external inviscid flow. There is a large transverse temperature gradient in the boundary layer, and heat conduction occurs between adjacent layers due to molecular thermal motion. The macroscopic motion of the fluid carries its contained heat, and the sum of these two parts of heat transfer is the convective heat transfer between the fluid and the solid wall.

Understanding the mechanism of convective heat transfer, it can be seen that the heat exchange between unit solid surface per unit time and fluid is related to the motion state of fluid, the temperature difference between fluid and solid wall, and the thermal properties of fluid, namely k and c_p, which are the three basic factors determining the convective heat transfer process.

2.2.2 Newton's formula and convective heat transfer coefficient

Because of the complicated shape of the object, the boundary layer of the fluid flowing through the solid is usually difficult to determine. Only the simplest object, such as flat plate, circular pipe and so on, can determine the temperature profile in the boundary layer by differential equations. In most cases, the heat exchange flux between the wall and the fluid can only be determined by experiment, so Newton's formula is very convenient in engineering calculation. Newton proposed that the heat exchange flux between the fluid and the wall is proportional to the temperature difference between the fluid and the wall. This relationship is written in the formula:

$$q = h(T_w - T_f) = h\Delta T(\mathrm{W/m^2}) \quad (2\text{-}13)$$

Where h is the scale factor, and its unit is $W/(m^2 \cdot K)$, it is called convective heat transfer coefficient or convective heat transfer coefficient. Of the three factors that determine convective heat transfer, all complexity is concentrated in the convective heat transfer coefficient h. h is determined by the geometry of the flow distribution, the flow state, and the thermophysical properties of the fluid, k and c_p. For natural convection, typical values of h are between 5 and $25W/(m^2 \cdot K)$ while for forced convection it is between 10 and $500W/(m^2 \cdot K)$.

【Example 2-2】 For a horizontally placed steam pipe, the outer diameter of the insulation layer is 583mm, and the measured average temperature of the outer surface is $T = 48℃$. The air temperature T_0 is 23℃, and the natural convection heat transfer coefficient h between the air and the outer surface is $3.42\ W/(m^2 \cdot ℃)$. Calculate the heat loss by natural convection per meter of pipe length.

【Solution】 When only natural convection is considered, according to equation (2-13):

$$q = h\Delta T = 3.42 \times (48 - 23) = 85.5(W/m^2)$$

Then the natural convection heat dissipation per meter of pipe length is:

$$q' = \pi dq = 3.14 \times 0.583 \times 85.5 = 156.5(W/m)$$

2.2.3 Boundary layer analysis and solution of convective heat transfer process

The heat transfer process occurs in the boundary layer near the surface of the solid wall, and its structure determines the value of h. First consider the speed of U_∞ an isothermal system in which an incompressible fluid flows through a flat plate parallel to it. As shown in Fig. 2-2, the flow velocity of the fluid on the wall is $U(0) = 0$, assume that the velocity distribution in the vertical direction is $U = U(y)$, and the velocity at infinity from the wall is $U(\infty) = U_\infty$. The thickness of the flow boundary layer is defined from the wall to the point where $U(y) = 0.99U_\infty$. For small values of x, i. e., area near the edge of the wall, the flow in the boundary layer is laminar. As x increases, the flow becomes fully turbulent after passing through a transition region. However, near the wall, there is always a "laminar inner layer". Like pipe flow, the flow properties are determined by the local Reynolds number $Re = xU_\infty \rho/\mu$. If $Re_x < 2 \times 10^3$ then the flow is laminar; if $Re_x > 3 \times 10^6$, then it is a turbulent flow. In between, it is a transition zone, which may be laminar or turbulent flow. However, in the case of direction problem, the critical Reynolds number is usually taken as 5×10^5.

Fig. 2-2 shows an isothermal flowing boundary layer system, with the thickness of the flowing boundary layer (δ_h) depends on the local Reynolds number Re and can be approximately expressed for laminar flow as:

$$\delta_h \approx l\left(\frac{8}{Re_l}\right)^{1/2} \tag{2-14}$$

Where l is the x value of corresponding to δ_h, Re_l is the local Reynolds number at the point $x = 1$.

If there is a temperature difference between the fluid and the plate, a "thermal boundary layer" is formed. See Fig. 2-3. The rate of heat transfer between the fluid and the solid wall

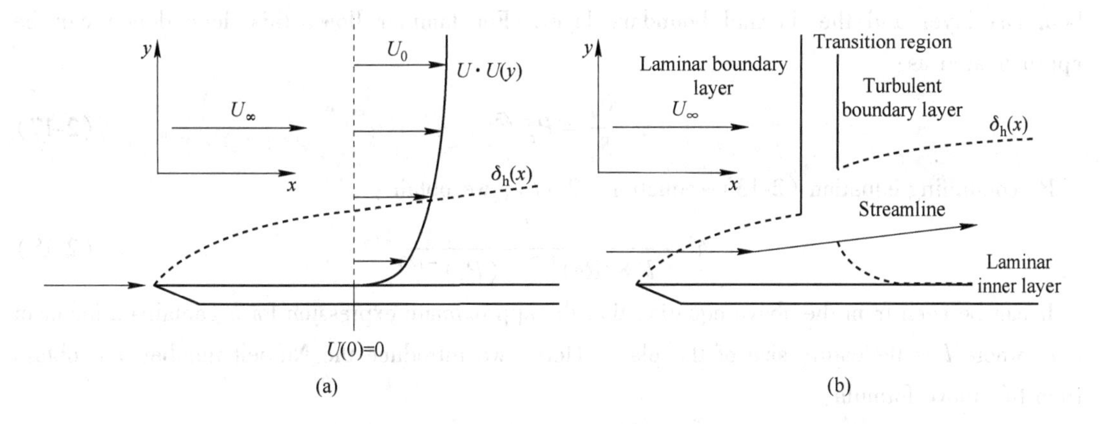

Fig. 2-2 Adiabatic flow boundary layer system on a flat plate

(a) The flow velocity of the fluid on the wall is $U(0)=0$; (b) The laminar and turbulent flow

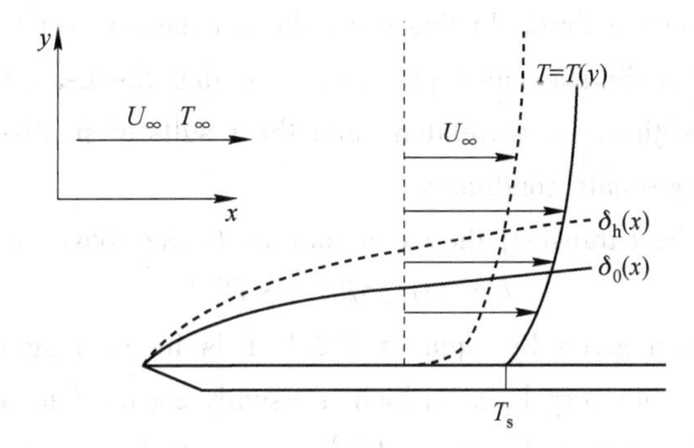

Fig. 2-3 Nonadiabatic flow boundary layer system on a flat plate

(The dashed line is the flow boundary layer and the solid line is the thermal boundary layer)

depends on the temperature gradient of the fluid at $y=0$. Applying Fourier's law of heat conduction, we have:

$$q''=-k\left(\frac{\partial T}{\partial y}\right)\bigg|_{y=0} \tag{2-15}$$

Where k is the thermal conductivity of the fluid. The above equation can be further approximated as:

$$q''\approx-\frac{k}{\delta_\theta}(T_\infty-T_S) \tag{2-16}$$

Where δ_θ is the thermal boundary layer thickness, T_S, T_∞ are respectively the incoming flow temperature and the solid wall surface temperature. The ratio of the thickness of the thermal to flow boundary layers (δ_θ/δ_h) depends on the Prandtl number $P_r=v/a$(of which $v=\mu/\rho$,μ called the kinematic viscosity coefficient of the fluid or the dynamic viscosity coefficient of the fluid, α is the thermal diffusion coefficient), that is, the ratio of the viscous dissipation to the thermal diffusion coefficient of the fluid, which respectively determine the structures of the flow boundary layer and the thermal boundary layer. For laminar flow, this dependence can be approximated as:

$$\frac{\delta_\theta}{\delta_h} = P_r^{-1/3} \tag{2-17}$$

By combining equation (2-13) ~ equation (2-17), we obtain:

$$h \approx \frac{k}{l(8/Re)^{1/2} \cdot (P_r)^{-1/3}} \tag{2-18}$$

It can be seen from the above equation that the approximate expression for h contains a factor of k/l, where l is thefeature size of the plate. Here, we introduce the Nusselt number and obtain from the above formula:

$$Nu = \frac{hl}{k} = 0.35Re^{1/2} \cdot P_r^{1/3} \tag{2-19}$$

Nu and B_i have the same form, but the meaning of k is different. Here in the fomula Nu、k is the thermal conductivity of the fluid. In this way, the convective heat transfer coefficient can be expressed in the form of a dimensionless parameter, so that the heat transfer coefficients under geometrically similar conditions are correlated, and the results of small-scale experiments can be used to determine the large-scale conditions.

Referring to books on heat transfer, the exact solution to the above problem is:

$$Nu = 0.35Re^{1/2} \cdot P_r^{1/3} \tag{2-20}$$

The approximate solution given by equation (2-19) is in good agreement with experiment. When the change of P_r is not very large, which is usually assumed to be 1 in many combustion problems, it can be further obtained that $h \propto U^{1/2}$ this conclusion is often used in the analysis of the response of fire detectors to heat from fire.

For turbulent flow, the temperature gradient at $y = 0$ is much larger than in the laminar case, and Nu is:

$$Nu = 0.0296Re^{4/5} \cdot P_r^{1/3} \tag{2-21}$$

For other expressions in the case of forced convection, see the relevant heat transfer references.

In the case of natural convection, the flow is driven by the buoyancy generated by the internal temperature difference, and the flow boundary layer and the thermal boundary layer are inseparable. From the analysis and derivation (omitted here, please refer to the books and literature on heat transfer), we can get a new dimensionless number: Grashof number, i. e. G_r, which represents the ratio of the buoyant force to the viscous force of the fluid, i. e.:

$$G_r = \frac{\beta g l^3(\rho_\infty - \rho)}{\rho v^2} = \frac{\beta g l^3 \Delta T}{v^2} \tag{2-22}$$

Where, G_r is the Gravitational acceleration and $\beta = 1/T$ (coefficient of volumetric thermal expansion). The above equation reflects the relative relationship between buoyancy force and viscous force due to the temperatures' difference of different parts of the fluid. The convective heat transfer coefficient can be expressed as functions of P_r and G_r. For the riser under laminar flow condition ($10^4 < G_r \cdot P_r < 10^9$):

$$Nu = \frac{hl}{k} = 0.59(G_r \cdot P_r)^{1/4} \tag{2-23}$$

Under turbulent flow condition($G_r \cdot P_r > 10^9$), there is:

$$Nu = \frac{hl}{k} = 0.13(G_r \cdot P_r)^{1/3} \tag{2-24}$$

Where the product of G_r and P_r is called the lagrange number: $Ra = G_r \cdot P_r$. Expressions in other cases can be found in the table above and in references on heat transfer.

Some commonly used dimensionless numbers are listed in Table 2-1.

Table 2-1 Common dimensionless numbers

Dimensionless number	Symbol	Expression	Meaning
Biot number	B_i	$=hl/k$	The ratio of the conductive thermal resistance inside a body to the convective thermal resistance on the surface of the body
Fourier number	F_o	$=\alpha t/l^2$	The ratio of the approximate penetration depth of a temperature wave to the characteristic size of an object at a given time
Lewis number	L_e	$=D/\alpha$	Ratio of mass diffusivity to thermal diffusivity
Reynolds number	R_e	$=U_\infty l/v=\rho U_\infty l/\mu$	Ratio of flow inertia force to viscous force
Prandtl number	P_r	$=\mu/\alpha=\mu c_p/k$	The ratio of the viscous dissipation to the thermal diffusivity of a fluid
Nusselt number	N_u	$=hl/k$	Ratio of convection heat transfer to heat conduction of fluid on solid wall surfaces
Grashof number	G_r	$=g\beta(T_S-T_\infty)l^3/v^2$	Ratio of buoyancy to viscous force of a fluid
Modified grashof number	G_r^*	$=G_r Nu=g\beta q_s l^4/kv^2$	
Lagrangian number	Ra	$=G_r P_r=g\beta(T_S-T_\infty)l^3 v\alpha$	
Standen number	St	$=Nu/(ReP_r)=h/\rho U_\infty c_p$	

The following is a summary of the methods of convective heat transfer calculation, and examples are given to illustrate the application of different empirical relations. According to the following simple steps, it is convenient to select the expression of convective heat transfer coefficient suitable for different situations.

(1) Firstly, make clear the geometric conditions of the flow, that is, whether the problem contains a flat plate, a sphere, or a cylinder. Because the form of the expression of the convective heat transfer coefficient is related to the geometric properties.

(2) Select the appropriate reference temperature, and then calculate the characteristic parameters of the fluid in the flow process according to it. The foregoing analysis has been carried

out under the assumption that thefeature parameters of the fluid are constant throughout the flow. If the conditions of the wall and the incoming flow change significantly, it is suggested that the membrane temperature, also known as the qualitative temperature, be used to determine the characteristic parameters. The arithmetic mean of the wall temperature and the incoming flow temperature ($T_f = (T_S + T_\infty)/2$).

(3) By calculating *Re*, and comparing with the critical *Re*, determine whether the flow state is laminar or turbulent. Such as a horizontal flow over a flat plate, $Re = 5\times10^6$, and critical $Re = 5\times10^5$ there is obviously a mixed boundary layer.

(4) Clarify whether it is the convective heat transfer coefficient of a certain point or the average convective heat transfer coefficient of the whole surface that needs to be calculated. The local *Nu* is used to determine the heat flux at a point on the surface, while the average *Nu* is used to determine the heat transfer rate across the surface.

【Example 2-3】 The outer surface temperature of the vertical wall, T_W is 333K, the outside air temperature, T_∞, is 293K, the wall height $l = 3$m, and the amount of heat transfer per hour through the free movement of the wall surface per square meter. ($K = 0.0276\mathrm{W/(m\cdot K)}$, $v = 16.96\times10^{-6}\mathrm{m^2/s}$, $P_r = 0.69$)

【Solution】 Assuming that the wall temperature is uniform and the air is still, the reference temperature is themembrane temperature. $T_f = \dfrac{T_S + T_\infty}{2} = 313\mathrm{K}$. It is known that $K = 0.0276\mathrm{W/(m\cdot K)}$, $v = 16.96\times10^{-6}\mathrm{m^2/s}$, $P_r = 0.69$, and $\beta = 1/T_f = 0.0032\mathrm{K^{-1}}$.

Analysis: derived from formula (2-24)

$$G_r = \frac{\beta g l^3 \Delta T}{v^2} = \frac{0.0032\times9.8\times3^3\times(333-293)}{(16.96\times10^{-6})^2} = 11.77\times10^{10}$$

then

$$G_r\cdot P_r = 11.77\times10^{10}\times0.699 = 8.23\times10^{10}$$

Because $Ra_L = G_{rL}\cdot P_r > 10^9$, the natural convection flow has been transformed into turbulent flow, so from equation (2-24),

$$Nu = 0.13(G_r\cdot P_r)^{1/3} = 0.13\times(8.23\times10^{10})^{1/3} = 435$$

And because $Nu = \dfrac{hl}{k} = 0.13(G_r\cdot P_r)^{1/3}$, therefore

$$h = \frac{Nuk}{l} = \frac{435\times0.0276}{3} = 4(\mathrm{W/(m^2\cdot K)})$$

The heat transfer rate of natural convection is:

$$q = h(T_S - T_\infty) = 4\times(333-293) = 160(\mathrm{W/m^2})$$

【Example 2-4】 As shown in Fig. 2-4, in order to prevent smoke and dust from entering the room, a glass door is generally placed in front of the fireplace, and its height is 0.7m, 1.02m wide, and maintain 232℃. If the indoor air temperature is 23℃, solve the convective heat transfer rate from the fireplace to the room. (Known $k = 33.8\times10^{-3}\mathrm{W/(m\cdot K)}$, $v = 26.41\times10^{-6}\mathrm{m^2/s}$, $P_r = 0.69$)

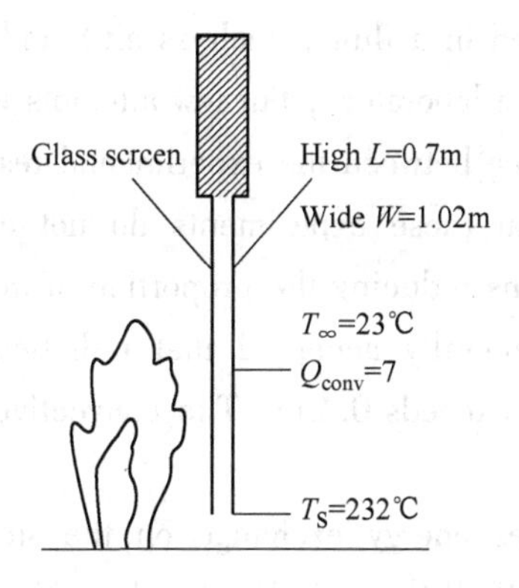

Fig. 2-4 Schematic diagram of example

【Solution】 Assume that the temperature of the glass door is uniform and the indoor air is still. Feature parameters: membrane temperature is selected as the reference temperature $T_f = \frac{T_S + T_\infty}{2} \approx 400K$, $p = 1atm$, known $K = 3.38 \times 10^{-2} W/(m \cdot K)$, $v = 26.41 \times 10^{-6} m^2/s$, $P_r = 0.69$, $\beta = 1/T_f = 0.0025K^{-1}$ ($T_S = 505K$, $T_\infty = 296K$).

After analysis, it can be concluded that:

$$Ra_L = G_{rL} \cdot P_r = \frac{g\beta(T_S - T_\infty)l^3}{\alpha v}$$

$$= \frac{9.8 \times 0.0025 \times (505 - 296) \times 0.7^3}{3.83 \times 10^{-5} \times 26.41 \times 10^{-6}} = 1.73 \times 10^9$$

Because $Ra_L = G_{rL} \cdot P_r > 10^9$, the natural convection flow has been transformed into turbulent flow, then

$$Nu = 0.13(G_r \cdot P_r)^{1/3} = 0.13 \times (1.736 \times 10^{10})^{1/3} = 337$$

Then

$$h = \frac{\overline{Nu}k}{L} = \frac{337 \times 33.8 \times 10^{-3}}{0.71} = 16(W/(m^2 \cdot K))$$

The heat exchange of natural convection is:

$$q = hA(T_S - T_\infty)$$

That is

$$q = 16 \times 1.02 \times 0.7 \times (505 - 296) = 2394(W)$$

It is worth pointing out that, in this case, the radiation heat transfer is greater than the natural convection heat transfer.

2.3 Heat radiation

Heat radiation is a kind of electromagnetic radiation emitted by an object due to its own temperature. It travels at the speed of light, and its corresponding wavelength range is 0.4 ~ 100μm (including visible light). When an object is heated and its temperature rises, it will lose heat both by convection (if placed in a fluid, such as air) and by radiation. In the past, many fire studies were carried out in the laboratory, but few attempts were made to simulate the real fire situation. There is a big difference between the experimental results and the real large-scale fire. The main reason for this is that these experiments do not consider heat radiation, because reducing the size of the fire means reducing the proportion of radiation compared with other ways of heat exchange. It is now generally accepted that radiation is the dominant mode of heat exchange when the flame height exceeds 0.2m. The convective heat transfer is more significant for smaller flames.

Heat radiation in fire includes energy exchange on the surface of objects, such as walls, ceilings, floors, furniture, etc. Of all the emissions or absorptions of many different gases and dust particles, water vapor and carbon dioxide are of practical value to fire engineering. They are in the most dominant thermal radiation range 1 to 100μm. it has strong endothermic and exothermic properties.

2.3.1 Basic concepts and laws

2.3.1.1 Radiation intensity and energy

The energy radiated by an object from a unit area in a unit time is called radiant energy. According to the Stefan-Boltzman equation, the radiant energy of a body is proportional to the fourth power of its temperature, i. e.

$$E = \varepsilon\sigma T^4 (\mathrm{W/m^2}) \tag{2-25}$$

In the formula, σ is the Stefan-Boltzman constant whose value is $5.667\times10^{-8}\mathrm{W/(m^2 \cdot K^4)}$. The temperature t is taken as Kelvin's. ε is the emissivity, which is a constant characterizing the surface properties of a radiating object and is defined as the ratio of the radiant energy of an object to the radiant energy of a Black Body at the same temperature, namely $\varepsilon = \dfrac{E}{E_b}$, for a Black Body, $\varepsilon = 1$. In practice, the emissivity of a material varies with different temperatures and wavelengths of the radiation.

In order to be able to calculate the radiant energy of an object in arbitrary directions, here we introduce the radiant intensity I_n, the energy radiated per unit time, per unit surface area and per unit solid angle in the normal direction. Thermal radiant energy through a surface in either direction can be replaced by spectral radiant energy:

$$q_v = \int_0^{4\pi} I_V nR\mathrm{d}\Omega = \int_0^{4\pi} I_V \cos\theta \mathrm{d}\Omega \tag{2-26}$$

In equation (2-26), Ω represents the solid angle ($\mathrm{d}\Omega = \sin\theta\mathrm{d}\theta\mathrm{d}\phi$). I_V is the radiation intensity, that is, the radiation energy per unit area and per unit solid angle. Radiation intensity is a valuable measure of thermal radiation because when a radiation beam propagates in a vacuum, its radiation intensity is constant.

[Example 2-5] There is one piece of steel plate at 27℃ whose blackness $\varepsilon = 0.8$. Try to calculate the radiant energy emitted by the unit area of the steel plate in unit time.

[Solution] According to the formula (2-27), the radiant energy emitted by the steel plate per unit area is:

$$E = \varepsilon\sigma T^4 = 0.8 \times 5.667 \times 10^{-8} \times (273 + 27)^4 = 367.2(\text{W/m}^2)$$

2.3.1.2 Planck's law

Radiant energy emitted from the surface of an object with a wavelength within range between 0 and ∞. The amount of radiant energy is expressed in terms of total radiant force or simply radiant force. The radiation force is defined as the amount of radiation emitted by a unit surface area of an object to the surrounding hemisphere in a unit time. The total radiated energy in the wavelength range within range between 0 and ∞ is expressed as $E(\text{W/m}^2)$. For a Black Body, the radiation force is denoted by E_b. In 1900, Planck revealed the distribution law of the monochromatic radiation force of a Black Body in vacuum at different temperatures based on the quantum theory of electromagnetic waves, which is called Planck's law. It reveals the relationship between the monochromatic radiation force of a Black Body and the wavelength, thermodynamic temperature T:

$$E_{b\lambda} = \frac{2\pi c_2 h\lambda^{5}}{\exp(ch/K\lambda T) - 1} \tag{2-27}$$

Where K——boltzmann constant, $K = 1.3806\times10^{-23}$ J/K;

c——speed of light in vacuum, $c = 2.998\times10^{8}$ m/s;

h——planck constant, $h = 6.624\times10^{34}$ J · s.

In the formula (2-27), $E_{b\lambda}$ at 0 ~ ∞ by integrating λ in the wavelength range between 0 and ∞, the radiation force of the Black Body can be obtained as:

$$E_b = \int_0^\infty E_{b\lambda} d\lambda = \int_0^\infty \frac{2\pi c^2 h\lambda^{-5} d\lambda}{\exp(ch/KT\lambda) - 1} = \sigma T^4 \tag{2-28}$$

The radiation force of a real surface is different from that of a Black Body, and the radiation force of a real surface is less than that of a Black Body. Definition ε_λ is the emissivity (blackness) of the true surface:

$$\varepsilon_\lambda = \frac{E_\lambda}{E_{b\lambda}}$$

ε_λ reflecting the difference between the radiation force of a real surface and the radiation force of a black body, $\varepsilon_\lambda < 1$.

For real surfaces, ε_λ is a function of wavelength. The concept of grey body is usually introduced for convenience. We call an object as a gray body if its monochromatic emissivity, ε_λ is wavelength independent, i.e. in the wavelength range between 0 and ∞, ε_λ is a constant less than 1. Thus, we can rewrite equation (2-28) into a formula suitable for calculating the radiation force on the surface of an actual object as:

$$E = \varepsilon\sigma T_4 \tag{2-29}$$

2.3.1.3 Kirchhoff law

If a fire is isolated, although the internal substances are different, the isolators with the same temperature can reach their own equilibrium state. The relationship is as follows:

$$\alpha_v + \rho_v + \tau_v = 1 \tag{2-30}$$

On the interface between substances, α, ρ and τ respectively represents the absorption, reflectivity, and transmission of energy. The assumption of thermodynamic equilibrium can be used to obtain more results and is widely used in the calculation of radiant heat exchanges. Kirchhoff's law states that in order to maintain balance, the absorptivity and emissivity of light will be inevitably related by the following equation.

$$\alpha_v = \frac{I_v}{I_{bv}} = \xi_v \tag{2-31}$$

More importantly, when equation (2-31) is applied to all substances,

$$\alpha_t = \xi_t \tag{2-32}$$

In special cases where the incident radiation is independent of the angle of incidence, equation (2-31) is equally valid and has the same spectral proportions as a Black Body, such as a Gray Body. This is exactly what happens when a substance participates in a fire in a radiant-heat-exchanging engineering model. Although the emissivity of a gas depends only on the properties of the gas, the absorptivity is still part of the origin of the temperature of the incident radiation beam. These may come from substances other than gases, such as the temperature of the walls.

2.3.2 Radiation from hot gases and non-luminous flames

For gases in which only the molecules are in dipole motion, the emission occurs over the entire thermal emission wavelength range of the spectrum from 0.4 to 100 (visible light). Gases with non-polar symmetric molecular structures, such as N_2, H_2, O_2, etc., are all substantially "transparent" to radiation at low temperatures. While gases having a polar molecular structure, such as CO_2, H_2O, CO, HCl, and hydrocarbons, its radiation (absorption) occurs in some inconsecutive narrow bands, showing inconsecutive radiance, and sometimes the amount of radiation is considerable. In addition, gas radiation is total integrated radiation. Thus, the radiation properties of a gas are related to its "thickness".

Consider that a beam of monochromatic light with wavelength λ passes through a gas layer with thickness L. When the beam passes through the thin layer dx, the reduction of its light intensity is proportional to the local light intensity, the thickness dx of the gas layer and the concentration C of the component absorbing light in the gas layer, that is:

$$dI_\lambda = K_\lambda C I_{\lambda x} dx \tag{2-33}$$

In the formula, K_λ is the constant of proportionality, the absorption coefficient of monochromatic light. Integrate the above equation in the region from $X = 0$ to $X = L$ to obtain:

$$I_{\lambda X} = I_{\lambda 0} \exp(-K_\lambda CL) \tag{2-34}$$

$I_{\lambda 0}$ is the incident light intensity at $X = 0$. The above equation is called Lambert-Beer law, so the absorption rate of monochromatic light is obtained as:

$$\alpha_\lambda = \frac{I_{\lambda 0} - I_{\lambda L}}{I_{\lambda 0}} = 1 - \exp(-K_\lambda CL) \tag{2-35}$$

According to Kirchhoff's law, it is equal to the radiance of monochromatic light at the same wavelength ε_λ. It can be seen from the above equation (2-35) that when $L \to \infty$, α_λ and ε_λ tends to 1.

Gas radiation is often involved in fire research and engineering calculation, but the calculation of gas radiation characteristics is very complex, and many quantities are difficult to measure and estimate. For this reason, Hottel and Egbertmanaged to find an empirical method to obtain the "equivalent gray body emissivity" of a certain volume of hot gas containing CO_2 and water vapor. This method operates basing on a series of precise measurements of radiant heat on CO_2 and water vapor (mixed and separated) at different partial pressures, different temperatures and different radiant gas geometries. Since the emissivity at a single wavelength is known to depend on the concentration of the radiation component and the "thickness" of the gas, Hotell defined the total effective emissivity of CO_2 and water vapor as a function of temperature over a range of values of $P_a L$. P_a is the partial pressure of the radiation component, and L is the average length of the ray path, which means that if there is a gas between two large parallel plates with diffuse radiation, the distance of radiation energy propagation through the gas is different. The distance of energy propagation in the direction of surface normal is equal to the distance between the plates, while in the direction of small angle, the radiation energy travels a longer distance in the gas. After carefully synthesizing and analyzing the experimental data from several sources, Hottel along with other researchers gave the average length of ray travel corresponding to several geometric states.

If the partial pressure of the radiation component and the geometrical state of the radiation gas are known, the average length (L) of the ray path and other equivalent Gray Body emissivities at certain temperatures can be obtained. If there is no data on the average length of the ray path, a satisfactory approximation will be obtained for a particular geometry using the following formula. Namely:

$$L = 3.6 \frac{V}{A} \tag{2-36}$$

Where, V is the total volume of the gas; A is the total surface area.

This method is applicable to gas mixtures with a total pressure of one atm. For the case where the total pressure is not one atm, a correction is required, and when CO_2 and water vapor are existing at the same time, an additional correction factor $\Delta\varepsilon$ is required. In this case, the equivalent Gray Body emissivity of the mixture shall be divided into the sum of the two radiations minus this correction factor, i. e. :

$$\varepsilon_g = C_{H_2O}\varepsilon_{H_2O} + C_{CO_2}\varepsilon_{CO_2} - \Delta\varepsilon \tag{2-37}$$

In the formula C_{CO_2}, C_{H_2O} are the pressure correction coefficients for CO_2 and water vapor emissivity respectively, and their determination methods can be referred to the relevant books on heat transfer. ε_{CO_2}, ε_{H_2O} are the emissivities of CO_2 and water vapor, respectively, at a total pressure of one atm. In most fire engineering applications, the pressure correction factor is 1.0 for the medium in large fires, and the frequency band overlap factor is approximately $\Delta\varepsilon \approx 1/2\varepsilon_{CO_2}$. So there is:

$$\varepsilon_g \approx \varepsilon_{H_2O} + \frac{1}{2}\varepsilon_{CO_2} \tag{2-38}$$

Although this result may be considered to be highly approximate, because the assumptions made did have some arbitrariness, it was not much different from the results measured by Rasbash et al. 1965. Experience has shown that the emissivities obtained by Hottel's method are acceptable up to a temperature of 1000℃. At temperatures higher than this, especially at larger flame "thicknesses", the emissivity obtained by this method is low, is shown as Table 2-2.

Table 2-2 Average length of gas ray travel under different geometric conditions

Gas geometry	Radiation status	Average length of gas ray travel	Correction factor
Sphere, diameter D	Radiation to the entire surface	0. 66D	0. 97
	Radiate to the bottom	0. 48D	0. 90
Cylinder, height H = 0. 5D	Opposite side radiation	0. 52D	0. 88
	Radiation to the entire surface	0. 50D	0. 90
Cylinder, height H = D	For the whole central radiation	0. 77D	0. 92
	Radiation to the entire surface	0. 66D	0. 90
	Radiate to the bottom	0. 73D	0. 82
Cylinder, height H = 2D	Opposite side radiation	0. 82D	0. 93
	Radiation to the entire surface	0. 80D	0. 91
Semi-infinite cylinder	Central radiation on the bottom	1. 00D	0. 90
Height $H \to \infty$	Radiation to the entire surface	0. 81D	0. 80
Infinite plate, spacing D	Area radiation of surface element	2. 00D	0. 90
	Radiate to both sides	2. 00D	0. 90
Cube, side length D	Radiate to either side	0. 66D	0. 90
	Radiate to either side	0. 90D	0. 91

2. 3. 3 Luminous flame and hot smoke radiation

With a few exceptions, such as formaldehyde and paraformaldehyde, liquids and solids burn with a yellow luminous flame. This characteristic yellow color results from the formation of tiny semi-smokeless carbon particles, of the order of 10 to 100nm in diameter, produced on the fuel side of the reaction zone inside the flame, which may be burned out during their passage through the oxidation zone of the flame and may also react and change further to form smoke which will be further distributed among the mixture of combustion products and air, forming hot flue gas. When these particles are in the flame or in the hot smoke, their temperature is relatively high, and each particle acts as a tiny Black Body or Gray Body, and its radiation spectrum is continuous, and the flame (hot flue gas) radiation depends on the temperature, the concentration of particles and the "thickness" of the flame and the average length of the ray run. The empirical formula for its emissivity is similar to Kirchhoff's law for the absorptivity of monochromatic light, i. e. :

$$\varepsilon = 1 - \exp(-KL) \tag{2-39}$$

Where K is the absorption coefficient, when the diameter of the carbon particle is much smaller than the wavelength of the radiation (most case $\lambda = 1\mu m$), it is proportional to the volume fraction of char particles in the flame (hot flue gas) and the radiation temperature, i. e. :

$$K = 3.72 \frac{C_0}{C_2} f_v T \tag{2-40}$$

In this formula, C_0 is a constant between 2 and 6; C_2 is Planck's second constant whose value is 1.4388×10^{-2} m · K; f_v is the volume fraction of carbon particles, i. e. the fraction of particles in the whole volume whose valuc is about 10^{-6} order of magnitude. T_s is the radiation temperature.

Generally speaking, the more carbon particles in the flame, the lower the average temperature of the flame. Rasbash has found that the average temperature of non-luminous mcthanol flame is 1200℃, while the average temperature of luminous flame formed by kerosene and benzene combustion is much lower at 900℃ and 921℃ respectively. This indicates that for a luminous flame, the heat loss of the flame due to the radiation of the char particles is relatively large, meanwhile the degree of completeness of combustion is relatively poor, and thus the flame temperature is low.

The radiation of the char particles greatly exceeds the molecular radiation produced by gases such as water vapor and CO_2. In the above empirical formula, we have generally pointed out the existence of flame radiation while gave non strict distinction between the two radiation sources of carbon particles and gases. In fact, the intermittent radiation produced by CO_2 and water vapor over the entire wavelength range of thermal radiation makes a peak in the intensity band of the consecutive radiation of the carbon particles in the "thin flame". However, for convenience, it is usually assumed that the luminous flame has the emission properties of a gray body, that is, the emissivity is independent of wavelength, and the gas radiation is ignored in the general calculation. If gas radiation must be taken into account. The following empirical formula gives a good approximation for the emissivity of an average mixture of char particles and hot gas:

$$\varepsilon = [1 - \exp(-KL)] + \varepsilon_g \exp(-KL) \tag{2-41}$$

In the formula, ε_g is the total emissivity of the gas, and L may be taken as the "thickness" of the flame (hot flue gas). For a more crude estimation, it is usually assumed that the "thick luminous flame" formed by the hydrocarbon fuel (L) is a Black Body, i. e. $\varepsilon = 1$.

Whether the emissivity is assumed or calculated, the equation $E = \varepsilon\sigma T^4$ (W/m^2) can be applied when the flame temperature is known to calculate the radiant energy. If the radiant heat flux at a point beyond the distance from one end of the flame, or the radiant heat transfer between two Black Bodys'surfaces, is to be estimated, angle factors must be used. It is generally assumed that the flame has a simple geometric shape, such as a rectangle with a height of 1 to 2 times the diameter of the fuel bed, and then the angle factor is determined by the aforementioned method. However, to calculate the radiation heat transfer between two non-Black-Body surfaces, it is necessary to go further.

We know that thermal radiation is an important factor in promoting the spread and development of fire, especially in confined spaces. In the process of indoor fire development, hot smoke accumulates under the ceiling to form a smoke layer, which has a strong thermal radiation to the lower part, which is very conducive to intensifying combustion and achieving full development of the fire.

2.4 Transfer of substances

When combustion occurs, the combustion products will continue to leave the combustion zone, and the fuel and oxidant will continue to enter the combustion zone, otherwise, the combustion can not continue. Here, the exit of products, the entry of fuel and oxidant, all have a problem of mass transfer. Mass transfer can be achieved by means of molecular diffusion, Stephen flow at the fuel interface, mass flow due to buoyancy, forced flow due to external forces, and mass mixing due to turbulent motion. Here, only the first three modes of mass transfer are described in this section.

2.4.1 Diffusion of substances

Suppose there is a stationary isothermal fluid B that seeps into another fluid A from one side and seeps fluid A out from the other side. As shown in Fig. 2-5, the concentration of A is different at different places in B. Because of the concentration difference, substance A will diffuse from the place where the concentration is high to the place where the concentration is low. Microscopically speaking, this diffusion is due to the intermixing of molecules due to their ceaseless thermal motion. So that the concentration of each component tends to be the same, thus causing the macroscopic diffusion phenomenon. It is found that in unit time, the material flow caused by the diffusion of fluid A per unit area is proportional to the concentration gradient of fluid A in B, that is,

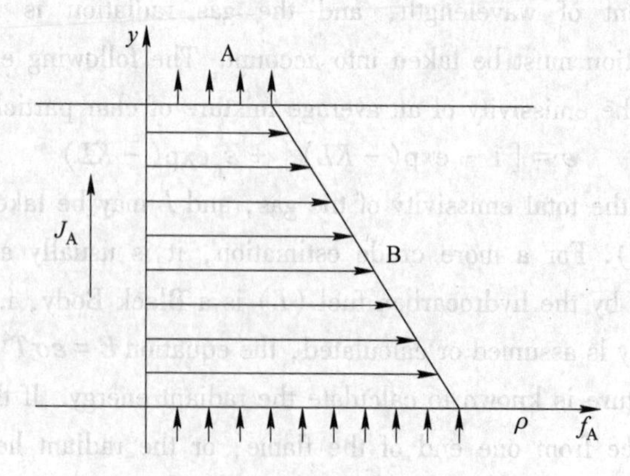

Fig. 2-5 Schematic diagram of diffusion law

$$J_A = -D_{AB}\frac{\partial \rho_A}{\partial y} \tag{2-42}$$

Where, J_A represents the mass flow resulting from the diffusion of fluid A per unit area in unit time; D_{AB} is called the diffusion coefficient of component A in component B. In general, the diffusion coefficient is composition-dependent, and the negative sign in equation (2-42) indicates that component A diffuses in the direction of decreasing concentration of A.

When considering the diffusion problem of a multi-component mixture with more than two components, the S component considered is often assumed to be one component, and the other component other than the S component is assumed to be another component, which is approximately treated as a two-component diffusion problem, so the diffusion equation can be written as:

$$J_S = -D_S\frac{\partial \rho_S}{\partial y} \tag{2-43}$$

At this time, the diffusion coefficient is related to the composition and concentration of each component, so it is often necessary to further simplify the specific calculation.

Equations (2-42) and (2-43) are known as the Fick diffusion law equations.

Equation (2-42) can also be written in terms of partial pressure gradient or mass fraction gradient. Make P, M, R, T and f respectively stand for the pressure, molecular weight, universal gas constant, temperature and mass fraction. (The subscripts A and B represent the components A and B, respectively, and amounts without subscripts represent the total mixture.) Assuming the gas is perfect, then:

$$p_A = \frac{\rho_A}{M_A}RT$$

That is:

$$\rho_A = \frac{P_A}{RT}M_A$$

Substituting the above equation into equation (2-44) gives:

$$J_A = -D_{AB}\frac{\partial \rho_A}{\partial y} \tag{2-44}$$

The mass fraction and partial pressure areconnected by Eq, and Eq, therefore:

$$J_A = -\frac{\rho D_{AB}}{M}\frac{\partial f_A M}{\partial y} \tag{2-45}$$

The molecular weight M of that mixture is a function of the mass fraction f_A and the molecular weight of the components A and B. Therefore, in Fig. 2-6, M varies with the value of y. However, in the case of combustion, M does not change much and can be regarded as a constant, so equation (2-45) can be changed to:

$$J_A = -\rho D_{AB}\frac{\partial f_A}{\partial y} \tag{2-46}$$

2.4.2 Stephen flow

In the problem of combustion, there is a phase interface between the high temperature gas flow and the adjacent liquid or solid substance. It is very important to understand the mass transfer at the phase separation interface for correctly writing the boundary conditions and accurately studying the combustion problems under various boundary conditions. In combustion problems, there is a normal flow at the phase interface. This is different from the single-component fluid mechanics problem. In general, a single component viscous fluid, when flowing over an inert surface, will form an adhesive layer at the surface if the gas pressure is not very low. However, under certain conditions, the multi-component fluid will form a certain concentration gradient at the surface, so it is possible to form a diffusion material flow normal to each component. In addition, if there is a physical or chemical process on the phase interface, then this physical or chemical process will also produce or consume a certain mass flow. Then, under the action of physical or chemical processes, a normal total material flow related to the diffusion material flow will be generated at the surface. This total mass flow is caused by the surface itself. This phenomenon was first discovered by Stephen when he studied the evaporation process of water surface, so it became Stephen Flow. It should be emphasized that this Stephen flow is due to a combination of diffusion and physico-chemical processes.

Below we use two examples to illustrate the conditions that produce Stephen flow and its physical essence.

The first example is Stephen's discovery of Stephen's flow while studying water surface evaporation process, as shown in Fig. 2-6. A-B is the water surface, and the space above the water surface is air. At this time, there are only two components, water vapor and air at the water-air phase interface. We use f_{H_2O} indicates the relative concentration of water vapor, using f_{air} indicates the relative concentration of air. Their distribution is shown in this figure. And there is:

$$f_{H_2O} + f_{air} = 1 \tag{2-47}$$

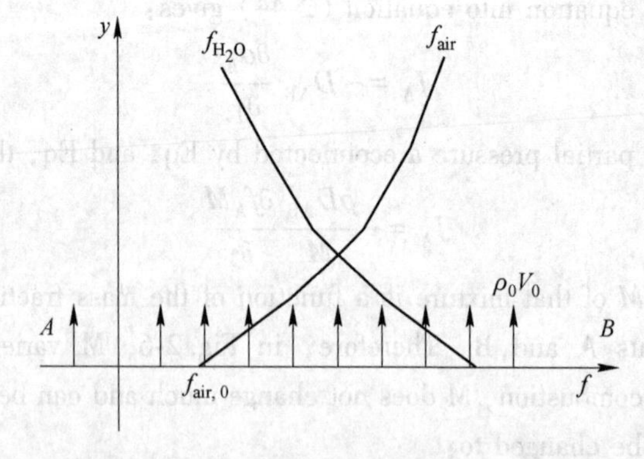

Fig. 2-6 Stephen flow with water evaporation

At this time, the diffusion flow of water vapor molecules at the phase separation interface is:

$$J_{H_2O,0} = -D_0\rho_0\left(\frac{\partial f_{H_2O}}{\partial y}\right)_0 \tag{2-48}$$

$$\because \quad \left(\frac{\partial f_{H_2O}}{\partial y}\right)_0 < 0, \therefore \; J_{H_2O,0} > 0.$$

At the same time, the air concentration gradient at the interface will also lead to the diffusion flow of air molecules.

$$J_{air,0} = - D_0\rho_0\left(\frac{\partial f_{air}}{\partial y}\right)_0 \tag{2-49}$$

Obtained from the formula (2-47):

$$\left(\frac{\partial f_{air}}{\partial y}\right)_0 = - \left(\frac{\partial f_{H_2O}}{\partial y}\right)_0$$

So

$$\left(\frac{\partial f_{air}}{\partial y}\right) > 0 \;,\; J_{air,0} < 0$$

That is to say, there is a diffusion flow of air flowing to the interface, but we know that the air will not be absorbed by the water surface, so where is the flow of air flowing to the interface? There is only one explanation here, that is, at thc phase separation interface, in addition to the diffusion flow, there must be an overall mass flow of the air-steam mixture opposite to the air diffusion flow, so that the total mass flow of air at the phase interface is zero. Assuming that the total mass flow of the gas mixture flows at the velocity V_0, the mass flow of each component can be divided into two parts: one part is the diffusion mass flow caused by the concentration gradient of the component, and the other part is the corresponding mass flow carried by the total mass flow of the gas mixture. So the following relation can be written:

$$g_{H_2O} = J_{H_2O,0} + f_{H_2O,0}\rho_0 V_0 = - D_0\rho_0\left(\frac{\partial f_{H_2O}}{\partial y}\right)_0 + f_{H_2O,0}\rho_0 V_0 \tag{2-50}$$

$$g_{air,0} = - D_0\rho_0\left(\frac{\partial f_{air}}{\partial y}\right)_0 + f_{air,0}\rho_0 V_0 = 0 \tag{2-51}$$

In that problem of water surface evaporation, $g_0 = g_{H_2O,0} + g_{air,0}$, $\because g_{air,0} = 0$, $\therefore g_0 = g_{H_2O,0}$, that is:

$$- D_0\rho_0\left(\frac{\partial f_{H_2O}}{\partial y}\right)_0 + f_{H_2O,0}\rho_0 V_0 = \rho_0 V_0$$

So there is:

$$- D_0\rho_0\left(\frac{\partial f_{H_2O}}{\partial y}\right)_0 = (1 - f_{H_2O,0})\rho_0 V_0 \tag{2-52}$$

It can be seen from this that in the problem of water surface evaporation, the Stephen flow (that is, the evaporation flow of water) is not equal to the diffusion material flow of water vapor, but is equal to the diffusion material flow plus the water vapor material flow carried by the overall movement of the mixed gas.

The second example is an analysis of the combustion of a carbon plate in pure oxygen. In this case, we assume that only the following reactions occur on the carbon surface:

$$C + O_2 \longrightarrow CO_2$$

At that point, there are two gas components of oxygen and carbon dioxide in the space above the carbon plate, so there is:

$$f_{O_2} + f_{CO_2} = 1$$

Differentiate the above equation with respect to y and multiply by, we get:

$$\rho_0 D_0 \left(\frac{\partial f_{CO_2}}{\partial y}\right)_0 = -\rho_0 D_0 \left(\frac{\partial f_{O_2}}{\partial y}\right)_0$$

That is:

$$g_{CO_2,0} = -g_{O_2,0} \tag{2-53}$$

Butwe also conclude from the reaction equation:

$$g_{CO_2,0} = -\frac{44}{32} g_{O_2,0} \tag{2-54}$$

By comparing the equations (2-53) and (2-54), it can be seen thatit is impossible to send out all the carbon dioxide on the surface of the carbon plate through only diffusion, so there must be an overall mass flow of the gas mixture in the same direction of the diffusive flow of carbon dioxide, so that the mass flow of carbon dioxide can meet the requirement of equation (2-54), that is, the CO_2 produced by the chemical reaction can be consecutively discharged from the carbon surface. This overall mass flow is the Stephen flow, i. e.

$$g_0 = g_{O_2,0} + g_{CO_2,0} = \rho_0 V_0$$

Or

$$g_0 = g_{O_2,0} - \frac{44}{32} g_{O_2,0} = -\frac{12}{32} g_{O_2,0} = g_c \tag{2-55}$$

The above equation shows that the Stephen flow at this time is the amount of carbon burned, that is, the burning rate of carbon.

From the above two examples, we can see that the condition for the generation of Stephen flowis that: there are both diffusion phenomena and physical or chemical processes at the phase interface. These two conditions are indispensable. It is very important to analyze the boundary conditions on the interface by using the concept of Stephen flow. This concept will be used.

2.4.3 Buoyancy movement caused by combustion

At the fire scene, the whole gas near the burning area is flowing, and this material flow is called the integral material flow. The oxygen and combustible gases required for combustion are carried into the combustion zone by this integral material flow, and the combustion products produced in the combustion zone are carried out by the material flow. This integral mass flow is caused by both forced and natural convection, i. e. buoyancy caused by combustion. Only the buoyant motion due to combustion is discussed here.

After an object in the room catches fire, the smoke generated by combustion will first fill the top of the room due to buoyancy, and the surrounding cold air will flow to the combustion area to supplement, and will be heated, expanding the volume, reducing the specific gravity and rising.

To further discuss the characteristics of this buoyancy-induced flow, now we assume this vertical duct of height H filled with air at a temperature T and with a specific gravity γ, the air temperature outside the tube is T_0, the specific gravity is γ_0; the plane of the lower end of the pipe is 1-1 plane, the upper end plane of the pipe is 2-2 plane; downward action on the pressure on the 2-2 plane is P_2, downward action on the plane 1-1 is P_1; upward action on the 1-1 plane is P, as shown in Fig. 2-7.

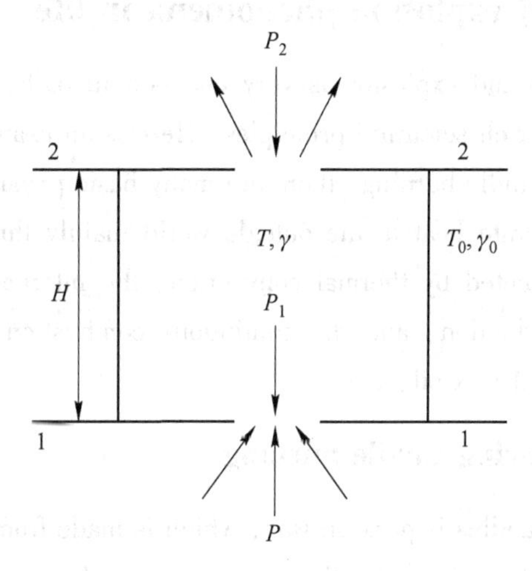

Fig. 2-7 Chimney effect

When the temperature inside the tube is equal to the temperature outside the tube, i. e. $T = T_0$, $\gamma = \gamma_0$. The fluid inside and outside the pipe is in a balanced state and does not flow. At this time, according to the fluid balance equation:

$$P = P_1 = P_2 + H\gamma = P_2 + H\gamma_0$$

If the temperature inside the tube is higher than that outside the tube, i. e. $T > T_0$, then because

$$P = P_2 + H\gamma_0 \qquad P_1 = P_2 + H\gamma \qquad \gamma < \gamma_0 \qquad P > P_1$$

So

$$P - P_1 = (P_2 + H\gamma_0) - (P_2 + H\gamma) = H(\gamma_0 - \gamma) \tag{2-56}$$

Then the cold air enters from the lower end of the pipe and the hot air flows out from the upper end of the pipe. This effect of promoting the upward flow of smoke and hot air due to gas convection in the vertical enclosure is called " chimney effect" . It can be seen from formula (2-56) that:

(1) The higher the pipeline H is, the greater the pressure difference ($P-P_1$) on the 1-1 plane at the lower end is, the more significant this effect is. This is the reason for the rapid upward progress of high-rise building fires through stairwells and elevator shafts.

(2) The greater the temperature difference between the inside and outside of the pipeline, the greater the specific gravity difference between the hot air and the cold air, the greater the pressure difference on the 1-1 plane at the lower end of the pipeline, the more significant the chimney effect is.

This chimney effect is particularly harmful to high-rise buildings in case of fire. In case of fire, stairways and elevator shafts will act as chimneys if fire prevention measures are not taken. According to the actual measurement, the vertical flow velocity of smoke in a fire can reach 2~4m/s, and buildings with dozens of floors will be filled with hot smoke in less than a minute.

2.5 Combustion and explosion phenomena in life

The phenomenon of burning and explosion is very common in daily life, but behind the simple phenomena, there are often rich scientific principles. Here is an example of the process of candle burning. In the process of candle burning, there are many basic physical knowledge of combustion and explosion: candles generate heat to the outside world mainly through thermal radiation, the shape of candle flame is affected by thermal convection, the internal heat transfer of candles is mainly through thermal conduction, and the continuous combustion of candles depends on the consecutive transfer of melted wax oil, etc.

2.5.1 Heat transfer during candle burning

The main raw material for candles is paraffin wax, which is made from waxy fractions of petroleum by cold pressing or solvent dewaxing. Candles are a mixture of several higher alkanes, mainly n-docosane ($C_{22}H_{46}$) and n-octacosane ($C_{28}H_{58}$), containing about 85% carbon and 14% hydrogen. The added auxiliary materials comprise white oil, stearic acid, polyethylene, essence and the like. Among them, stearic acid ($C_{17}H_{35}COOH$) is mainly used to improve softness. Candles are easy to melt, less dense than water and insoluble in water. They will melt into liquid when heated, appearing to be colorless and transparent, and volatile when slightly heated, and observers can smell the unique odor of paraffin. The candle burning we see is not the burning of paraffin solid, but the ignition device ignites the cotton core, releasing heat to melt the paraffin solid, and then vaporizes it. Paraffin vapor is generated, and the paraffin vapor is combustible.

When a candle burns, the products of combustion are carbon dioxide and water. Chemical expression is:

$$\text{Paraffin} + O_2 \xrightarrow{\text{Ignition}} CO_2 + H_2O \tag{2-57}$$

Paraffin wax contains only two elements, one is hydrogen and the other is carbon. In the process of combustion, paraffin vapor reacts with oxygen in the air. Hydrogen combines withoxygen to produce water, while carbon combines with oxygen to produce carbon dioxide. When the combustion is insufficient, carbon monoxide or carbon particles may also be produced. The chemical equation for full combustion is:

$$C_xH_y + \left(x + \frac{y}{4}\right) O_2 = \frac{y}{2} H_2O + xCO_2 \tag{2-58}$$

In fact, when paraffin is ignited, it does not directly produce carbon dioxide and water, but has a very obvious but not correctly recognized process. In fact, paraffin is ignited by hydrogen in the combustion process, and hydrogen actually releases atomic energy in the combustion process, which is the conversion from gaseous state to ionic state. The electrons released by the hydrogen pass through the carbon atoms and emit light and heat, which is the flame that people see. The flame core is mainly composed of paraffin vapor with the lowest temperature; the inner flame is a process in which electrons released by hydrogen emit light and heat through carbon atoms, and therefore the brightest part; the outer flame is a process in which carbon atoms react with oxygen in the air to form carbon dioxide and release chemical energy, as well as hydrogen ions in high energy states, so its temperature is the highest. As carbon dioxide and hydrogen ions rise to high altitude, the temperature gradually decreases, and hydrogen changes from ionic state to atomic state, and combines with oxygen in the air to form water molecules.

Each of us can easily draw the shape of a candle flame on paper with a pen, but how many of us have asked ourselves, "Why is the flame of a candle in the shape of an elongated droplet"? Why is the flame of a candle always upward, whether it is upright or horizontal? If we ask these questions, can we give the answers easily? Regarding the shape of the flame. Faraday's report "The Story of the Candle" has been discussed. Because of gravity, hot air is light, less dense, and tends to rise; cold air is heavy, and of high density. There is a tendency to sink. Cold and hot air flow to form convection, as shown in Fig. 2-8.

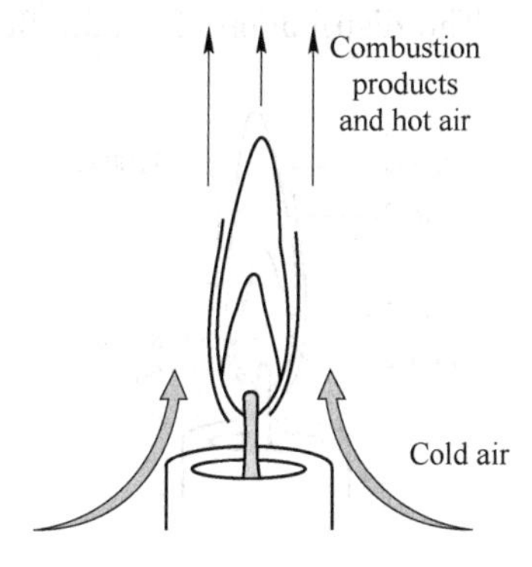

Fig. 2-8 Effect of convection on flame shape

The hot gases from the burning of the candle, as well as the heated air around the flame, will flow upward; at the same time, the cold air around them will be replenished, thus convection is formed. It is convection that keeps the flame always upward and elongates the flame of the candle. In addition, the added cold air cools the periphery of the top of the candle body to form the shape of a cup. The wax oil is stored in the candle cup and does not flow down easily, so that it can be continuously transported up the candle wick and burned. When the candle burns, if the airflow formed is irregular, it will not be able to form a flat cup mouth, which will lead to the continuous outflow and pouring of wax liquid, which is both wasteful and polluting.

2.5.2 Mass transfer in candle combustion

How is the wax oil continuously sent to the candle flame as fuel? Here is an interesting phenomenon called "capillarity", one is the attraction of a liquid surface to a solid surface. The line in the middle of the candle is called the wick, which absorbs the liquid wax. When candles are first produced, there is already solid wax on the wick, and when you light the candle with fire, the solid wax oil is heated, first melted and then vaporized, and then the wax oil vapor is ignited by the fire, so that the candle is ignited. When the candle burns, it releases heat to melt the solid wax oil below, which is sucked up by the candle wick and continues to be heated, melted, vaporized and ignited by the fire above, thus forming a cycle.

There is a reason why the candle flame cannot completely ignite the wick: the melted wax extinguishes it and prevents it from spreading. If a lighted candle is turned upside down, head up, bottom up, and the wax is allowed to run down the wick, the candleis extinguished. This is because the flame does not have time to heat the rapidly flowing wax oil to the point where it can burn.

When the candle is lit, the initial burning flame is small and gradually becomes larger. The flame is divided into three layers (outer flame, inner flame and flame core). The flame core is mainly candle steam with the lowest temperature. The paraffin wax in the inner flame does not burn fully, and the temperature is higher than that in the flame core. Because of some carbon particles, the outer flame is in full contact with the air, the flame is the brightest, burning fully, and the temperature is the highest. The distribution of candle flames is shown in Fig. 2-9.

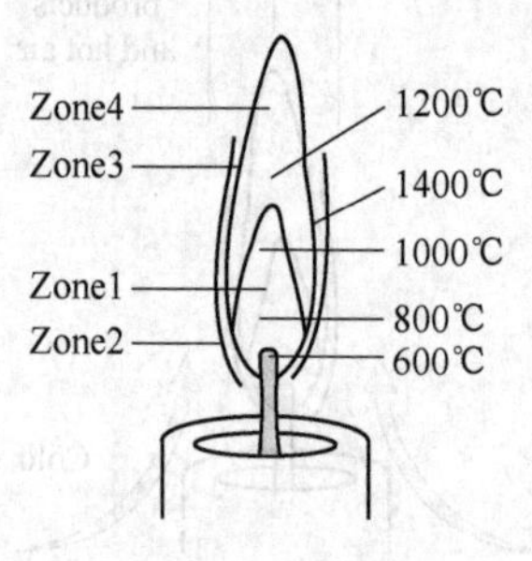

Fig. 2-9 Temperature distribution of candle flame

Compared with modern candles, ancient candles have many disadvantages. Li Shangyin, a poet of the Tang Dynasty, wrote a poem entitled "When exactly can we cut the candle in the west window together?" Why do poets cut candles? At that time, the candle wick was made of cotton thread and stood upright in the center of the flame. Because it could not be burned out and charred, the end of the remaining wick had to be cut off with scissors from time to time. This is undoubtedly troublesome. In 1820, Champaschelle, a Frenchman, invented a Candlewick made of three cotton threads, which loosened naturally when burning, and the end of the Candlewick was just outside the flame, so that it could burn completely. Another benefit of the complete combustion of the wick is that the carbonized wick is prevented from contaminating the molten pool to the stable combustion of the chimneys. At the same time, the black smoke produced by the incomplete combustion of the candle wick can be prevented from polluting the air.

Exercises

1. What are the ways of heat transfer? Try to give examples to illustrate the characteristics of these heat transfer modes.
2. The ambient temperature around the heat shield is 20℃, 30mm thick insulation on the wall. The thermal conductivity of the insulating material is 0.2W/(m·K). The surface temperature of the insulating layer in contact with the flat wall is 230℃, while the outer surface temperature is 40℃. Solve the convection heat transfer coefficient between the outer surface of the insulation layer and the medium.
3. There are two black body surfaces parallel and very close to each other, if: (1) the surface temperatures are 1000K and 800K, respectively; (2) the surface temperatures are 400K and 200K, respectively. Find the ratio of the radiant heat transfer in these two cases.
4. One steel plate with thermal conductivity of 380W/(m·K) has a thickness of 3mm, and its two sides are closely attached to the stainless steel plate with a thermal conductivity of 16W/(m·K) and a thickness of 2mm. The outer wall temperatures of the stainless steel plate are 400℃ and 100℃ respectively. Solve the temperature distribution inside the steel plate and the density of heat flux flowing through the flat wall.
5. Try to solve Steady-State Heat Conduction Equation in Single-Layer Cylindrical Wall by Fourier's Law.
6. According to the Buoyancy Movement Caused by Combustion, try to analyse the formation process and harm of "chimney phenomenon" in the fires of high-rise Building.
7. What is diffusion? What is the relationship between diffusive material flow and its concentration gradient distribution?
8. Give examples to illustrate the characteristics of Stephen flow and its generation conditions.

Chapter 3 Ignition Theory

扫码获得
数字资源

3.1 Ignition classification and ignition conditions

3.1.1 Classification of fire

The ignition modes of combustibles are generally divided into the following categories:

(1) Chemical spontaneous combustion: for example, a match catches fire by friction; an explosive explodes by impact; spontaneous combustion of metallic sodium in air Spontaneous combustion of bituminous; bituminous coal spontaneously ignited due to excessive accumulation, etc. This kind of ignition phenomenon usually does not require external heating, but occurs at room temperature according to its own chemical reaction, so it is customarily called chemical spontaneous combustion.

(2) Thermal spontaneous combustion: if the mixture of combustible and oxidant is uniformly heated in advance, with the increase of temperature, when the mixture is heated to a certain temperature, it will automatically ignite (at this time, the ignition occurs in the whole volume of the mixture). This ignition mode is customarily called thermal self-ignition.

(3) Light up (Or called forced ignition): refers to the ignition of the combustible mixture due to the strong heating of the local area by the energy obtained from external energy sources, such as electric heating coils, electric sparks, hot particles, ignition flames, etc. At this time, the flame will be initiated near the ignition source, and then spread to the whole combustible mixture by means of the combustion wave. This way of ignition is customarily called smoldering. Most fires are caused by smoldering.

It must be pointed out that the above three fire classification methods do not properly reflect the relationship and difference between them. For example, both chemical spontaneous combustion and thermal spontaneous combustion have the effect of both chemical reaction and heat, while the difference between thermal spontaneous combustion and ignition is only the difference between overall heating and local heating it is by no means the difference between "automatic" and "forced". In addition, fire is sometimes called explosion, and thermal spontaneous combustion is also called thermal explosion. This is because the characteristics of ignition at this time are similar to those of explosion, the chemical reaction rate increases sharply with time, and the reaction process is very rapid.

3.1.2 Ignition conditions

Generally speaking, the so-called ignition refers to the process in which the reaction of the mixture in intuition is automatically accelerated, and the temperature is automatically increased so as to cause a part of the space and eventually a flame appears at a certain time. This process reflects an important sign of combustion reaction, that is, from one part of space to another. Or the phenomenon that there is a sudden jump in the number of chemical reactions from one moment of time to another, which can be represented by Fig. 3-1.

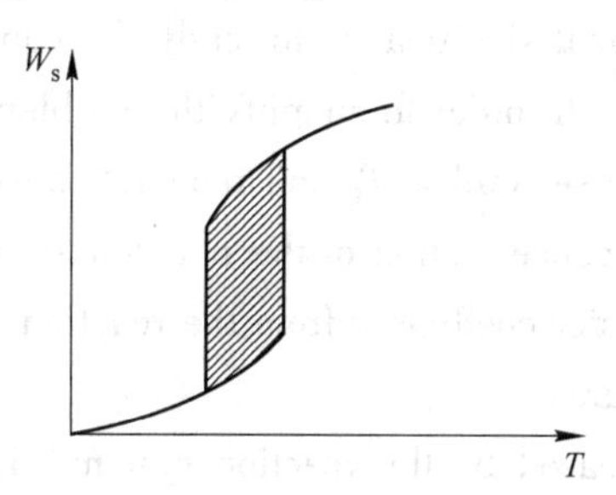

Fig. 3-1 External signs of the ignition process

Fig. 3-1 shows that the ignition condition is: if under certain initial conditions, the system will not be able to maintain a slow reaction state at a low temperature level throughout the time period, but will have a violent and accelerated transition process, so that the system will reach a high temperature reaction state at a certain moment, that is, to reach the combustion state, then this initial condition is the ignition condition.

It should be noted that: (1) The fact that the system has reached the ignition conditions does not mean that it has been ignited, but only that the system has met the ignition conditions; (2) The phenomenon of ignition is based on the initial state of the system, and its critical property cannot be wrongly interpreted as the property that the chemical reaction rate jumps with the change of temperature. Therefore, the temperature represented by the abscissa in Fig. 3-1 is not the temperature of the reaction, but the initial temperature of the system; (3) The ignition condition is not a simple initial temperature condition, but a comprehensive reflection of chemical kinetics parameters and fluid mechanics parameters. For a certain kind of combustible premixed gas, the ignition conditions can be expressed by the following functional relationship under the closed condition:

$$f(T_0, h, p, d, u_\infty) = 0$$

Where, T_0 represents the ambient temperature; h is the convective heat transfer coefficient; p is the premixed gas pressure; d is the diameter of the container; u_∞ is the ambient air velocity.

3.2 Semyonov thermal spontaneous combustion theory

3.2.1 Semenov's theory of thermal spontaneous combustion

Combustible mixed gas in any reaction system, on the one hand, it will slowly oxidize and release heat to raise the temperature of the system, and at the same time, the system will dissipate heat outward through the wall to lower the temperature of the system. According to the theory of thermal self-ignition, ignition is the result of the interaction between exothermic factors and heat dissipation factors. If the exothermic reaction is dominant, the system will accumulate heat. As the temperature increases, the reaction accelerates and spontaneous combustion occurs; On the contrary, if the heat dissipation factor is dominant and the temperature of the system drops, it cannot spontaneously combust.

Therefore, it is of great practical significance to study the conditions of fuel self-ignition under the condition of heat dissipation. In order to simplify the problem for study, suppose:

(1) The temperature of the vessel wall is T_0 and remains constant;

(2) The temperature and the concentration of the reaction system are uniform;

(3) The convective heat transfer coefficient from the reaction system to the vessel wall is h and does not vary with the temperature;

(4) The quantity of heat released by the reaction system (i. e., the heat of reaction at this stage) Q is constant (J/mol).

If the volume of the reaction vessel is V and the reaction rate is W (the change in the amount of substance per unit volume per unit time), then the amount of heat released by the reaction system per unit time q_1 for:

$$q_1 = QVW \tag{3-1}$$

According to the theory of chemical reaction rate and Arrhenius law, for a general second-order reaction, the reaction rate can be expressed by the following formula within the time to ignition:

$$W = K_0 C_A C_B e^{-\frac{E}{RT}} \tag{3-2}$$

Where, K_0 is the Arrhenius reaction rate constant; C_A and C_B are the molar concentrations of fuel and air molecules respectively, and the heat release of the system is obtained by substituting the W value into the former formula:

$$q_1 = K_0 QVC_A C_B e^{-\frac{E}{RT}} \tag{3-3}$$

The amount of heat Q_2 lost through the wall per unit time can be expressed by the following equation (when the temperature is not high, the radiation loss is negligible):

$$q_2 = hA(T - T_0) \tag{3-4}$$

Where, h is the convective heat transfer coefficient through the wall; A is the heat transfer area of the wall; T is the reaction system temperature; T_0 is the vessel wall temperature.

At the initial stage of the reaction, C_A and C_B are very close to the initial concentrations C_{A0} and C_{B0} before the start of the reaction, and Q, V, and K_0 are all constant, so the exothermic rate q_1 and the mixture temperature T is an exponential function, that is, $q_1 \sim e^{-\frac{E}{RT}}$, as shown by the curve values in Fig. 3-2. As the pressure (or concentration) of the mixture increases, the curve shifts to the upper left (q_1).

And the relationship between heat dissipation rate and mixture temperature is a linear function, as shown in line q_2 in Fig. 3-2. When the temperature T_0 of the container wall increases, the straight line moves to the right, for example q_2''. When the heat release rate is less than the heat dissipation rate ($q_1 < q_2$), the temperature of the reactants will gradually decrease, and it is obviously impossible to cause ignition. Conversely, if the rate of heat release is greater than the rate of heat dissipation ($q_1 > q_2$), there is always a chance that the mixture will catch fire. For example, when the pressure of the gas mixture is increased, the rate of the exothermic reaction increases from q'_1, while the vessel wall temperature remains at T_0, at this time, the rate of heat dissipation is much lower than the rate of heat release, so that the mixture can heat itself and ignite at any time.

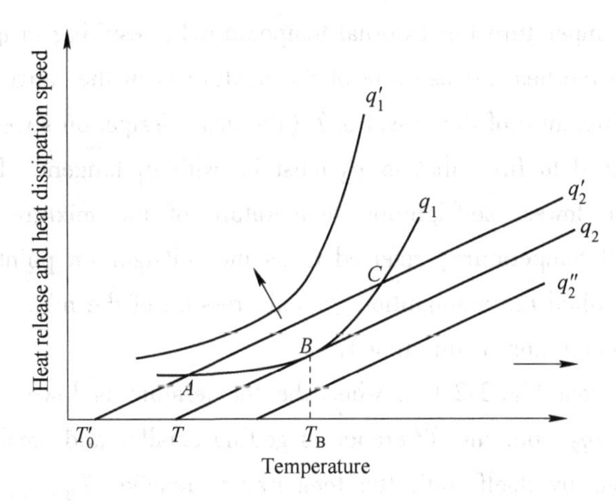

Fig. 3-2 Heat release and dissipation rate of mixed gas in the container

From the above analysis, it can be seen that the transition of the reaction from impossible ignition to possible ignition must pass through one point, namely $q_1 = q_2$, that's what it takes to catch fire. But $q_1 = q_2$ is not yet a sufficient condition for ignition, as can be seen from the following. For example, the pressure of the mixed gas is reduced so that the exothermic rate of the reaction increases from q_1 while the temperature of the vessel wall is maintained T'_0. At this time q_1 and heat dissipation speed q'_2, intersect at points A and C. Satisfied on both points $q_1 = q_2$ conditions, but they are not yet the ignition point. Point a indicates that the system is in a stable thermal equilibrium state. If the temperature rises slightly, the heat dissipation rate exceeds the heat release rate, and the temperature of the system will automatically decrease and return to the stable state of point A. If the temperature increases from point A is slightly lowered, at which point $q_1 > q'_2$, the temperature of the system will rise and return to point A. As a result, the system will react isothermally at point A for a long time, and it is impossible to cause ignition. In contrast, point C indicates that the system is in an unstable state of thermal equilibrium: as long as there is a slight decrease in temperature, the rate of heat release from the system q_1 is less than the heat dissipation speed q'_2, resulting the system cools down and returns to point A; if there is a slight increase in temperature , then $q_1 > q'_2$, the temperature of the system keeps rising, resulting in a fire. But this is not the ignition temperature at all, because if the initial temperature of the system is T'_0, it is not possible for it to heat itself past point A to point C. Unless an external energy source heats the system and raises the temperature of the system to point C, the system is always in a stable state at point A. Therefore, point C is not the automatic ignition temperature of the mixture, but the forced ignition temperature of the mixture. In that adiabatic compression of the mixture, this situation can be encountered, for example, in a diesel engine, where the cylinder wall temperature is not high, but the mixture is strongly compressed and heated to the forced ignition temperature.

From the above, it can be seen that under a certain pressure (or concentration), a certain mixed gas reaction system can only change from slow reaction to rapid automatic heating at a certain container wall temperature (or external temperature), resulting in ignition. It can be seen from Fig. 3-2 that when the heat release rate of the mixture is in the range of q_1. The curve runs only when the wall temperature of the vessel is T_0 (the heat dissipation speed according to q_2) can be automatically converted to fire, that is, q_2 must be with q_1 tangent. The temperature at the tangent point B is the lowest self-ignition temperature of the mixture at this pressure (or concentration) and wall temperature, referred to as the self-ignition point. The pressure of the mixture at this time is called the autoignition critical pressure of the mixture (or the mixture is at the natural critical concentration at this time).

It can also be seen from Fig. 3-2 that when the temperature is lower than T_B and gradually heated, the mixture $q_1 > q_2$, but the difference is getting smaller and smaller. At this time, the mixture is slowly heated by itself until the temperature reaches T_B, q_1 still greater than q_2. However, with the increase of temperature, the difference between the two becomes larger and larger, which promotes the violent reaction, and the reaction gradually turns into an explosion.

From this point of view, the definition of ignition temperature includes not only the condition that the heat release rate and the heat dissipation rate of the exothermic system are equal, but also the condition that the two rates of change with temperature should be equal, that is,

$$q_1 = q_2 \tag{3-5}$$

$$\frac{dq_1}{dT} = \frac{dq_2}{dT} \tag{3-6}$$

This is the condition under which the reaction changes from slow to ignition under the condition of heat dissipation.

It can be seen that the ignition temperature of the mixture is not a constant, but varies with the nature of the mixture, the pressure (concentration), the temperature and thermal conductivity of the container wall, and the size of the container. In other words, the ignition temperature depends not only on the reaction rate of the mixture, but also on the heat dissipation rate of the surrounding medium. When the property of the mixed gas is not changed, reducing the surface of the container and improving the insulation of the container can reduce the autoignition temperature or the critical pressure of the mixture. Two related issues are discussed below.

3.2.2 Relationship between ignition temperature and vessel wall temperature

According to the conditions at the time of the fire $q_1 = q_2$, it is known that:

$$K_0 Q V C_A C_B e^{-\frac{E}{RT_B}} = hA(T_B - T_0) \tag{3-7}$$

According to $\frac{dq_1}{dT} = \frac{dq_2}{dT}$, derivate the above equation to obtain (expect T_B, the rest are known):

$$K_0 QVC_A C_B \frac{E}{RT_B^2} e^{-\frac{E}{RT_B}} = hA \tag{3-8}$$

Divide equation (3-8) by equation (3-7) to obtain:

$$T_B - T_0 = \frac{RT_B^2}{E} \tag{3-9}$$

The value of T_B can be found by solving the quadratic equation:

$$T_B = \frac{E}{2R} + \frac{E}{2R}\sqrt{1 - \frac{4RT_0}{E}} \tag{3-10}$$

The sign before the square root in the formula should be negative, otherwise the result is too large and does not conform to the actual situation. Expand the square root in the formula (3-10) in series to obtain:

$$T_B = \frac{E}{2R} - \frac{E}{2R}\left(1 - \frac{2RT_0}{E} - \frac{2R^2T_0^2}{E^2} + \Lambda\right) - T_0 + \frac{RT_0^2}{E} + \Lambda \tag{3-11}$$

Considering that $E = 209350\text{kJ/mol}$ in general, when the error is within 0.5%, the following items can be ignored. Thus

$$\Delta T = T_B - T_0 \approx \frac{RT_0^2}{E} \tag{3-12}$$

Equation (3-12) shows that, under ignition conditions, the ignition temperature of the mixture T_B. The difference from the initial temperature T_0 (wall temperature) adapted to the ignition conditions is small.

3.2.3 Relationship between mixture pressure and other parameters during ignition

In the formula (3-11), T_B substitute the value of into the formula (3-7) to obtain:

$$K_0 QVC_A C_B e^{-\frac{E}{R\left(T_0 + \frac{RT_0^2}{E}\right)}} = hA\left(\frac{RT_0^2}{E}\right) \tag{3-13}$$

Because $\frac{RT_0}{E} = \frac{\Delta T}{T_0} \ll 1$, in the above formuls

$$T_0 + \frac{RT_0^2}{E} = T_0\left(1 + \frac{RT_0}{E}\right) \approx T_0$$

Thus, equation (3-13) can be written as:

$$\frac{K_0 QVC_A C_B E}{hART_0^2} \cdot e^{-\frac{E}{RT_0}} = 1 \tag{3-14}$$

Let the total molar concentration of the reactants be C, $C = C_A + C_B$, x_A represents the mole fraction of the fuel, x_B represents the mole fraction of air (oxygen), then

$$C_A = Cx_A$$
$$C_B = Cx_B$$

At the same time, under the ignition condition, according to the ideal gas state equation:

$$C = \frac{P_C}{RT} \tag{3-15}$$

In the formula P_C is the critical pressure of the mixture (i. e. the pressure of the mixture at ignition).

Replace $C_A C_B$ in equation (3-14) with the function of pressure and temperature to obtain

$$\frac{K_0 QVEP_C^2 x_A x_B}{hAR^3 T_0^4} \cdot e^{-\frac{E}{RT_0}} = 1 \tag{3-16}$$

This is the relationship between mixture pressure and temperature and other parameters under ignition conditions. When other conditions are known, the pressure of the mixture is less than that in equation (3-16). P_C value, the mixture cannot ignite; if it is greater than this value, it can ignite. This formula is the basic formula for the ignition condition. It can also be written in logarithmic form as

$$\ln\frac{P_C}{T_0^2} = \frac{E}{2RT_0} + \frac{1}{2}\ln\frac{AhR^3}{K_0 QVEx_A x_B} \tag{3-17}$$

Under certain conditions of container and gas mixture composition, the above formula can also be written as:

$$\ln\frac{P_C}{T_0^2} = \frac{A}{T_0} + B \tag{3-18}$$

Where A and B are constants.

The above equation is called Semenov equation, which is one of the basic formulas of ignition conditions. From this equation, we can see some basic laws of ignition.

(1) When the composition of the mixture and the shape of the container are constant, the higher the external temperature T_0 (container temperature), the smaller the critical pressure required for ignition. In equations (3-16) to (3-18), P_C value follows $e^{-\frac{E}{RT_0}}$ change in one term far exceeds the change in T_0 in the denominator. Accordingly, as that value of T_0 increase, P_C always decrease. In the actual experiment, the relationship between the critical pressure of various hydrocarbons and the temperature of the container is shown in Fig. 3-3. At the lower left of the curve in Fig. 3-3 (no explosion zone), the fuel does not ignite; at the lower right of the curve (explosion area) can cause fire. It can be seen from the figure that when the temperature of the container decreases, the pressure of the mixture required to cause ignition increases; When the pressure of the mixture decreases, the outside temperature must be raised to ensure ignition. The ignition range of the mixture shrinks no matter whether the pressure or the temperature decreases. On the contrary, increasing the pressure or temperature of the mixture is beneficial to the ignition of the fuel.

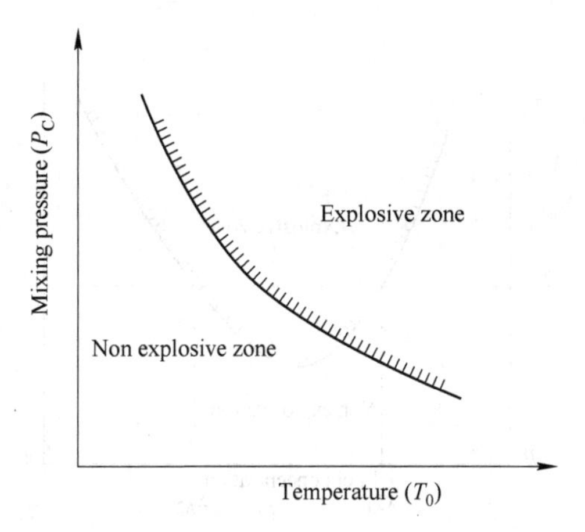

Fig. 3-3 Relationship between ignition critical pressure and vessel temperature

When the flight altitude of the aircraft engine increases, the combustion performance deteriorates, such as the decrease of restart performance, incomplete combustion, etc. The main reason is that the pressure and temperature in the combustion chamber decrease, which slows down the reaction speed and increases the heat dissipation speed, thus making it difficult for the fuel to ignite. This is also the main reason why it is particularly difficult to start a car when driving in the plateau or cold region.

(2) The composition of the mixture is closely related to the ignition. When the temperature is constant and the fuel concentration (mole fraction x_A) begins to decrease, P_C will gradually decrease, but after exceeding a certain value, P_C will gradually increase due to the increase of air concentration x_0. The relationship between them is shown in Fig. 3-4. When the pressure is constant, the relationship between mixture concentration and ignition temperature is the same. As shown in Fig. 3-5.

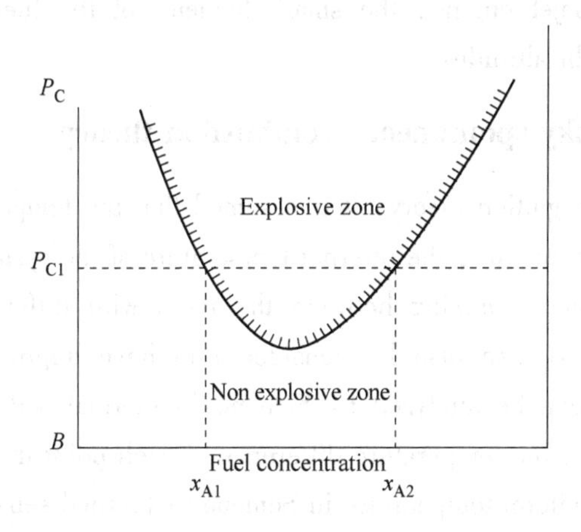

Fig. 3-4 Relationship between mixture composition and ignition critical pressure

It can also be seen from Fig. 3-4 and Fig. 3-5 that under a certain pressure and external temperature, not all components of the mixture can cause ignition, but only the mixture within a

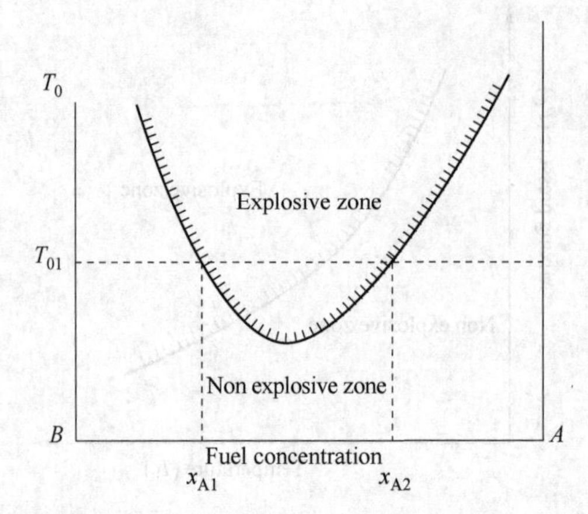

Fig. 3-5 Relationship between mixture composition and ignition temperature

certain concentration limit can cause ignition. For example, at the ambient temperature T_{01} in Fig. 3-5, the mixture composition can only cause ignition between x_{A1} and x_{A2}. This range is called the combustion limit or explosion limit of the mixture at this temperature. When the temperature or pressure of the mixture increases, the explosion limit also increases. On the contrary, when T_0 and P_C decrease, the explosion limit also decreases. Below certain values of T_0 and P_C, no mixture can ignite. Generally, the value of P_C is the smallest when the fuel is mixed with air at a stoichiometric ratio.

(3) The ratio of the volume of the combustion chamber to the heat dissipation area of the vessel also has an effect on the critical pressure of ignition. It can be seen from the formula (3-16) that the larger the combustion chamber volume (mixed gas volume V) or the smaller the container wall area, the lower the critical pressure P_C of mixture ignition, that is, the more conducive to ignition. Thus, in a turbojet engine, the small diameter of the fuel cell is not conducive to operation at very high flight altitudes.

3.3 Frank-Kamenetsky spontaneous combustion theory

In Semenov's thermal self-ignition theory, it is assumed that the temperature of each point in the system is equal. For a gas mixture, the internal temperature of the system can be considered to be uniform due to the convective mixing between the parts with different temperatures. A solid substance with a smaller B_i can also be considered to have approximately the same internal temperature. Both cases can be analyzed by Semenov's thermal self-ignition theory. However, when B_i is large ($B_i>10$), the temperature difference of each point in the system is large. In this case, the assumption of uniform temperature in Semenov's thermal self-ignition theory is obviously not valid, as shown in Fig. 3-6.

Therefore, it is necessary to establish a new theoretical model to analyze the material system under the large B_i number, which is the Frank-Kamenetsky thermal spontaneous combustion theory. The theory takes into account the inhomogeneity of temperature distribution in the material

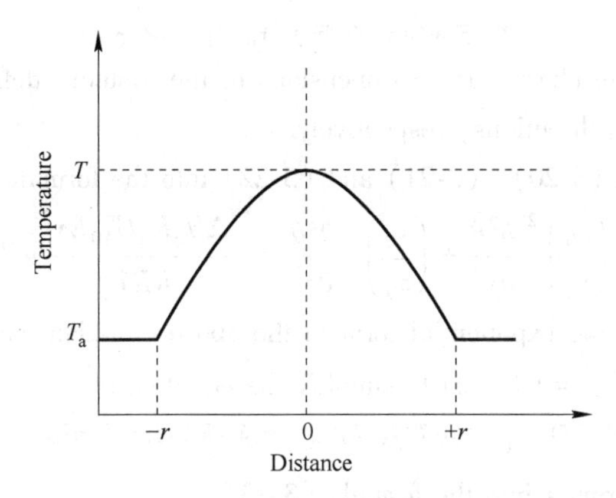

Fig. 3-6 Temperature profile in the reaction system of Frank-Kamenetsky theory

system under the condition of large B_i number, and takes whether the system can finally obtain a steady-state temperature distribution as the criterion for judging spontaneous combustion. A steady-state analysis method for thermal self-ignition is proposed.

3. 3. 1 Frank-ramenetskil's thermal spontaneous combustion theory

According to this theory, when combustible substances are piled up, the oxygen in the air will have a slow oxidation reaction with them. On the one hand, the heat released by the reaction will increase the internal temperature of the object, on the other hand, it will dissipate to the environment through the boundary of the accumulation body. If the system does not have the conditions for spontaneous combustion, the internal temperature will gradually increase from the time of material accumulation. And after a period of time, the internal temperature distribution of the material tends to be stable, when the heat released by the chemical reaction is equal to the heat lost by the boundary heat transfer. If the system has the conditions for spontaneous combustion, the system will catch fire after a period of time (called the ignition delay period) from the beginning of material accumulation. Obviously, in the latter case, before the autoignition of the system, a time-dependent unsteady temperature distribution appears inside the substance. Therefore, whether the system can reach the steady-state temperature distribution becomes the basis for judging whether the material system can spontaneously combust.

When the system does not meet the self-ignition conditions, the steady-state temperature distribution equation is obtained:

$$\frac{\partial^2 T}{\partial x^2} + \frac{\partial^2 T}{\partial y^2} + \frac{\partial^2 T}{\partial z^2} + \frac{q}{K} = 0 \qquad (3\text{-}19)$$

In that formula, Q is the exothermic rate of the reaction system.

$$Q''' = \Delta H_c K_n C_{A0}^n \exp(-E/RT) \qquad (3\text{-}20)$$

For the convenience of analysis, dimensionless temperature and the dimensionless distances x_1, y_1, z_1 are introduced:

$$\theta = (T - T_0)/(RT_0^2/E) \qquad (3\text{-}21)$$

$$x_1 = x/x_0, y_1 = y/y_0, z_1 = z/z_0 \tag{3-22}$$

Here x_0, y_0, z_0 are the characteristic dimensions of the system, defined as the length of the system in the x, y, and z directions, respectively.

Substitute the formulas (3-20), (3-21) and (3-22) into the formula (3-19).

$$\frac{\partial^2\theta}{\partial x_1^2} + \left(\frac{x_0}{y_0}\right)^2 \frac{\partial^2\theta}{\partial y_1^2} + \left(\frac{x_0}{z_0}\right)^2 \frac{\partial^2\theta}{\partial z_1^2} = -\frac{\Delta H_c K_n C_{A0}^n E x_0^2}{KRT_0^2} e^{-E/(RT)} \tag{3-23}$$

Since $(T-T_0) \ll T_0$, the exponential term in the above equation can be expressed as follows when z is small, $(1+z)^{-1} = (1-z)$ to simplify the equation of,

$$e^{-E/(RT)} = e^{-E/R(T+T_0-T_0)} = e^{-[E/(RT_0)][1+(T-T_0)/T_0]^{-1}}$$

Substitute the above formula into the formula (3-23).

$$\frac{\partial^2\theta}{\partial x_1^2} + \left(\frac{x_0}{y_0}\right)^2 \frac{\partial^2\theta}{\partial y_1^2} + \left(\frac{x_0}{z_0}\right)^2 \frac{\partial^2\theta}{\partial z_1^2} = -\delta\exp(\theta) \tag{3-24}$$

Where

$$\delta = \frac{\Delta H_c K_n C_{A0}^n E x_0^2}{KRT_0^2} e^{-E/(RT_0)} \tag{3-25}$$

The corresponding boundary conditions are: at the boundary surface $z_1 = f_1(x_1, y_1)$ go $\theta = 0$; at maximum temperatures

$$\frac{\partial\theta}{\partial x_1} = 0, \frac{\partial\theta}{\partial y_1} = 0, \frac{\partial\theta}{\partial z_1} = 0$$

It is clear that the solution of equation (3-24) is completely controlled by x_0/y_0, x_0/z_0 and δ, that is, the steady-state temperature distribution inside the object depends on the shape of the object and the size of δ. When the shape of the object is determined, its steady-state temperature distribution depends only on the value of δ.

According to the formula (3-25), δ characterize that relative magnitude of the chemical heat release inside the body and the outward heat transfer through the boundary. Therefore, when δ greater than a certain critical value δ_{cr}, equation (3-24) has no solution, that is, the steady-state temperature distribution cannot be obtained inside the body. Obviously, δ_{cr} depends only on the shape of the system, as can be seen from the analysis of equation (3-24).

When $\delta = \delta_{cr}$, the parameters related to the system are all critical parameters, and the ambient temperature at this time is called critical ambient temperature $T_{a,cr}$, from equation (3-25):

$$\delta_{cr} = \frac{\Delta H_c K_n C_{A0}^n E x_{0c}^2}{KRT_{q,cr}^2} e^{-E/(RT)_{q,cr}} \tag{3-26}$$

If substances are stacked in simple shapes such as infinite flat plate, infinite cylinder, sphere and cube, the internal heat conduction can be summarized as one-dimensional heat conduction, and the corresponding steady-state heat conduction equation is established as shown in Fig. 3-6:

$$\frac{d^2T}{dx^2} + \frac{\beta}{x}\frac{dT}{dx} + \frac{Q'''}{K} = 0 \tag{3-27}$$

Where $\beta = 0$ for a flat plate of thickness $2x_0$, $\beta = 1$ for an infinite cylinder of radius x_0, $\beta = 2$

for a sphere of radius x_0 and β = 3.28 for a cube of side length $2x_0$.

The corresponding nondimensionalization of equation (3-27) yields:

$$\frac{\partial^2\theta}{\partial x_1^2} + \frac{\beta}{x_1}\frac{d\theta}{dx_1} = -\delta\exp(\theta) \tag{3-28}$$

The expression of δ is the same as equation (3-25). For these simple shapes, the critical self-ignition criterion parameters δ_{cr} are obtained by mathematical solution: for an infinite flat plate, $\delta_{cr} = 0.88$; for an infinite cylinder, $\delta_{cr} = 2$; for spheres, $\delta_{cr} = 3.32$; for versus cubics, $\delta_{cr} = 2.52$. When the system $\delta > \delta_{cr}$, the system spontaneously combusts.

3.3.2 Solution of critical criterion parameter δ_{cr} for spontaneous combustion

For a substance with a simple geometrical shape, δ_{cr} can be solved mathematically. Where, an infinite flat plate is taken as an example to illustrate the solution process of δ_{cr}.

The dimensionless heat conduction equation and boundary conditions are:

$$\frac{\partial^2\theta}{\partial x_1^2} + \partial e^{\theta} = 0 \tag{3-29}$$

$$\text{When} \quad x_1 = 1,\ \theta = 0 \tag{3-30}$$

$$\text{When} \quad x_1 = -1,\ \theta = 0 \tag{3-31}$$

Solving equation (3-29) yields:

When $x_1 \geqslant 0$ $$x_1 = -\frac{1}{\sqrt{2a\delta}}\ln\frac{1-\sqrt{1-e^{\theta}/a}}{1+\sqrt{1-e^{\theta}/a}} + b$$

When $x_1 < 0$ $$x_1 = \frac{1}{\sqrt{2a\delta}}\ln\frac{1-\sqrt{1-e^{\theta}/a}}{1+\sqrt{1-e^{\theta}/a}} + b$$

Where a and b are constants of integration to be determined.

Applying the boundary condition equations (3-30) and (3-31), we obtain:

$$1 = -\frac{1}{\sqrt{2a\delta}}\ln\frac{1-\sqrt{1-e^{\theta}/a}}{1+\sqrt{1-e^{\theta}/a}} + b$$

and

$$-1 = \frac{1}{\sqrt{2a\delta}}\ln\frac{1-\sqrt{1-e^{\theta}/a}}{1+\sqrt{1-e^{\theta}/a}} + b$$

The two expressions are subtracted and sorted out to obtain:

$$\delta = \frac{1}{2a}\left(\ln\frac{1-\sqrt{1-1/a}}{1+\sqrt{1-1/a}}\right)^2 \tag{3-32}$$

Fig. 3-7 below shows the δ as a function of a, it can be seen from the figure that there is a maximum value of δ, when δ is greater than this maximum, there is no solution to a, and correspondingly there is no solution to the steady-state heat conduction equation, so this maximum is the required critical criterion parameter for self-ignition δ_{cr}, in the figure, $\delta_{cr} \approx 0.88$.

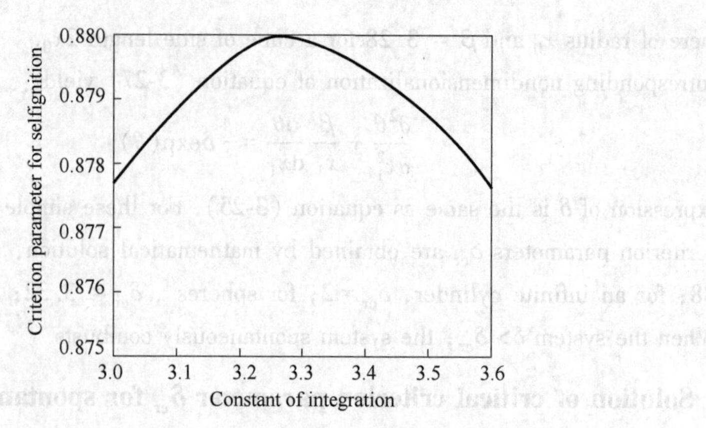

Fig. 3-7 δ variation with *a*

3. 3. 3 Theoretical application

The conditions of spontaneous combustion of various material systems can be studied by using the frank karmenetsky theoretical model of thermal spontaneous combustion and some experimental means. This is undoubtedly meaningful for preventing spontaneous combustion of substances and determining the cause of fire. Arrange the relation (3-26) and take logarithms on both sides to get:

$$\ln\left(\frac{\delta_{cr} \cdot T_{a,cr}^2}{x_{0c}^2}\right) = \ln\left(\frac{E\Delta H_c K_n C_{A0}^n}{KR}\right) - \frac{E}{RT_{a,cr}} \tag{3-33}$$

This shows that for a particular substance, the first term on the right is a constant, and the left term is a linear function of $1/T_{a,cr}$. For many systems, this linear relationship holds. For a given geometry of the material, the relationship between $T_{a,cr}$ and x_{0c} (the characteristic dimensions of the specimen) can be determined by experiment. The following example illustrates the application of the *F-K* autoignition model to predict the likelihood of autoignition of a substance.

【Example 3-1】 That data of the cubic stack activate carbon obtained by the experiment are as follow. It is known that when the material is stacked in the form of an infinite flat plate, the minimum stacking thickness has the risk of spontaneous ignition at 40℃.

x_0 (cubic stack side length) (mm)	25.40	18.60	16.00	12.50	9.53
$T_{a,cr}$ (critical temperature) (K)	408	418	426	432	441

【Solution】 The following table is made according to the test data provided:

$\ln(2.52T_{a,cr}^2/x_{0c}^2)$	6.47	7.15	7.49	8.01	8.59
$1000/T_{a,cr}$	2.45	2.39	2.35	2.35	2.27

Use the above table with $1000/T_{a,cr}$ as the horizontal axis, with $\ln(2.52T_{a,cr}^2/x_{0c}^2)$ as the vertical axis for Coordinate system.

It follows from Fig. 3-8 that $T_{a,cr}=40+273=313\text{K}$, $\ln(\delta_c T_{a,cr}^2/x_{0c}^2)=-2.2$. For the "semi-infinite plate" stacking mode, $\delta_c=0.88$

So

$$\ln(0.88\times 313^2/x_{0c}^2)=-2.2$$

From which we can get

$$x_{0c}=839(\text{mm})$$

That is to say, when the ambient temperature is 40℃, in order to avoid spontaneous combustion, the thickness of activated carbon stacked in the form of "semi-infinite plate" should not be greater than $2x_{0c}=1.678\text{m}$.

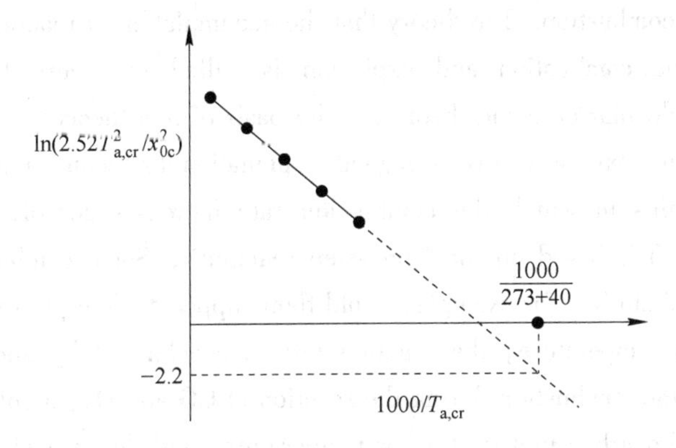

Fig. 3-8 Relationship between the self-ignition data of cubic reactor activated carbon

For combustible gas-air mixtures, the delay rarely exceeds 1 second. For solid piles, the auto-ignition delay can be several hours or several days or even several months, depending on the amount of material stored and the ambient temperature. The experimental results show that the autoignition delay time is 68 hours for the cubic stack of activated carbon with a side length of 1.2m. Therefore, for the larger size of the packed solid, the delay period of spontaneous combustion is longer, even if the experimental conditions and funds permit, people are not willing to spend such a long time to do the experiment. Therefore, the Frank-Kamenetsky model of spontaneous combustion provides a good way for people. With this method, we can determine the conditions of spontaneous combustion of a large number of stacked solids through small-scale experiments. It provides a solid theoretical basis for preventing the accumulation of solid spontaneous combustion and determining the cause of spontaneous combustion fire.

3.4 Chain reaction

3.4.1 Thermal deflagration theory and its limitations

According to the molecular collision theory of chemical reaction and Arrhenius law, the reaction rate is determined by the number of molecules of the activation energy required to reorganize the molecular structure of the reactants when the kinetic energy collides between the molecules in the system, and the average velocity determining the kinetic energy of the molecules is a function of temperature. Therefore, the temperature of the reaction system not only affects the amount of molecular kinetic energy, But also affects the collision frequency between molecules.

When the reactants of combustion reaction are gathered together and under certain temperature conditions, some molecules can complete exothermic reaction due to collision between molecules and release combustion heat. If the reaction system is adiabatic, this part of combustion heat will increase the temperature of the whole reaction system, and the temperature increase will accelerate the reaction rate between reactants, and the heat release rate will also increase. When the temperature of the system is further increased, the reaction system seems to be in a process of positive feedback heating and accelerating the reaction until the reaction rate increases to infinity, which is explosion or combustion. The theory that the accumulation of reaction heat accelerates the reaction and even the combustion and explosion is called the thermal deflagration theory. Arrhenius law and molecular collision theory are the basis of this theory.

However, this theory can only give a logical explanation for some simpler combustion rate-temperature relationships in which the combustion rate increases sharply with the increase of reaction temperature. It is based on the " one-step reaction". Some combustion phenomena can not be explained satisfactorily, for example, a cold flame appears when phosphorus and ether vapor are oxidized at a low temperature, the reaction rate is accelerated by adding water vapor that cannot participate in the combustion during the reaction of CO and O_2, a cold flame appears when the combustion of hydrocarbon fuel is at a low temperature, and the explosion limit of the reaction of H_2 and O_2 is the so-called " burning peninsula phenomenon". The above phenomena can not be explain by that simple thermal deflagration theory, it forces people to make further research on the reaction and mechanism.

3.4.2 Concept of chain reaction

As mentioned above, Semenov's theory and thermal explosion theory show that spontaneous combustion and explosion are mainly caused by the fact that the heat released by the chemical reaction of the system is greater than the heat dissipated from the system to the surrounding environment, resulting in the accumulation of heat and the automatic acceleration of the reaction rate, which can explain many related combustion and explosion phenomena. It has been found in practice that the ignition is not in all cases an automatically accelerated reaction due to the accumulation of exothermic heat. In order to explain some special combustion and explosion phenomena, it is necessary to study the chemical reaction mechanism of combustion and explosion, and then the chain ignition theory should be used. According to this theory, the automatic acceleration of the reaction does not necessarily depend only on the accumulation of heat, but also on the rapid increase of activation centers through the branching of the chain reaction to accelerate the reaction until it catches fire and explodes. Chain reactions can be divided into two categories according to the chain initiation and chain transfer processes: straight-chain reactions (such as H_2+Cl_2) and branched-chain reactions (such as H_2+O_2). The former does not branch during its development. The latter will give rise to branched chains.

The reason why the chain reaction process can proceed at a very fast speed is that each elementary reaction or each step in the chain reaction produces one or more active centers. These activated centers then react with the reactants in the reaction system. The reaction activation energy of these elementary reactions is very small, generally below 4×10^4 kJ/kmol. It is much smaller than the activation energy of the usual molecule-to-molecule combination (such as 16×10^4 kJ/kmol). When ions, free radicals and atoms combine with each other, the activation energy is even smaller, almost close to zero.

3.4.3 Straight chain reaction

Only one activation center is produced at each step in the chain reaction process, and this activation center reacts with the reactants in the reaction system to produce products and new activation centers. This continues until the reaction is complete or the chain reaction is interrupted.

Usually, the chain reaction starts with a certain amount of energy input from the outside world (such as light, heat, impact, etc.), so that the molecules of the reactants differentiate into activation centers, which is called "chain initiation". This activated center reacts rapidly with other molecules and begins a chain reaction. Each step in the subsequent chain reaction process consumes an activated center and then produces another activated center for the next reaction; these successive chain reaction processes are "Propagation of chains". In the process of chain reaction, many activation centers are absorbed when they collide with the wall of the reaction vessel. The collision of the activation centers results in the formation of molecules with poor activity, which interrupts the chain reaction. Therefore, it is called chain scission, chain termination or chain interruption. Such as the reaction of chlorine with hydrogen, Chlorine molecules can first be decomposed into two chlorine atoms by the action of photons under illumination:

$$H_2 + Cl_2 \longrightarrow 2HCl$$

$$(1)\quad M + Cl_2 \longrightarrow 2Cl^* + M \quad \text{(Chain initiation)}$$

$$(2)\quad Cl^* + H_2 \longrightarrow HCl + H^* \quad \text{(Chain transmission)}$$

$$(3)\quad H^* + Cl_2 \longrightarrow HCl + Cl^*$$

$$(4)\quad H^* + HCl \longrightarrow H_2 + Cl^*$$

$$(5)\quad M + 2Cl^* \longrightarrow Cl_2 + M \quad \text{(Chain termination)}$$

From the above chain reaction process, the mechanism of HBr formation from H_2 and Br_2 can be understood, in which (4) and (5) are chain inhibition. Because the active centers H and Br in the two-step reaction do not react with the reactant H_2 or Br_2 in the system, but react with HBr (product) to regenerate the original reactant, and also produce H and Br that can carry out chain propagation. The reaction rate of the reaction process is not interrupted in these links but only inhibited, so this reaction is called the inhibition of the chain.

For such a linear reaction process, the rate of formation of HBr cannot be determined by its overall reaction formula:

$$H_2 + Br_2 \longrightarrow 2HBr \tag{3-34}$$

The reaction rate is calculated directly and must be analyzed according to the chain reaction mechanism.

3.4.4 Branch chain reaction (or branch chain reaction)

Branching chain reactions are characterized by the fact that each radical consumed in the reaction chain (elementary reaction) produces more than one new radical, which proceeds to the next chain-ring reaction. In this way, the speed of reaction increases geometrically, and the reaction can quickly reach the level of explosion.

The multiplication factor α is often used to express the multiplication of free radicals in the course of a chain reaction cycle. In the process of branching chain reaction, some steps can produce two free radicals, but some steps can only produce one free radical. Usually, the multiplication factor of chain reaction is $1 \leqslant \alpha \leqslant 2$. When $\alpha = 1$, it is a straight chain reaction, and when $\alpha > 1$, it is a branched chain reaction.

That is to say, after a radical H^* participates in the reaction, three H^* are produced at the same time as the final product H_2O is formed through a chain transfer, and these three H^* begin to form three other chains, and each H^* will produce three more. Thus, as the reaction proceeds, the number of H^* increases. So the reaction is accelerating. The increase in the number of H^* is shown in Fig. 3-9.

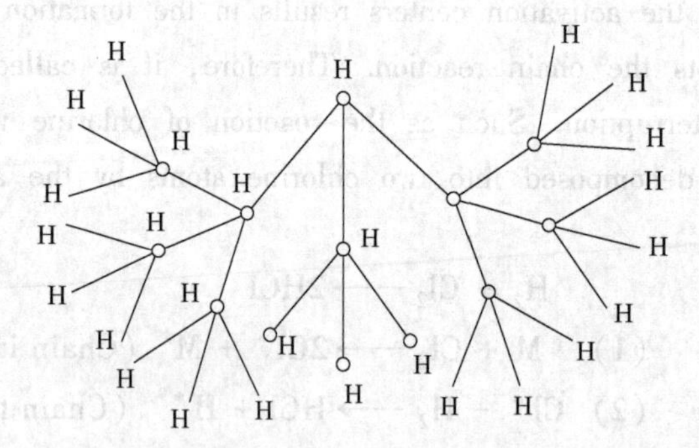

Fig. 3-9 Schematic diagram of hydrogen atom number increase

3.4.5 Ignition conditions of chain spontaneous combustion (development conditions of chain branching reaction)

The reaction rate of simple reactions decreases with time due to the continuous consumption of reactant concentration, but in some complex reactions, the reaction rate increases automatically with the increase of product concentration. The chain reaction is of the latter type, in which the rate of reaction is influenced by the concentration of some intermediate unstable product. After some external energy causes the reaction to produce activated centers, the propagation of the chain continues, the number of activated centers increases due to branching, and the reaction rate accelerates sharply, leading to ignition and explosion. However, in the course of the chain reaction, there are not only reactions leading to the formation of activated centers, but also reactions leading to the elimination of activated centers and the interruption of the chain. Whether the speed of the chain reaction can be increased to cause ignition and explosion also depends on the relationship between the two, that is, the rate of increase in the concentration of activated centers.

There are two factors for the increase of the concentration of activated centers in the chain reaction: one is the result of thermal motion, and the other is the result of chain branching. In addition, at any time of the reaction, there is a possibility that the activated centers will be destroyed, and its speed is also proportional to the concentration of the activated centers themselves.

3.4.5.1 Chemical reaction velocity in chain reaction

According to the chain reaction theory, the automatic acceleration of the reaction does not necessarily depend on the accumulation of heat, but can also be accelerated by the gradual accumulation of free radicals in the chain reaction until ignition. Whether the number of free radicals in the system can be accumulated is the result of the interaction between the factors of free radical growth and the factors of free radical destruction in the chain reaction process. Free radical accumulation occurs in systems where free radical growth factors are dominant.

In the process of chain initiation, the reaction molecules are decomposed into free radicals due to the initiation factors. The rate of radical formation is represented by W_1, which is generally small because the initiation process is a difficult process.

In the chain transfer process, the number of free radicals will increase for the branch reaction due to the branching of the branch reaction. For example, H^* in the hydrogen-oxygen reaction generates three from one during chain transfer. Obviously, the higher the concentration of H^* is, the faster the number of free radicals increases. Let that growth rate of a free radical in the chain transfer process be W_2, $W_2 = fn$, where f is the reaction rate constant for the generation of a free radical in the branched chain. Since the branching process is the decomposition of stable molecules into free radicals, which requires the absorption of energy, the temperature has a great influence on f. With the increase of temperature, the value of f increases, that is, the percentage of activated molecules increases, and W_2 also increases. In the process of chain transfer, the radical propagation velocity W_2 due to the dissociation of branched chains plays a decisive role in the increase of the number of radicals.

In the process of chain termination, free radicals collide with the wall or recombine with each other to lose energy and become stable molecules, and the free radicals themselves are destroyed. Let the radical destruction rate be W_3. The greater the radical concentration n is, the more the collision chance is, and the greater the destruction velocity W_3 is. That is, W_3 is proportional to n, written as $W_3 = gn$, where G is the rate constant of the chain termination reaction. Since the chain termination reaction is a compound reaction, there is no need to absorb energy (in fact, a small amount of energy is released). Under the ignition condition, G is smaller than f, so it can be considered that the influence of temperature on G is small, and G is approximately regarded as independent of temperature.

The relationship between the number of free radicals and time in the whole chain reaction is as follows:

$$\frac{dn}{dt} = W_1 + W_2 - W_3$$
$$= W_1 + fn - gn$$
$$= W_1 + (f - g)n \tag{3-35}$$

Let $f-g=\varphi$, then the above formula can be written as:

$$\frac{dn}{dt} = W_1 + \varphi n \tag{3-36}$$

Suppose $t=0, n=0$, set integrate the above formula to obtain:

$$n = \frac{W_1}{\varphi}(e^{\varphi t} - 1) \tag{3-37}$$

If a represents the number of molecules of a free radical participating in the reaction to produce the final product in the chain transfer process (for example, in the chain transfer process of the hydrogen-oxygen reaction, one H^* is consumed to produce two H_2O molecules), then the reaction rate, that is, the production rate of the final product, is:

$$W_{\text{produce}} = aW_2 = afn = af\frac{W_1}{\varphi}(e^{\varphi t} - 1) \tag{3-38}$$

3.4.5.2 Ignition condition of chain reaction

The rate of radical formation during chain initiation is very small and can be neglected. The main factors causing the change of the number of free radicals are the growth rate of free radicals W_2 caused by chain branching in the process of chain transfer and the destruction rate of free radicals W_3 caused by chain termination. W_2 is closely related to temperature, but W_3 is not. It is not difficult to understand that W_2 becomes larger and larger as the temperature increases. Free radicals are more likely to accumulate. The system is more likely to catch fire. The relative relationship between W_2 and W_3 at different temperatures is analyzed below to find out the ignition conditions.

When the system is at low temperature, W_2 is very small, and W_3 is large relative to W_2, so $\varphi = f-g<0$. According to the formula (3-38), the reaction rate is

$$W_{\text{produce}} = af\frac{W_1}{-|\varphi|}(e^{-|\varphi|t} - 1) = af\frac{W_1}{-|\varphi|}\left(\frac{1}{e^{|\varphi|t}} - 1\right) \tag{3-39}$$

For

$$t \to \infty , \frac{1}{e^{|\varphi| t}} \to 0$$

So

$$W_{\text{produce}} \to af\frac{W_1}{|\varphi|} = \text{constant} = W_0 \tag{3-40}$$

This shows that, in φ. In the case of < 0, the number of free radicals can not be accumulated, the reaction rate will not automatically accelerate, but can only tend to a certain value, so the system will not catch fire.

As the temperature of the system increases, W_2 increases, and W_3 can be regarded as not changing with temperature, which may lead to the situation that $W_2 = W_3$. According to equation (3-36), the reaction rate will increase linearly with time.

Because

$$\frac{dn}{dt} = W_1 , \ n = W_1 t$$

So

$$W_{\text{produce}} = aW_2 = afn = afW_1 t \tag{3-41}$$

Since the reaction rate increases linearly, not at an accelerated rate, the system will not catch fire.

If the system temperature further increases and W_2 further increases, then $W_2 > W_3$, that is, $\varphi = f - g > 0$. According to equation (3-38), the reaction rate W_{produce} will increase exponentially with time, and the system will ignite.

If the above three conditions are plotted on the $W_{\text{produce}}-t$ diagram, it is easy to find the ignition conditions. As shown in Fig. 3-10, only when $\varphi > 0$, the growth rate of the free radicals W_2 formed in the branched chains is greater than the destruction rate of the free radicals W_3 in the chain termination process, the system may catch fire.

$\varphi = 0$ is the critical condition, and the corresponding temperature is the auto-ignition temperature. Above this autoignition temperature, the system will ignite spontaneously as long as chain initiation occurs.

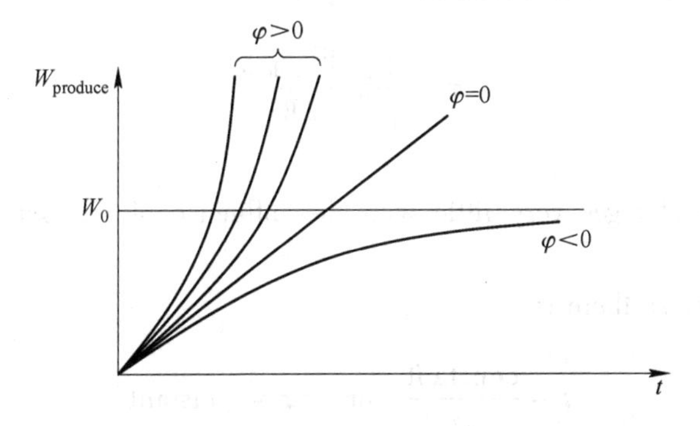

Fig. 3-10 Reaction rates under different φ values

3.4.5.3 Ignition induction period in chain reaction theory

The ignition induction period in the chain reaction has three cases:

(1) $\varphi<0$, the chemical reaction rate of the system tends to be a constant, the chemical reaction rate of the system will not automatically accelerate, the system will not catch fire, $\tau=\infty$.

(2) $\varphi>0$, it can be seen from the formula (3-38) that the reaction rate will increase exponentially with time, but because W_1 is very small, the reaction will be very slow in the initial period, that is, during the ignition delay period τ. After the delay period, due to the accumulation of activated centers, the reaction rate automatically accelerates and ignites, which is called chain spontaneous combustion. The variation of the reaction rate with time is shown in Fig. 3-11. In this case, the automatic acceleration of the reaction mainly depends on the automatic accumulation of activation centers in the system.

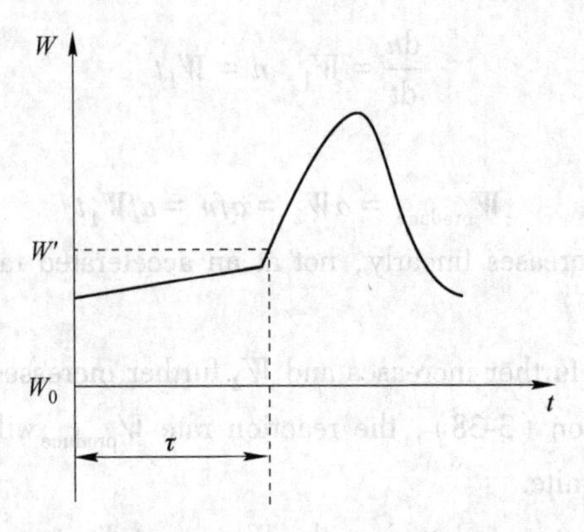

Fig. 3-11 Chain Spontaneous Combustion Diagram

The size of the ignition induction period τ can be obtained from the following relationship:

$$W_{\text{produce}}=\frac{f_a W_1}{\varphi}(e^{\varphi\tau}-1)$$

When φ is large, $\varphi\approx f$, and correspondingly, 1 in the above formula can be omitted. If the above formula is taken as the logarithm, we can get:

$$\tau=\frac{1}{\varphi}\ln\frac{W_{\text{produce}}}{aW_1}$$

Actually $\ln\frac{W_{\text{produce}}}{aW_1}$ changes very little with the influence of the outside world and can be regarded as a constant, so there is

$$\tau=\frac{\text{constant}}{\varphi}\quad\text{or}\quad\tau\varphi=\text{constant} \tag{3-42}$$

(3) $\varphi=0$ is a limiting case where the ignition induction period is the time when $W_{\text{produce}}=W_0$.

3.4.6 Burning peninsula phenomenon

Stoichiometrically mixed H_2 and O_2 were placed in a container at 1 atmosphere pressure and immersed in a constant temperature hot bath at 500℃. When a vacuum of a few millimeters of mercury is then drawn inside the container, H_2 and O_2 explode. When the pressure in the vessel is gradually increased to 0.01 ~ 0.13MPa, the mixture of hydrogen and oxygen cannot explode. When the pressure of the mixed gas is further increased to 0.2MPa, the explosion can occur again. It can be seen that even the hydrogen-oxygen mixture which is very easy to explode can only explode under certain conditions such as temperature and pressure. The pressure and temperature conditions are the explosion limits of the combustible mixture. Fig. 3-12 shows the explosion limit of the stoichiometric hydrogen-oxygen mixture, which is Peninsular, so it is also called the burning peninsula of the hydrogen-oxygen mixture. It can be seen from the figure that the explosion limit curve divides the pressure and temperature range of the gas mixture into an explosion zone and a non-explosion zone. In the non-explosive region, although there is a sufficiently high temperature, the mixture cannot explode due to the effect of pressure on the reaction rate. To find out the reason, we must study the reaction mechanism.

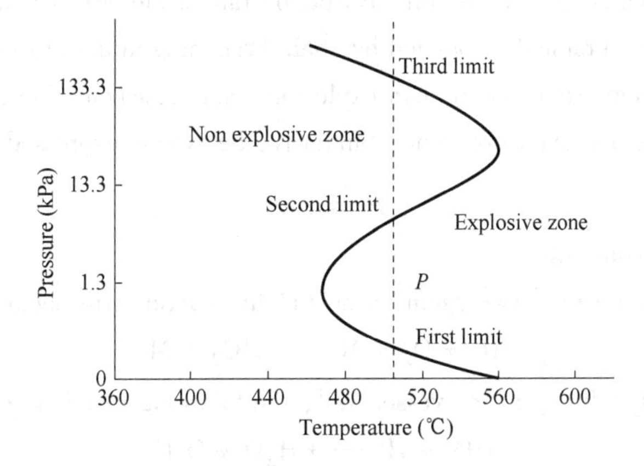

Fig. 3-12 Phenomenon of hydrogen-oxygen ignition peninsula

It can be seen from Fig. 3-12 that there are three ignition limits for the hydrogen-oxygen reaction from the relationship between the critical ignition temperature and the critical ignition pressure of the hydrogen-oxygen mixture, which is simply explained by the chain reaction ignition theory.

Suppose there is a point P in the explosion zone between the first limit and the second limit, keep the system temperature constant and reduce the system pressure, and point P moves vertically downward. At this time, due to the low pressure of the hydrogen-oxygen mixture gas, the free radicals diffuse quickly, and the hydrogen free radicals are easy to collide with the wall, and the destruction of free radicals mainly occurs on the wall. The lower the pressure, the greater the rate of radical destruction. When the pressure drops to a certain value, the destruction rate of free radicals may be greater than the growth rate of free radicals produced by chain branching in the process of chain transfer, so the system changes from explosion to non-explosion, and the first limit appears between the explosion zone and the non-explosion zone. If an inert gas is added to the mixed gas, the diffusion of hydrogen radicals to the vessel wall can be prevented, resulting in a downward movement of the lower limit. As can be seen from Fig. 3-12 of the ignition peninsula, if the temperature of the mixture is increased, the critical ignition pressure can be reduced, that is, the two are inversely proportional. Semyonov summed up this relationship as follows:

$$P_i = Ae^{\frac{B}{T_i}} \tag{3-43}$$

In the formula, A and B are constants, and their values are related to the properties of the activation center, the reacting substance and the incombustible additive, as well as the shape and size of the vessel wall. In fact, the above formula is the expression of the first limit of ignition.

If the temperature of the system is kept constant and the pressure of the system is increased, point P moves upward and vertically. At this time, due to the high pressure of the mixed gas of hydrogen and oxygen, the free radicals collide with a large number of stable molecules in the gas during the diffusion process and consume their own energy. The free radicals combine into stable molecules. Therefore, the free radicals are mainly destroyed in the gas phase. The gas mixing pressure increases, the free radical gas destruction speed increases. When the pressure of the mixed gas increases to a certain value, the destruction rate of the free radicals may be greater than the growth rate of the free radicals produced by chain branching in the process of chain transfer, so the system changes from explosion to non-explosion, and a second limit appears between the explosion zone and the non-explosion zone. Similarly, Semyonov expressed the limit as:

$$P_i = A'e^{\frac{B'}{T_i}} \tag{3-44}$$

Where A' and B' are constants.

When the pressure increases again, a new chain reaction will occur, namely:

$$H' + O_2 + M \longrightarrow HO_2 + M$$

Before the HO_2 diffuses to the vessel wall, the following reaction occurs to generate OH':

$$HO_2' + H_2 \longrightarrow H_2O + OH'$$

As a result, the growth rate of free radicals increases, and then explosion can occur, which is the third limit of explosion. At this time, the heat release of the limit is greater than the heat dissipation, which belongs to a thermal explosion and completely follows the law of thermal spontaneous combustion theory. Therefore, the third limit in the "fire peninsula" phenomenon is essentially the thermal self-ignition limit.

Experiments show that the above two equations can be used to calculate and analyze not only the "ignition peninsula" phenomenon of hydrogen-oxygen mixtures, but also the "ignition peninsula" phenomenon of $CO + O_2$ mixtures.

At present, a third ignition theory, the chain reaction thermal explosion theory, has been proposed, which holds that the initial stage of the reaction may be a chain reaction, but as the reaction proceeds, it releases heat and automatically heats up, and finally becomes a pure thermal explosion.

3.4.7 Chain reaction of hydrocarbon oxidation*

In addition to the induction period, the high temperature gas phase oxidation of hydrocarbons often shows obvious stages, that is, the phenomenon of cold flame often occurs before ignition (compared with the hot flame during ignition, the temperature is lower, the glow is weaker, and the heat produced is less, so it is called cold flame). This phenomenon is one of the characteristics of gas phase oxidation of hydrocarbons.

The oxidation process or mechanism of various hydrocarbons at high temperature in the gas phase is very complex. According to modern theory, the oxidation process of hydrocarbons is essentially a series of chain reactions through free radicals.

A free atom or radical refers to an atom or group of atoms with free electrons, such as a hydrogen free atom (H), a hydroxyl radical (OH), a hydrocarbon radical (R), etc. Usually a free radical is produced by the action of light radiation, heat, electricity, or other (e.g., chemical initiator) energy on a molecule. For example, a hydrocarbon molecule RH may collide with an inert, energy-rich molecule M. Obtaining sufficient vibrational energy to dissociate is called thermal dissociation.

$$RH + M \longrightarrow R + H + M \qquad (3\text{-}1)$$

According to the modern chain reaction theory, free radicals are the active particles that cause chain development, so they are also called active centers. This is because free radicals are more reactive than molecules, especially molecules with saturated bonds. The experimental results show that the activation energy required for the chemical reaction between free radicals and molecules is generally not more than 30 ~ 40kJ, and a few are 40 ~ 80kJ. When the saturated bond molecules react with each other, the activation energy required is several hundred kilojoules, and the difference between the two is significant.

Because the free radical has free valence, it can use its chemical force to interact with the bonding electrons of the molecules it meets, that is to say, in this case, there is a competition between the free radical and a group in the molecule for a bonding valence electron of the molecule at the same time. The result of competition is often the destruction of this bond in the molecule. For example, the reaction of a hydrocarbon peroxide radical with a hydrocarbon breaks the bond between RH to produce a hydrocarbon radical and a hydrocarbon peroxide:

$$ROO + RH = ROOH + R \qquad (3\text{-}2)$$

The R produced will react with the oxygen in the air to regenerate the hydrocarbon peroxide radical:

$$R + O_2 = ROO \qquad (3\text{-}3)$$

Another characteristic of free radical reactions is that when a monovalent free radical reacts with a saturated valence molecule, the free valence will not disappear, and a new monovalent free radical will be produced in the reaction product, and sometimes another free radical with two free valences will be produced, such as the above reactions (3-2) and (3-3). And hydrogen oxidation at low pressure (temperature about 500. degree. C.) reactions (3-4), (3-5), (3-6) as follows:

$$H + O_2 \Longrightarrow OH + O \quad (3\text{-}4)$$

$$O + H_2 \Longrightarrow OH + H \quad (3\text{-}5)$$

$$OH + H_2 \Longrightarrow H_2O + H \quad (3\text{-}6)$$

It can be seen from the above that in a system in which a chemical change can occur, if the first free radical appears, it will quickly react with one to form a new free radical, which in turn will react to form another free radical. In this way, as it goes on, the free radical can easily form a long chain to produce chemical changes, which is the chain reaction. The first few reactions (3-1) to (3-6) are like this. The chain of this chain reaction is interrupted only when the free radicals disappear.

Like other chain reactions, the oxidation chain reaction of hydrocarbons consists of three basic steps:

3.4.7.1 Initiation of the chain

It refers to the formation of the first atoms or free radicals from the raw material molecules. The production of free radicals depends on the breaking of bonds in the molecule, so the activation energy required for it is equal to the bond energy in the saturated valence molecule. Free radicals can be formed when molecules absorb more energy than the bond breaking energy of the light quantum. It is usually where the bond energy of the molecule is low that the first break occurs to form a free radical.

3.4.7.2 Development of the chain

Once free radicals occur, they can automatically develop and form long chains, that is, they act alternately in turn until the raw materials are exhausted or the free radicals are eliminated. Due to the difference of raw materials and conditions, the chain development can be divided into four types: (1) straight chain; (2) continuous branched chain; (3) rare branched chain; (4) degenerate branched chain.

3.4.7.3 Termination of the chain

This is the reaction of the disappearance of free radicals. For example, when two free radicals interact with each other or with an inert molecule M, the energy is transferred to the inert molecule, causing the free valence to saturate each other, thus causing the free valence to disappear. This phenomenon is also known as gas phase destruction.

$$H + OH + M \Longrightarrow H_2O + M$$

$$H + H + M \Longrightarrow H_2 + M$$

In addition, chain scission also occurs when the reaction chain touches the wall of the reaction vessel. According to research, when free radicals hit the wall, they are adsorbed by the wall and form a weak compound on the surface of the wall. The free radicals no longer react with the raw material molecules, but can interact with the free radicals from the container. The colliding particles fly into the container as neutral, inactive molecules. Therefore, the wall can affect the slowing down of the chain reaction, and its effect is similar to that of the inhibitor. This phenomenon is called wall destruction.

The oxidation of hydrocarbons is discussed in further detail below. According to Bax-Engler's process oxidation theory, the oxidation of hydrocarbons is carried out by destroying one bond of oxygen, not two bonds of oxygen, because it takes 489.879kJ to destroy two bonds of oxygen at the same time, while only 293.09 ~ 334.96kJ is needed to destroy one bond. Therefore, the oxidation of hydrocarbons is the first generation of hydrocarbon peroxide or peroxide radical ROO (namely R—O—O), and peroxide will also decompose into free radicals. With the generation of free radicals, the reaction has the nature of a chain reaction and thus can be automatically continued and automatically accelerated by the occurrence of branching. The whole oxidation process before combustion is a series of chain reactions with the participation of free radicals. Some features of the reaction in the first and second induction phases are discussed below with respect to the phase prior to the onset of the cool flame (induction period τ_1). The first is the separation of free radicals from hydrocarbon molecules by irradiation, thermal dissociation, or other action, as in a reaction

$$RH + M = R + H + M$$

Hydrocarbon molecules can also react directly with oxygen to produce free radicals, such as

$$RH + O_2 = R + HO_2$$

HO_2 is a less active free radical, which can continue to react with hydrocarbons to form H_2O_2 and R, so that the chain continues to develop.

$$RH + HO_2 = R + H_2O_2$$

The generated hydrocarbon radical R can be combined with molecular oxygen to generate a peroxide radical:

$$R + O_2 = ROO$$

At this time, oxygen is added to the carbon atom with free valence in the hydrocarbon. Because the carbon atom in the hydrocarbon molecule is covered by hydrogen, the first attack of oxygen molecule is not the weaker C—C bond in the hydrocarbon molecule but the stronger C—H bond. Generally, the average bond energy of C—C bond is 346.98kJ/mol, and the average bond energy of C—H bond is 413.47kJ/mol. In fact, the C—H bond in different molecules and the energy of different C—H bonds in molecules are different.

Among the carbon atoms in various positions, the carbon atom with the weakest C—H bond is the most vulnerable to attack. Thus, a tertiary carbon atom (directly associated with three carbon atoms) are most readily formed on the carbon atoms to which the atoms are attached, secondary carbon atoms (directly attached to two carbon atoms to which atoms are attached), primary carbon atoms (those directly connected to only one carbon atom and those in methane) is the least reactive. In the low-temperature cold flame range, the probability of the generation of the three is roughly 33 : 3 : 1.

Hydrocarbyl peroxide radicals inperiod τ_1 continues to react with hydrocarbon molecules to produce monohydrocarbyl peroxides and free radicals, so that the chain reaction continues to develop:

$$ROO + RH = ROOH + R$$

The hydrocarbon peroxide generated at this time is easy to break due to the weak —O—O— chain (the chain energy is only 125.61 ~ 167.48kJ), and can be decomposed into different products with different hydrocarbon structures, such as

$$RCH_2OOH = RCH_2O + OH$$
$$RCH_2O = R + HCHO$$
$$OH + RH = R + H_2O$$

The above reaction has the characteristics of degenerate branching, that is, the decomposition of peroxide is slow and does not always develop along the branches, which makes the oxidation of hydrocarbons have a period of gentle pressure rise, that is, the induction period.

Formaldehyde is produced by the above reaction, but alcohol products can also be produced by the following reaction:

$$RCH_2O + RH = RCH_2OH + R$$

Therefore, aldehydes, alcohols, peroxides, and the like are always analyzed in the initial products of the oxidation of the alkane, and in addition, alkanes can be cracked at a relatively high temperature to produce olefins, and aldehydes can be further oxidized to produce CO, CO_2, H_2O, and the like.

To sum up, the oxidation reaction is characterized by the generation of peroxide in τ_1, which decomposes under the catalysis of formaldehyde to generate multiple free radicals and branch the reaction. The reaction has an induction period due to the nature of the degenerate branch. At the same time, there is a sudden increase in pressure after the induction period due to the generation of branch chain reaction, and at the same time, there is a cold flame due to the activation of formaldehyde. When the temperature rises, the reaction rate is accelerated, and the decomposition of the peroxide is accelerated, so that the induction period τ_1 shorten. When the pressure increases, the concentration of the reactant increases, and the reaction rate also increases, which will also increase the induction period τ_1 shorten. The addition of peroxide to the reactant causes the decomposition of the peroxide to proceed to the right,

$$ROOH = RO + OH$$

Cause that branching reaction to proc faster, thereby causing τ_1 shorten.

The period from cold flame to spontaneous combustion (induction period τ_2) are less studied. It is generally believed that the main feature of the reaction at period τ_2 is that the decomposition of the peroxide radical prevents the formation of peroxide. This is because the higher temperature during the period τ_2 is beneficial to the decomposition reaction of peroxide free radicals.

The decomposition of the peroxide radicals during the period generally proceeds in the following manner:

$$RCH_2 + O_2 = RCH_2OO$$

$$RCH_2OO = RCHO + OH$$
$$RCH_2OO = H_2O + RCO$$
$$RCO = R + CO$$
$$RCH(OO)R = RCHO + RO$$
$$RCH(OO)CH_2R = RCHO + RCH_2O$$
$$RCH_2O = R + HCHO$$

In the above reaction, the peroxide radical is decomposed to generate various aldehydes (formaldehyde, etc.) and free radicals. These radicals continue to interact with oxygen molecules or hydrocarbons to propagate the chain in a straight chain.

In the reaction, some free radicals will decompose into olefins, which will further react with hydrocarbons to form hydrocarbon free radicals to continue the chain.

$$C_3H_7 + O_2 = C_3H_7OO$$
$$C_3H_7OO = C_3H_6 + HO_2$$
$$C_3H_8 + HO_2 = C_3H_7 + H_2O_2$$

Anyway, the straight chain reaction plays a controlling role in τ_2. It should bc noted that the free radical decomposition reaction of the above-mentioned peroxide can also occur and compete with peroxide formation reactions in period τ_1. When the temperature is not high, the formation of peroxide is dominant. Only when the temperature reaches about 400℃, the decomposition reaction of peroxide free radicals gradually gains advantages. It will thus be appreciated that when other conditions are unchanged and the temperature is increased, the reaction will pass through a region of highest branching rates (marked by a cool flame) and then reach a region of weak branching rates. In this way, there is an induction period τ_2 with slower pressure rise after the cold flame.

The speed of the reaction at the beginning of τ_2 is related to the concentration of the intermediate remaining at the end of τ_1, especially to the concentration of formaldehyde, because it can catalyze the decomposition reaction of the period τ_2. When the initial temperature of the system is higher, due to the branching reaction of the period τ_1 is strong, and the remaining formaldehyde is reduced, so that the induction period τ_2 was prolonged. But, in the period τ_2, the decomposition of peroxide free radicals continues to produce formaldehyde, which increases the concentration of formaldehyde and releases a large amount of heat energy, and eventually accelerates the reaction automatically and leads to spontaneous combustion. When the pressure of the system is increased, since the reaction speed is increased, τ_2 and τ_1 are both going to get shorter. With regard to the recurrence of cold flames, it has been speculated that it is related to the inhibition of formaldehyde during the period τ_1. According to research, although formaldehyde is right, the reaction in the period τ_2 has a catalytic effect, however, there is an obvious inhibitory effect in the period τ_1. For example, the induction period can be prolonged by adding an appropriate amount of formaldehyde to the mixture of pentane and oxygen or hexane and oxygen. Formaldehyde can react with free radicals to form the inactive formaldehyde free radical CHO, which causes the reaction to break the chain, for example:

$$HCHO + OH = H_2O + CHO$$

$$HCHO + CH_3O = CHO + CH_3OH$$

τ_1 formaldehyde is continuously generated by the branching reaction during the period, the branching reaction in period T keeps producing formaldehyde, and when a cold flame with strong branching characteristics appears, the concentration of formaldehyde also increases, so that the cold flame is extinguished. When the temperature continues to rise, the cold flame can be repeated with the development of the branching reaction.

3.5 Forced fire

3.5.1 Characteristics of forced ignition

Forced ignition, also known as ignition, generally refers to the use of hot high-temperature objects to ignite the flame, so that a small part of the mixture ignites to form a local flame core, and then the flame core ignites the adjacent mixture, thus causing the flame to spread layer by layer, so that the whole mixture burns. All combustion devices and combustion equipment need to go through the ignition process before they can start working, so it is of great practical significance to study the ignition problem. The following is an analysis of the different characteristics of forced ignition and spontaneous ignition.

First, forced ignition occurs only locally in the gas mixture (near the ignition source), while spontaneous ignition occurs throughout the gas mixture space.

Secondly, spontaneous ignition is caused by the fact that all the mixed gases are surrounded by the ambient temperature T_0, and the temperature of all the combustible mixed gases is gradually increased to the self-ignition temperature due to the automatic acceleration of the reaction. During forced ignition, the mixture is at a lower temperature. In order to ensure that the flame can propagate in the colder mixture, the ignition temperature is generally much higher than the self-ignition temperature.

Thirdly, whether the combustible mixture can be ignited depends not only on whether the local mixture in the boundary layer of the hot object can be ignited, but also on whether the flame can propagate itself in the mixture. Therefore, the forced ignition process is much more complex than the spontaneous ignition process.

Both forced ignition process and spontaneous ignition process have the common characteristics of self-heating and autocatalysis driven by thermal reaction and (or) chain reaction, both need the initial excitation of external energy, and also have the problems of ignition temperature, ignition delay and ignition flammability limit. However, their influence factors are different, and the influence factors of forced ignition are more complex than those of spontaneous ignition. Besides the chemical properties, concentration, temperature and pressure of the combustible mixture, it is also related to the ignition method, ignition energy and the flow properties of the mixture.

3.5.2 Common ignition methods

Commonly used ignition methods in engineering are: hot object ignition, flame ignition, electric spark ignition, etc. No matter which ignition method is used, its basic principle is to make the mixture locally ignited by external heat.

3.5.2.1 Ignition of a hot object

Resistance wires, rods or plates heated by electric current are all hot objects, which can be used to ignite premixed combustible gas (or heat radiation can be used to heat refractory bricks or ceramic rods to form various hot objects, which can be ignited in combustible mixture). The advantage of this ignition device is its simple and compact structure, but its disadvantage is that it is prone to oxidation and ablation. The ignition mechanism of a hot object is illustrated by taking a high-temperature particle as an example.

It is assumed that, as shown in Fig. 3-13, in an infinite flammable mixture (whose temperature is T_0, less than T_w) has a hot metallic particle (whose temperature is T_w). As a result of the temperature difference, a particle loses heat to an adjacent gas mixture, and the rate of heat flow is a function of the flow and thermal properties of the gas mixture. In the thin boundary layer around the particle, the temperature of the mixture changes from T_w down to T_0. For the combustible mixture, the temperature distribution curve in the thermal boundary layer is higher than that in the non-combustible mixture due to the heating of the mixture by the exothermic reaction. This temperature distribution is shown in Fig. 3-13, where curve a indicates that the mixture is non-flammable and curve b indicates that the mixture is flammable. According to the temperature gradient of the wall, when the reaction of the gas is exothermic, the heat flow transferred from the wall to the gas mixture is lower than that when the gas mixture is inert.

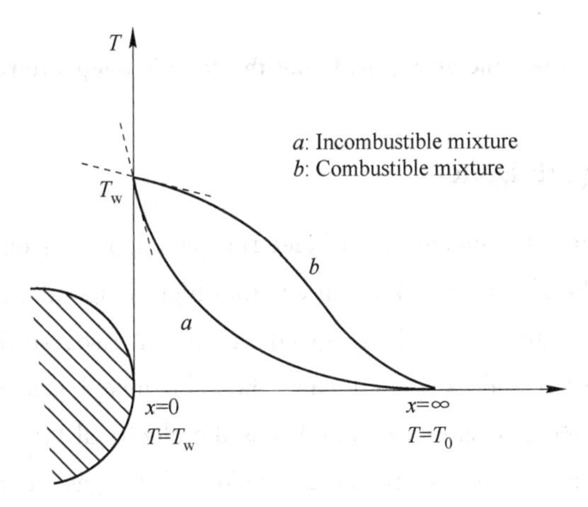

Fig. 3-13 Temperature distribution near hot particles in

If a higher particle temperature is selected than before, the difference between the reactant gas and inert gas temperature distributions becomes more pronounced. The higher the reaction gas particle temperature is, the smaller the heat flow from the wall is. As shown in Fig. 3-14, at the critical particle temperature T_c, the heat flux from the wall to the reaction mixture is equal to zero, and the slope of curve b at $x=0$ is zero. At this point, All the heat released by the chemical reaction in the thermal boundary layer is transferred to the cold mixed gas outside. When the particle temperature is slightly higher than T_c, the chemical reaction rate increases to such an extent that the rate of heat release from the chemical reaction is greater than the rate of heat transfer outward from the thermal boundary, so the temperature in the thermal boundary layer increases, and its maximum temperature occurs at a small distance from the particle surface. The heat flow is then partially transmitted to the particle. Under such conditions, a stable temperature distribution becomes impossible because the temperature maximum is constantly moving away from the surface of the particle. When the temperature gradient at the particle surface is equal to zero, the gas reaction layer (i. e., flame) begins to propagate toward the unburned mixture. The onset of such flame propagation is considered a criterion for forced ignition.

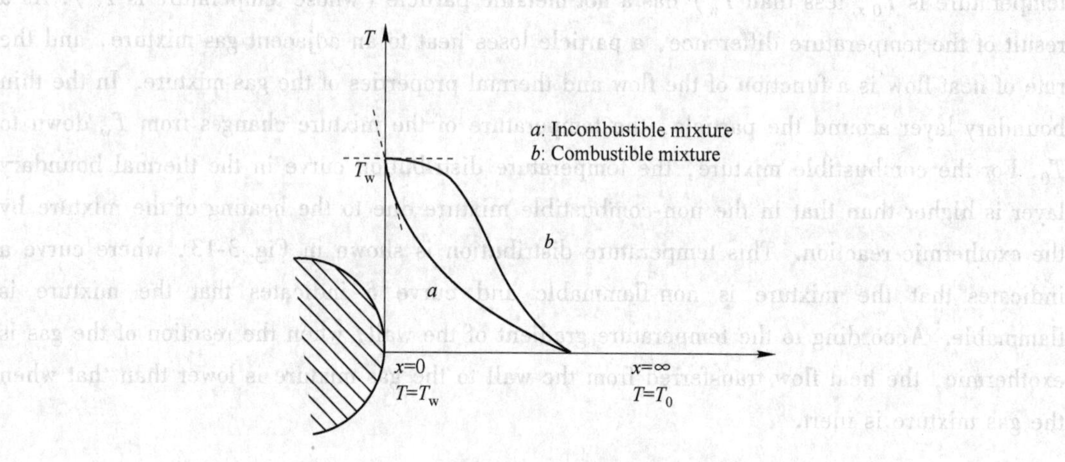

Fig. 3-14 The same as Fig. 3-13, but the particle temperature is higher

3.5.2.2 An electric spark ignites

There are two theories about the mechanism of electric spark ignition: one is the thermal theory of ignition, which regards the electric spark as an external high-temperature heat source, because its existence makes the temperature of the local mixed gas near it rise, so that it reaches the critical working condition of ignition and is ignited, and then the mixed gas in the whole container is ignited and burned by flame propagation; the other is the electrical theory of ignition, it considers that the ignition of the mixture is due to the ionization of the gas near the spark part and the formation of active centers, which provides the conditions for the chain reaction, and the mixture burns as a result of the chain reaction. The experiments show that the two mechanisms exist at the same time. Generally, the ionization is dominant at low temperature, but when the voltage is increased, the thermal effect is dominant. It is a widely used ignition method in engine combustion chamber, so it is of great practical significance to study spark ignition.

The characteristic of spark ignition is that the energy required is not large. For example, the energy required for spark ignition of stoichiometric hydrogen-air mixture is only 2.01×10^{-5} J. Ignition, as shown in Fig. 3-15, is achieved by an electric spark discharge between two electrodes placed in the combustible mixture. The electrodes may be flanged or unflanged and are usually made of stainless steel.

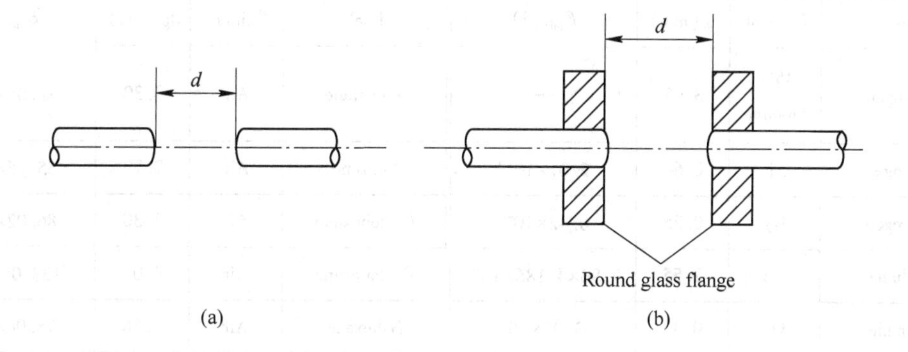

Fig. 3-15 Geometric arrangement of flange electrode for spark ignition research

(a) Flangeless electrode; (b) Flanged electrode

3.5.2.3 Flame ignition

On many industrial combustion devices and aero-engines, because the fuels used are difficult to ignite (such as pulverized coal, heavy oil, etc.) or have poor ignition conditions, it is required to use ignition devices with higher energy to achieve the ignition process. At this time, the above-mentioned ignition methods such as hot objects and electric sparks can not meet the requirements. In this case, a multi-stage ignition mode is generally adopted. That is to say, the gas fuel or light fuel oil is ignited by electric spark to form an ignition flame with large energy, and then the flame is used to ignite pulverized coal, heavy oil or other fuels to achieve combustion conditions.

That is to say, the so-called flame ignition is to ignite the flammable mixture in the combustion chamber by other methods to form a stable small flame, which is used as a heat source to ignite the mixture that is difficult to ignite. Its greatest advantage is that it has greater ignition energy.

3.5.3 Minimum energy for ignition

The experiment shows that when the gas mixture ratio, temperature and pressure in the electrode gap are constant, the electrode discharge energy must have a minimum value in order to form the initial flame center. If the discharge energy is greater than the minimum value, the initial flame center may be formed; if it is less than the minimum value, the initial flame center cannot be formed. This minimum discharge energy is the minimum energy for ignition, as shown in Table 3-1 below. It can be seen from Table 3-1 that the minimum ignition energy E_{min} required for different mixed gases is different. For a given mixture, when the mixture pressure and initial temperature are different, the minimum ignition energy E_{min} not the same.

Table 3-1 Quenching distance and ignition energy of the mixture stoichiometry at room temperature and 0.1MPa

Fuel	Oxidant	dp(mm)	E_{min}(J)	Fuel	Oxidant	dp(mm)	E_{min}(J)
Hydrogen	45% bromine	3.65	—	N-pentane	Air	3.30	82.05×10^{-5}
Hydrogen	Air	0.64	2.01×10^{-5}	Benzene	Air	2.79	55.05×10^{-5}
Hydrogen	O_2	0.25	0.42×10^{-5}	Cyclohexane	Air	3.30	86.02×10^{-5}
Methane	Air	2.55	$7.90\times4.186\times10^{-5}$	Cyclohexane	Air	4.06	138.05×10^{-5}
Methane	O_2	0.30	33.07×10^{-5}	N-hexane	Air	3.56	95.06×10^{-5}
Acetylene	Air	0.76	3.01×10^{-5}	1-hexane	Air	1.87	21.93×10^{-5}
Acetylene	O_2	0.09	0.04×10^{-5}	N-heptaneisooctane	Air	3.81	115.07×10^{-5}
Ethylene	Air	1.25	11.09×10^{-5}	Isooctane	Air	2.84	57.39×10^{-5}
Ethylene	O_2	0.19	0.25×10^{-5}	N-alkyl	Air	2.06	30.18×10^{-5}
Propane	Air	2.03	30.52×10^{-5}	1-alkyl	Air	1.97	27.63×10^{-5}
Propane	Ar+air	1.04	7.7×10^{-5}	N-butylphenzene	Air	2.28	$8.84\times4.186\times10^{-5}$
Propane	He+air	2.53	45.33×10^{-5}	Oxyethylene	Air	1.27	37×10^{-5}
Propane	O_2	0.24	0.42×10^{-5}	Oxypropylene	Air	1.30	19×10^{-5}
1-3 Butadiene	Air	1.25	23.53×10^{-5}	Methacrylate	Air	1.65	62.04×10^{-5}
Isobutane	Air	2.20	34.41×10^{-5}	Diethylether	Air	2.54	49.02×10^{-5}
Carbon disulfide	Air	0.51	1.51×10^{-5}				

3.5.4 Electrode flameout distance

A large number of experiments show that when other conditions are given, the minimum ignition energy E_{min} is related to the distance d between electrodes, as shown in Fig. 3-16 below. It can be seen from Fig. 3-16 that when the electrode distance d is too small, the gap is too small, and the initial flame transfers too much heat to the electrode, so that the heat transferred to the surrounding premixed combustible gas is reduced accordingly, resulting in the flame can not propagate. When d is less than a certain critical value d_p, no matter how large the ignition energy is, the mixed gas can not be ignited. The maximum distance d_p between the electrodes that cannot ignite the mixed gas is called the electrode flameout distance, also known as the quenching distance. The relationship between d_p and E_{min} is as follows:

$$E_{min} = Kd_p^2 \tag{3-45}$$

Where K is a proportionality constant, and for most hydrocarbons, the K value is approximately $7.02 \times 10^{-3} J/cm^3$.

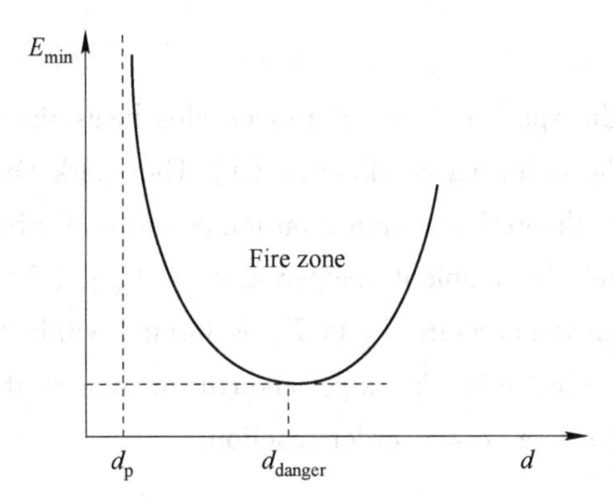

Fig. 3-16 relationship between minimum ignition energy E_{min} and electrode flameout distance

When $d>d_p$, as d increases, the ignition energy decreases until E_{min}; however, the ignition energy will increase with the increase of d. That is, under given conditions, the electrode distance has a most dangerous value d_{danger}, and when the electrode distance is greater than or less than the most dangerous value, the minimum ignition energy increases. When the electrode distance is equal to the most dangerous value, the minimum ignition energy is the smallest.

The minimum ignition energy E_{min} and the electrode flameout distance (quenching distance) d_p mainly depend on the parameters such as the nature, composition, pressure, temperature, flow conditions of the mixed combustible gas and the electrode shape. Fig. 3-17 shows the relationship between E_{min} and the fuel composition x_f in the premixed combustible gas. It shows that E_{min} is the lowest in the vicinity of chemical equivalent ratio; when the fuel is too rich or too lean, E_{min} goes to infinity, it means that the premixed combustible gas cannot be ignited. When the ignition energy of the electric spark decreases, the ignition range x_{f1} to x_{f2} of the premixed combustible gas becomes narrow, and there is an ignition limit.

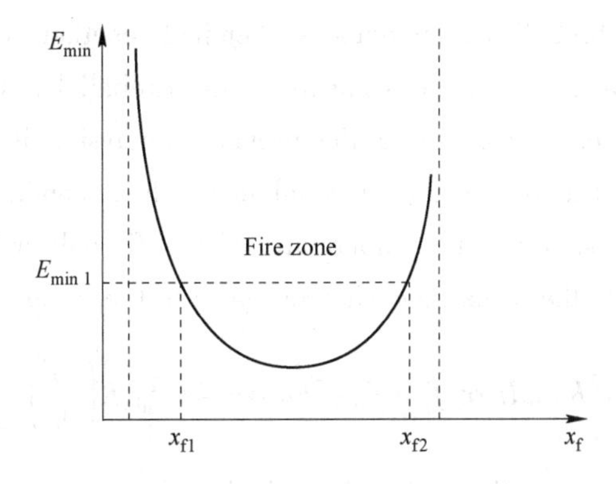

Fig. 3-17 Relationship between minimum ignition energy E_{min} and x_f

3.5.5 Semi-empirical formula for minimum ignition energy of electric spark in static air mixture

3.5.5.1 Assumptions

In the static gas mixture, the spark between the electrodes heats the gas. The physical model is shown in Fig. 3-18 under the following conditions: (1) The spark heating zone is spherical, the maximum temperature is the theoretical combustion temperature of mixed gas T_m, the temperature distribution is uniform, and the ambient temperature is T_∞; (2) During ignition, a linear temperature distribution from temperature T_m to T_∞ is formed within the flame thickness δ; (3) The distance between the electrodes is large enough to ignore the extinction effect of the electrodes; (4) The reaction is a second-order reaction.

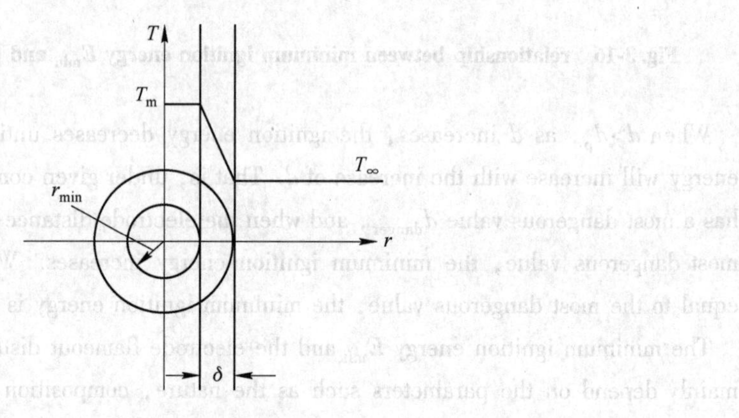

Fig. 3-18 Electric spark ignition model

3.5.5.2 Minimum fireball radius

When the fireball radius reaches the minimum fireball radius, the corresponding energy is the minimum ignition energy, and the ignition is successful. Now find the minimum fireball radius.

The gas mixture in the fireball area undergoes chemical reaction and releases heat under the heating of the electric spark, and at the same time, the fireball loses heat to the unburned gas mixture through the surface. According to the previous analysis, if ignition occurs, the heat released by the chemical reaction should be equal to the heat conducted by the fireball at the beginning of propagation, so that the temperature of the fireball will rise and form a stable temperature distribution. At the same time, it propagates to the unburned mixture, that is,

$$\frac{4}{3}\pi r^3 K_{0s}\Delta H_c \rho_\infty^2 f_F \cdot f_{Ox} e^{-\frac{E}{RT_m}} = -4\pi r_{min}^2 K\left(\frac{dT}{dr}\right)_{r=r_{min}} \tag{3-46}$$

The temperature gradient on the right side of the above equation can be approximately simplified as:

$$-\left(\frac{dT}{dr}\right)_{r=r_{min}} = \frac{T_m - T_\infty}{\delta} \tag{3-47}$$

In the formula δ is the laminar flame front thickness. If it is further assumed

$$\delta = Cr_{min} \tag{3-48}$$

In the formula, C is a constant, which is obtained by substituting the formulas (3-46) and (3-47) into (3-48)

$$r_{min} = \left[\frac{3K(T_m - T_\infty)e^{\frac{E}{RT_m}}}{CK_{0s}\Delta H_c\rho_\infty^2 f_F \cdot f_{0x}}\right]^{1/2} \tag{3-49}$$

3.5.5.3 Minimum energy of electric spark ignition E_{min} Formula

For the air mixture in the fireball of radius r_{min}, the temperature rises from the initial temperature T_∞ to the theoretical combustion temperature T_m, the energy is supplied by the electric spark, and this energy is the minimum ignition energy E_{min}

$$E_{min} = K_1\left(\frac{4}{3}\pi r_{min}^3\right)\rho_\infty \bar{c}_p(T_m - T_\infty) \tag{3-50}$$

Where, K_1 is an empirical correction factor, because the minimum ignition energy E_{min} provided by the spark is not necessary to raise the temperature of the mixed gas to the theoretical combustion temperature T_m, but it is often higher than T_m, so the assumption made by K_1 correction is: $\bar{c}_p$ is the average isobaric heat capacity of the gas mixture. Substitute equation (3-49) into equation (3-50).

$$E_{min} = K'\bar{c}_p\lambda^{\frac{3}{2}}\Delta H_c^{-3/2} f_F^{-3/2} f_{0x}^{-3/2}\rho_\infty^{-2}(T_m - T_\infty)^{5/2}e^{\frac{3E}{2RT_m}} \tag{3-51}$$

The above formula is the semi-empirical formula for the minimum energy of electric spark ignition, where K' is a constant,

$$K' = K_1\frac{4}{3}\pi\left(\frac{3}{cK_{0s}}\right)^{3/2} \tag{3-52}$$

3.5.5.4 Main factors affecting the ignition of electric spark

When the mixed gas is ignited by electric spark, the electrode distance must be greater than the flameout distance d_p, and the discharge energy must be greater than a certain minimum ignition energy, otherwise the electric spark cannot ignite the mixed gas.

The minimum ignition energy and the electrode flameout distance of the electrode are different for different premixed combustible gases. For a given mixture, the minimum ignition energy is different when the mixture ratio, mixture pressure and mixture temperature are different.

(1) The larger the Heat capacity $\bar{c}_p$, the larger the minimum ignition energy E_{min}, the mixed gas is not easy to ignite, because the heat capacity is large, the mixed gas absorbs more heat when it warms up.

(2) The larger the thermal conductivity K is, the lager the minimum ignition energy E_{min} is, the mixture is not easy to ignite. Because the spark energy is quickly conducted out, the temperature of the mixed gas in contact with the spark is not easy to rise.

(3) The combustion heat ΔH_c is large, the minimum ignition energy E_{min} is small, and the mixed gas is easy to ignite.

(4) High mixed gas pressure, that is, density ρ_∞ is large, and the minimum ignition energy E_{min} is small, indicating that the mixed gas is easy to ignite.

(5) The initial temperature of mixed gas T_∞ is high, the minimum ignition energy E_{min} is small, and the mixed gas is easy to cause.

(6) Activation energy of gas mixing E is large, the minimum ignition energy E_{min} is large, and the mixed gas is not allowed.

3.6 Spontaneous combustion of white phosphorus experiment

This experiment is to introduce the method of demonstrating spontaneous combustion of white phosphorus with red phosphorus. The purpose of the experiment is to observe the process and phenomenon of spontaneous combustion. The experimental principle is: when the red phosphorus is heated in isolation air to reach its sublimation temperature of 416℃, it becomes phosphorus vapor, and the phosphorus vapor condenses in cold to form white phosphorus. If the phosphorus vapor is dispersed and condensed on the scrip, when the scrip is in contact with the air, because a large number of small particles of white phosphorus are oxidized, the heat generated makes the scrip with white phosphorus burn.

3.6.1 Experimental methods

(1) Add 0.1~0.2g red phosphorus to the dry test tube. Then fix the tube on the steel platform and tilt the nozzle downward slightly, as shown in Fig. 3-19.

(2) Cut the white paper for writing into strips of paper about 20cm long and 2cm wide. Stretch the strip of paper along the wall of the test tube into the tube so that the tip of the strip is about 3cm from the red phosphorus. Then plug the mouth of the test tube with a ball of cotton soaked in copper sulfate solution.

(3) Preheat the test tube evenly with an alcohol lamp, and then fix it at the bottom of the test tube to heat. The red phosphorus gradually vaporizes and condenses onto the scrip. Stop heating after the red phosphorus is completely vaporized.

(4) When the tube cooled, the scrip from the tube out (do not break the scrip, do not shake off the white phosphorus attached to the strip), the white phosphorus attached to the small particles, oxidized by the air, burning immediately.

(5) If you want to make the scrip out after not burning can be in the scrip before pulling out, with a long eyedropper alcohol drops on the scrip with white phosphorus (eyedropper do not contact with white phosphorus), the scrip adsorption alcohol and then pull out. In this way, the

paper moistened with alcohol will not burn immediately. After 3 ~ 5 minutes, the white phosphorus will burn until the alcohol is completely volatilized.

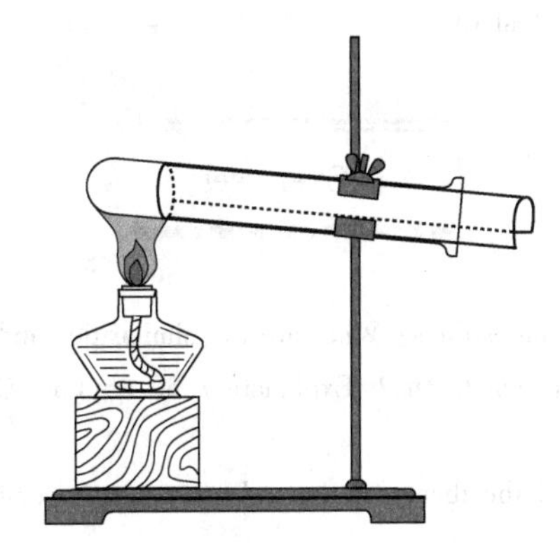

Fig. 3-19 Experimental diagram of spontaneous combustion of phosphorus

3.6.2 Precautions

(1) After repeated experiments, 0.1 ~ 0.2 grams of red phosphorus is the appropriate dosage. Too much not only lengthens the experiment time, but also may make the white phosphorus attached to the slip fall off when the slip is pulled out, which leads the burning white phosphorus to other places or burns the clothes and skin of the experimenter. Too little will not reach the temperature of oxidation combustion, combustion can not be ignition.

(2) In order to make the strip out of the tube, after a few minutes to start burning; The amount of alcohol (industrial alcohol) you drop is important. The amount of alcohol, complete evaporation takes a long time, the start of combustion time is also long. The opposite is shorter. The amount of alcohol drops is determined by not dropping alcohol when the strip is pulled out.

(3) The carbon disulfide and white phosphorus used in the experiment are toxic substances, so attention should be paid to safety. Carbon disulfide vapor, like hydrogen cyanide, is toxic to humans in excess of 10 milligrams per cubic meter of air. Therefore, it is necessary to control the dosage of white phosphorus carbon disulfide solution, and pay attention to ventilation.

White phosphorus is a serious drug, 0.1 grams of white phosphorus into the human body can kill. In order to prevent phosphorus vapor from escaping in the air, it is necessary to plug a group of cotton immersed in copper sulfate concentrated solution at the mouth of the test tube, so that the escaped phosphorus vapor and copper sulfate act to produce non-toxic and volatile phosphoric acid.

(4) If the instrument used in the experiment is accompanied by white phosphorus (such as test tube), it should also be soaked in 5% copper sulfate solution before cleaning, so that the white phosphorus can be transformed into non-toxic and not easy to self-ignite compounds. Do not use the tube brush directly with white phosphorus, in case the white phosphorus on the tube brush combustion hazard.

Absorption of white phosphorus surface water filter paper, unburned strips of paper dripping with white phosphorus carbon disulfide, absorption of white phosphorus vapor cotton and other things, must be timely concentrated to destroy.

Exercises

1. What are the ignition modes of combustibles? What are the similarities and differences between them? How to correctly understand the ignition conditions? Explanation of Ignition Conditions by Semenov's Thermal Spontaneous Combustion Theory.
2. What is the basic starting point of the theory of thermal spontaneous combustion? What is the mechanism of thermal self-ignition?
3. What is the rate of heat release? What is the speed of heat dissipation? How to express them in a formula? The critical condition of ignition in Semenov's thermal self-ignition theory is analyzed and explained by using the position relationship between the exothermic curve and the heat dissipation curve.
4. What is the induction period of thermal spontaneous ignition? When the initial temperature is much higher than the ignition temperature, is there still an induction period? Why?
5. What are the measures to prevent thermal spontaneous combustion?
6. Are ignition and flameout reversible processes? Please explain why.
7. How is the critical condition for chain self-ignition defined?
8. With the explosion processes of H_2 and O_2 indicate the development of chemical explosion.
9. Explanation of Combustion Peninsula Phenomenon of Hydrogen and Oxygen Mixture by Branch Chain Reaction Theory.
10. Why does the hydrogen-oxygen reaction have the phenomenon of ignition peninsula?
11. What are the common ignition methods?
12. What is the minimum ignition energy for spark ignition?

Chapter 4 Combustion and Explosion of Combustible Gases

During the production processes in petrochemical factories, a variety of combustible gases will be produced, and combustible gases are often used as raw materials in the production process of these enterprises. In people's daily life, combustible gases can be seen everywhere. The combustion of combustible gas can cause explosion and detonation under specific conditions, which can cause serious damage to building facilities, industrial equipment and so on and simultaneously endangers people's safety. Therefore, it is of great significance to study the law of gas combustion and explosion for the prevention of such accidents and disaster relief after the accidents.

4.1 Laminar premixed flame propagation mechanism

If a chemical reaction occurs somewhere in the stationary combustible mixture, the reaction will propagate in the mixture with the time going on. According to the different reaction mechanisms, this can be divided into two forms: slow combustion and detonation. The normal propagation of flame depends on heat conduction and molecular diffusion to increase the temperature of unburned mixture and cause chemical reaction in the reaction zone. One that the combustion wave is continuously pushed into the unburned mixture. The velocity of this form of propagation is generally not higher than 1 to 3m/s. This propagation is stable, and the speed is a constant under certain physical and chemical conditions (such as temperature, pressure, concentration, mixing ratio, etc.). The propagation of detonation wave does not occur through heat and mass transfer. It relies on the compressive effect of shock waves to make the temperature of unburned mixture rise consecutively, which causes chemical reaction and makes the combustion wave advance to the unburned mixture consecutively. The speed of spread of this form is very high, often more than 1000m/s, which is in sharp contrast to the normal flame spreading speed, and its propagation process is also stable.

We will further clarify this case from the point of view of chemical fluid mechanics.

To study its basic features, we consider the simplest case of a plane wave in a one-dimensional steady flow, that is, the mixture flow (or the propagation speed of the combustion wave) is assumed to be a one-dimensional steady flow, viscous and body forces neglected, and supposing the mixture to be a complete gas. Its specific heat capacity at constant pressure before and after combustion, c_p is constant, and its molecular weight also remains constant. The reaction zone is small compared to the characteristic dimensions of the tube (e.g., tube diameter), and it has no friction and no heat exchange with the tube wall. In the process of analysis, we do not analyze the propagation of the combustion wave in the static combustible mixture, but when the combustion wave is stationary and the mixture continues to flow towards the combustion wave, the velocity of the combustion wave relative to the combustible mixture at infinity, u_∞, which is the propagation speed of combustion wave, whose physical model is shown in Fig. 4-1.

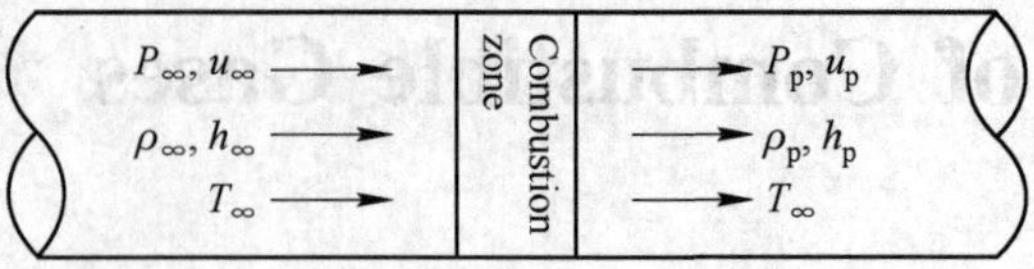

P_∞, u_∞ → | Combustion zone | → P_p, u_p
ρ_∞, h_∞ → | | → ρ_p, h_p
T_∞ → | | → T_∞

Fig. 4-1 Combustion process diagram

According to the above assumptions, the following conservation equation can be obtained:

(1) Equation of continuity:

$$\rho_p u_p = \rho_\infty u_\infty = m = \text{const} \tag{4-1}$$

Where the subscript "∞" is a parameter representing the combustible mixture at infinity upstream of the combustion wave. The subscript "p" denotes the parameters of the combustion products at infinity downstream of the combustion wave.

(2) Momentum equation: Since viscous and body forces are neglected, the momentum equation is:

$$P_p + \rho_p u_p^2 = P_\infty + \rho_\infty u_\infty^2 = \text{const} \tag{4-2}$$

(3) Energy equation: Due to the neglect of viscous and body forces and the absence of heat exchange, the energy equation can be reduced to:

$$h_p + \frac{u_p^2}{2} = h_\infty + \frac{u_\infty^2}{2} = \text{const} \tag{4-3}$$

Equation of state (complete gas):

$$pV = \rho RT$$

Or

$$P_p = \rho_p R_p T_p \quad P_\infty = \rho_\infty R_\infty T_\infty$$

(4) Heat equation of a certain state: the heat equation for a constant heat capacity is:

$$h_p - h_{p*} = c_p(T_p - T_*),\ h_\infty - h_{\infty *} = c_p(T_\infty - T_*) \tag{4-4}$$

Where h_* is the enthalpy (including chemical enthalpy) at the reference temperature T_*.

From the formulas (4-3) and (4-4):

$$c_p T_p + \frac{u_p^2}{2} - (\Delta h_{\infty p})_* = c_p T_\infty + \frac{u_\infty^2}{2} \tag{4-5}$$

Where $(\Delta h_{\infty p})_* = h_{p*} - h_{\infty *} = Q$ (heat of reaction of combustible mixture per unit mass), so the formula (4-5) can be rewritten as:

$$c_p T_p + \frac{u_p^2}{2} - Q = c_p T_\infty + \frac{u_\infty^2}{2} \tag{4-6}$$

From equations (4-1) and (4-2):

$$P_\infty + \frac{m^2}{\rho_\infty} = P_p + \frac{m^2}{\rho_p} \tag{4-7}$$

Or

$$\frac{P_p - P_\infty}{\dfrac{1}{\rho_p} - \dfrac{1}{\rho_\infty}} = -m^2 = -\rho_\infty^2 u_\infty^2 = -\rho_p^2 u_p^2 \tag{4-8}$$

This equation(4-8) is a straight line with a slope of $-m^2$ shown in Fig. 4-2 $P \sim 1/\rho$ (or specific volume $V=1/\rho$). This straight line is called Rayleigh line, and it shows the relationship that is to be satisfied between process end states P_P and ρ_p under the given case of the initial states P_∞ and ρ_∞.

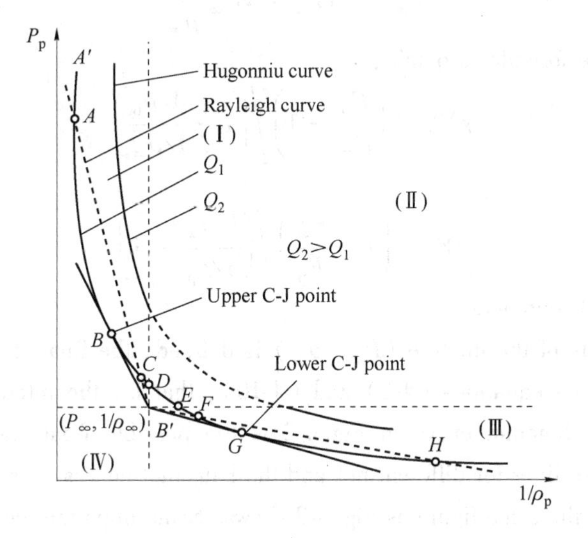

Fig. 4-2 Combustion state diagram

On the other hand, from equation (4-4) and equations (4-6) and (4-8):

$$h_p - h_\infty = c_p T_p - c_p T_\infty - Q = \frac{u_\infty^2}{2} - \frac{u_p^2}{2} = \frac{m^2}{2}\left(\frac{1}{\rho_\infty^2} - \frac{1}{\rho_p^2}\right)$$

$$= \frac{m^2}{2}\left(\frac{1}{\rho_\infty} - \frac{1}{\rho_p}\right)\left(\frac{1}{\rho_\infty} + \frac{1}{\rho_p}\right) = \frac{1}{2}(P_p - P_\infty)\left(\frac{1}{\rho_\infty} + \frac{1}{\rho_p}\right) \tag{4-9}$$

Using the equation of state and the following equation (γ is the specific heat ratio, which is an important parameter describing the thermodynamic properties of gas, and is defined as the ratio of the specific heat c_p at constant pressure to the specific heat c_V at constant volume):

$$c_p/R = \frac{\gamma}{\gamma - 1}$$

Eliminate the temperature parameter in the formula,

$$\frac{\gamma}{\gamma - 1}\left(\frac{P_p}{\rho_p} - \frac{P_\infty}{\rho_\infty}\right) - \frac{1}{2}(P_p - P_\infty)\left(\frac{1}{\rho_\infty} + \frac{1}{\rho_p}\right) = Q \tag{4-10}$$

This equation is called the Hugoniot equation, and its curve in Fig. 4-2 is the Hugoniot curve, which is obtained in a given initial state P_∞, ρ_∞ and the heat of reaction Q after the parameter m is eliminated, which reveals the relationship between the final state P_p and ρ_p.

In addition, from equation (4-8):

$$u_\infty^2\left(\frac{1}{\rho_\infty}-\frac{1}{\rho_p}\right)=\frac{P_p-P_\infty}{\rho_\infty^2}$$

Namely:

$$u_\infty^2=\frac{1}{\rho_\infty^2}\frac{P_p-P_\infty}{(1/P_\infty-1/\rho_p)}$$

Because the speed of sound c_∞ can be written as:

$$c_\infty^2=\gamma RT_\infty=\gamma P_\infty\frac{1}{\rho_\infty}$$

Substitute the above formula to obtain:

$$\gamma M_\infty^2=\left(\frac{P_p}{P_\infty}-1\right)\Bigg/\left(1-\frac{1/\rho_p}{1/\rho_\infty}\right) \tag{4-11}$$

Or

$$\gamma M_p^2=\left(1-\frac{P_\infty}{P_p}\right)\Bigg/\left(\frac{1/\rho_\infty}{1/\rho_p}-1\right) \tag{4-12}$$

Where Ma is the Mach number.

Once the initial state of the mixture (P_∞, ρ_∞) is defined, the final state (P_p,ρ_p) must meet the requirements of both equations (4-8) and (4-10), that is, the intersection of the Rayleigh straight line and the Hugoniot curve in Fig. 4-2 is the possible final state. Now we draw the Rayleigh lines (a set of lines for different m) and the Hugoniot curves (a set of curves for different Q) simultaneously on the same figure as Fig. 4-2 shows. Some important conclusions can be drawn from the analysis of Fig. 4-2.

(1) Fig. 4-2 (P_∞, $1/\rho_\infty$) is the initial state, the parallel lines of the axes (that is, the quantity dotted lines perpendicular to each other in the figure), are made respectively by crossing the point (P_∞, $1/\rho_\infty$) P_p shaft, $1/\rho_p$ then the plane(P_∞, $1/\rho_\infty$) is divided into four regions (Ⅰ, Ⅱ, Ⅲ, Ⅳ). The final state of the process can only occur in region (Ⅰ) and (Ⅲ), but not in region (Ⅱ) and (Ⅳ). This is because, as can be seen from the equation (4-8), the slope of the Rayleigh line is negative, and therefore the two dotted straight lines crossing the point (P_∞, $1/\rho_\infty$) are the limit conditions of the Rayleigh straight line, so the DE segment (shown by the dotted line) in the Hugoniot curve has no physical meaning, so the whole (Ⅱ) and (Ⅳ) regions have no physical meaning, and the final state cannot fall into these two regions.

(2) Intersections A, B, C, D, E, F, G, H, etc. are possible final state. Region (Ⅰ) is the detonation region, while region (Ⅲ) is the slow burn region. Because in region (Ⅰ), $1/\rho_p<1/\rho_\infty$, $P_p>P_\infty$ meaning the gas is compressed and slowed down after passing through the combustion wave. Secondly, from equation (4-11), we can see that the value of the numerator on the right side of the equation is much greater than 1, while the denominator is less than 1, so the value on the right side of the equation must be much greater than 1. 4. If we take $\gamma=1.4$, we get $M_\infty>1$. It can be seen that the combustion wave propagates in the mixture at supersonic speed. Therefore, the region (Ⅰ) is a detonation region. In contrast, in the (Ⅲ) region, $1/\rho_p>1/\rho_\infty$, $P_p<P_\infty$

which means after the combustion wave, the gas expands and its velocity increases. At the same time, it can be seen from equation (4-11) that the absolute value of the numerator on the right side of the equation is less than 1, while the absolute value of the denominator is greater than 1, so the value on the right side of the equation will be less than 1, making $M_\infty < 1$, so the combustion wave propagates in the mixture at subsonic speed, and this zone is called the slow combustion zone.

(3) The Rayleigh and Hugoniot curves are tangent to B and G respectively. Point B is called the upper Chapman-Jouguet point, or upper C-J point for short, and the wave with end point B is called the C-J detonation wave. The AB section is called strong detonation and the BD section is called weak detonation. Under most experimental conditions, C-J detonation waves are spontaneously produced, but artificial supersonic combustion can cause strong detonation waves. The EG section is a weak slow burning wave, the GH section is called the strong slow burning wave. Experiments show that most of the combustion processes are close to isobaric processes, so the strong slow combustion wave cannot occur, and the weak slow combustion wave in eg section will be of practical significance with $M_\infty \approx 0$.

(4) When $Q=0$, the Hugoniot curve passes through the initial state (P_∞, $1/\rho_\infty$). This is an ordinary aerodynamic shock wave.

4.2 Laminar premixed flame propagation velocity

4.2.1 Definition of flame propagation speed

4.2.1.1 Flame front (wavefront)

If a container is filled with a uniform mixture of gases, a part of the mixture is burned and a flame is formed when a spark or other heating method is used, after which, the energy is transferred to the cold mixture layer adjacent to the flame by heat conduction, which increases the temperature of the mixture and causes chemical reaction, forming a new flame. In this way, a layer of mixed gas is ignited in turn, in which way, a thin chemical reaction zone begins to propagate from the point of ignition to the unburned mixture, forming a clear boundary between the burned area and the unburned area, and this thin layer of chemical reaction luminous area is called the flame front.

Experiments show that the thickness of the flame front is quite thin compared to the characteristic size of the system, so it is often regarded as a geometric surface in the analysis of practical problems.

4.2.1.2 Flame displacement velocity and flame normal propagation velocity

The flame displacement velocity is the forward velocity of the flame front in the unburned mixture compared to the rest coordinate system, and the normal direction of the flame front points to the unburned gas. If the displacement of the flame front in the time interval from t to $t+\mathrm{d}t$ is $\mathrm{d}n$, then the displacement velocity u is:

$$u = \lim_{\Delta t \to 0} \frac{\Delta n}{\Delta t} = \frac{dn}{dt} \tag{4-13}$$

The normal propagation velocity of the flame is the velocity of the flame facing the unburned mixture at infinity in its normal direction. If the displacement velocity of the flame front is u and the velocity of the unburned mixture is w, its fractional velocity in the direction normal to the flame front is w_n, the flame normal propagation speed S_1 is:

$$S_1 = u \pm w_n \tag{4-14}$$

The sign is taken negative when the displacement velocity u is in the same direction as the velocity of the gas flow. Conversely, then take the positive sign. When the flow velocity $w = 0$, $S_1 = u$, the velocity of flame movement observed is the flame propagation velocity.

4.2.2 Flame-front's structure

Imagine a flat flame front in a circular tube (in fact, the flame front is parabolic when the flame propagates in the tube), the flame front is stationary in the tube, and the premixed flammable mixture flows along the tube toward the flame front at a velocity of S_1 (check Fig. 4-3). Experiments show that the flame front is a very narrow region, with a width of only hundreds or even tens of microns, which separates the burned gas from the unburned gas. The processes of chemical reaction, heat conduction and material diffusion are completed in this very narrow width (area with width of δ). Fig. 4-3 shows the variation of reactant concentration, temperature, and reaction rate within the flame front. Since the width and surface curvature of the flame front are very small, it can be considered that the temperature and concentration in the flame front are only functions of the coordinate x. It can be concluded from the figure that within the width of the front, the temperature gradually rises from the initial temperature T_0 of the original premixed gas to the combustion temperature T_f, while the concentration C of the reactant gradually decreases from close to C_0 on the o-o section to close to zero on the a-a section strictly speaking, the initial state of the premixed gas, $T = T_0$, $C = C_0$, $W = 0$, should correspond to the situation at the cross section at $x \to -\infty$; while the final state of the burned gas $T = T_f$, $C = 0$, $W = 0$, should be equivalent to the situation at the cross section at $x \to +\infty$. In that flame front, only 95% to 98% of the fuel is actually reacting. The width of the flame front changes very little, but the temperature and concentration change greatly in this width, and the maximum temperature gradient dt/dx and concentration gradient dc/dx appear. Therefore, there is a strong heat flow and diffusion flow in the flame. The direction of the heat flow is from high temperature flame to low temperature, fresh mixture, while the direction of diffusion flow is from high concentration to low concentration. For example, the molecules of fresh mixture diffuse from o-o section to a-a section. On the contrary, the molecules of combustion products, such as free radicals and activation centers (such as OH, H, etc.) in burned gas, diffuse to fresh mixture. The migration of molecules in the flame is therefore not only affected by mass flow (directional flow of gas), but also by diffusion. This results in intensive mixing of the combustion products with the fresh mixture over the entire width of the flame front.

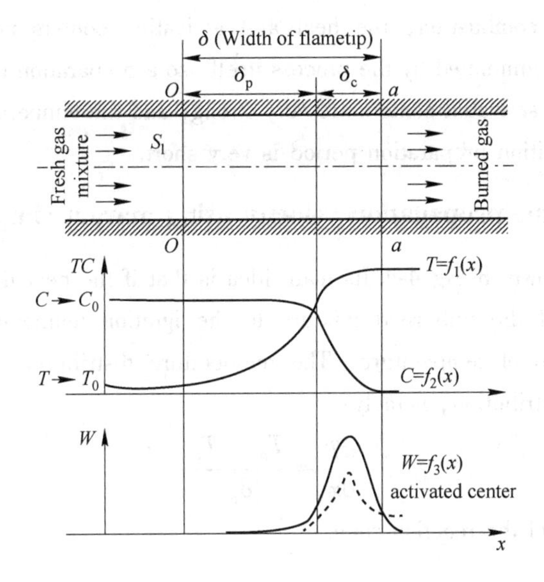

Fig. 4-3 Schematic diagram of flame front structure

(Width of flame tip; Fresh gas mixture; Burned gas; Activated center)

The variation of the chemical reaction rate can also be seen in Fig. 4-3. In the initial large width δ_p, the chemical reaction rate is very small, which can be generally ignored, and the changes in temperature and concentration are mainly due to heat conduction and diffusion, so this part of the flame front width is collectively called the "preheating zone", where the fresh mixture is heated. There after, with the increase of temperature, the chemical reaction rate increases exponentially, light and heat are emitted at the same time, and the temperature rises quickly to the combustion temperature T_f. As the temperature increases and the reactant concentration decreases, the temperature at which the chemical reaction rate reaches its maximum is slightly lower than the combustion temperature T_f, but close to the combustion temperature. As can be seen, the chemical reactions in the flame are always carried out at high temperatures close to the combustion temperature (this is very important as it is the basis of the thermodynamic theory of flame propagation speed). The faster the chemical reaction, the faster the flame propagation, and the shorter the residence time of the gas in the flame front. But this short time is sufficient for chemical reactions at high temperatures. Most of the combustible mixture (about 95% ~ 98%) reacts at a high temperature close to the combustion temperature, so the flame propagation speed corresponds to this temperature. These changes occur in a very narrow region δ_c of the remaining flame front width, where the reaction rate, temperature, and concentration of active centers reach their maxima. This region is commonly referred to as the "reaction zone" or "combustion zone" or the "chemical width" of the flame front. The chemical width of the flame front is always less than its physical width (or the flame front width δ_c), i.e. $\delta_c < \delta_p$. Another characteristic of the chemical reaction occurring in the flame front is that the ignition delay (induction period) is very short, or there is even not one. This is different from the spontaneous combustion process. In the process of spontaneous combustion, the heat and activation centers needed to accelerate the chemical reaction are accumulated by the process itself, so a preparation time is needed. The heat flux and the diffusion of activated centers are very strong, and the temperature of the premixed gas rises rapidly, so the ignition preparation period is very short.

4.2.3 Laminar flame propagation velocity using marant simplified analysis

Its physical model is shown in Fig. 4-4. Its main idea is that if the heat derived from zone Ⅱ can raise the temperature of the unburned mixture to the ignition temperature T_i, the flame can maintain the propagation of temperature. The temperature distribution in the reaction zone is assumed to be linear distribution, namely:

$$\frac{dT}{dx} \approx \frac{T_m - T_i}{\delta_c} \tag{4-15}$$

Where, δ_c is the width of the reaction zone.

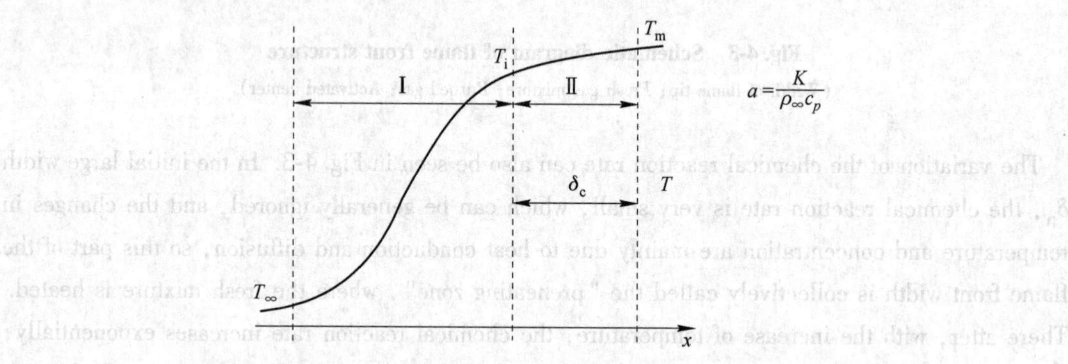

Fig. 4-4 Marant's temperature distribution model in the flame front

Ⅰ—preheating zone; Ⅱ—Reaction zone

Therefore, the heat balance equation is:

$$Gc_p(T_i - T_\infty) = FK\frac{T_m - T_i}{\delta_c} \tag{4-16}$$

Where, G is the mass flow; F is the cross-sectional area of the pipe; K is the thermal conductivity.

Because

$$G = \rho F u = F\rho_\infty S_l \tag{4-17}$$

So:

$$\rho_\infty S_l c_p(T_i - T_\infty) = K\frac{T_m - T_i}{\delta_c}$$

Or:

$$S_l = \frac{K}{\rho_\infty c_p} \cdot \frac{T_m - T_i}{T_i - T_\infty} \cdot \frac{1}{\delta_c} = a \cdot \frac{T_m - T_i}{T_i - T_\infty} \cdot \frac{1}{\delta_c} \tag{4-18}$$

Where $a = K/\rho_\infty c_p$, is called thermal diffusivity.

Plus:

$$\delta_c = S_l \tau_c = S_l \frac{\rho_\infty f_{s\infty}}{w_s} \tag{4-19}$$

Where τ_c is the chemical reaction time; ρ_∞ is the initial mass concentration of the mixture; $f_{s\infty}$ is the initial relative mass concentration of the mixture; w_s is reaction velocity of the combustible mixture.

Substitute equation (4-19) into equation (4-18).

$$S_l = \sqrt{a \cdot \frac{T_m - T_i}{T_i - T_\infty} \cdot \frac{W_s}{\rho_\infty \cdot f_{s\infty}}} \tag{4-20}$$

Equation (4-20) shows that the laminar flame propagation speed, S_l, is proportional to the thermal diffusivity, a, and the square root of the chemical reaction rate, W_s. This conclusion has been proved to be correct by experiments.

And because

$$W_s = K_{0s} \rho_\infty^n f_{s\infty}^n e^{\frac{-E}{RT_m}}, a = \frac{K}{\rho_\infty c_p}$$

So

$$S_l = \sqrt{\frac{K(T_m - T_i) K_{0s} \rho_\infty^{n-2} f_\infty^n e^{\frac{-E}{RT_m}}}{c_p(T_i - T_\infty)}} \tag{4-21}$$

According to the $p \propto \rho$ relation, we can get:

$$S_l \propto \sqrt{\rho_\infty^{n-2}} \propto \sqrt{p^{n-2}} = p^{\frac{n}{2}-1}$$

Where, n is the order of the reaction. Equation (4-21) implies that for a second-order reaction, the flame propagation speed, S_l, will be independent of pressure. The reaction order of most hydrocarbons with oxygen is close to 2, so the flame propagation speed S_l has little relationship with pressure, which is also confirmed by experiments.

4.2.4 Factors affecting flame propagation speed

4.2.4.1 The ratio of the amount of fuel to oxidizer's effect

Figs. 4-5 and 4-6 show the effects of the CO/air and hydrogen/air ratios on the flame propagation speed. It is very clear from these figures that when the mixture is too thick or too thin, the flame cannot propagate, which forms the upper and lower limits of the combustion limit, where the combustion rate decreases rapidly. For most mixtures, the maximum flame propagation speed occurs when the components are at stoichiometric ratios.

4.2.4.2 Influence of fuels' structure

Gerslein, Levine and Wong measured the burning rates of a large number of hydrocarbon and air mixtures to understand the effect of fuel molecular structure onits burning rate.

The range of the combustion limit becomes narrower as the molecular weight increases. Fig. 4-7 shows the effect of varying the number of carbon atoms in the three hydrocarbon fuel molecules on the rate of combustion. For saturated hydrocarbons (such as ethane, propane, butane, pentane, hexane, etc. in paraffins), the combustion rate ($\approx$70cm/s) is almost independent of the number of carbon atoms n_c. For unsaturated hydrocarbons, u_0 increases when n_c decreases. When $n_c<4$, u_0 decreases sharply with the increase of n_c, and then decreases very slowly with the further increase of carbon molecules. When $n_c>8$, the burning rate reaches its saturation value and does not change.

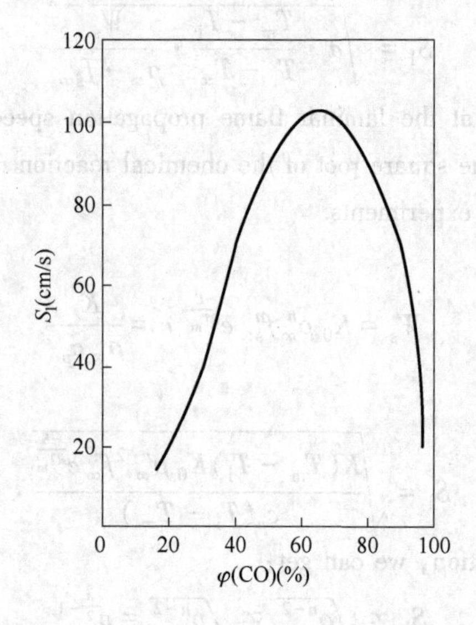

Fig. 4-5 Effect of CO/air ratio on combustion rate

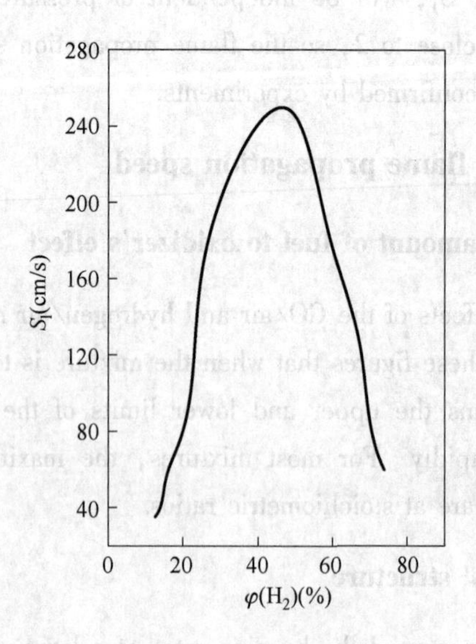

Fig. 4-6 Relationship between hydrogen/air concentration and combustion rate

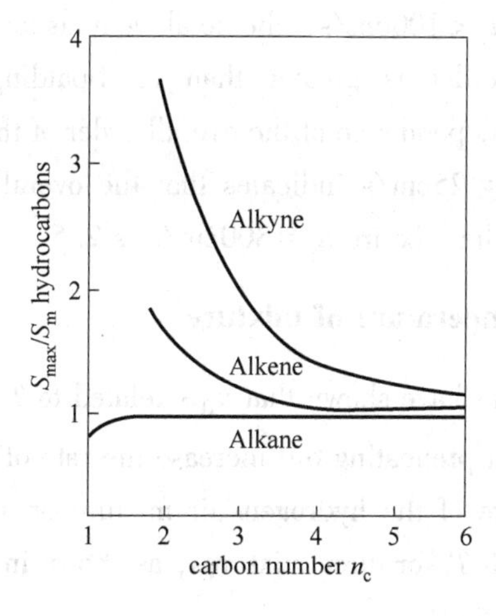

Fig. 4-7 Effect of different carbon atom numbers on *u*

4. 2. 4. 3 The effects of stress

Lewis has studied the effect of pressure on the combustion rate of different $CH/O_2/N_2/He$ mixtures by constant volume combustion. He determined the n values of different mixtures according to the principle that the proportional relationship of u_0p^n is established. Fig. 4-8 shows the pressure index n for different hydrocarbon flames. In general, when the combustion rate is low (e. g. , <50cm/s) , the combustion rate increases with decreasing pressure. In the range of 50 ~ 100cm/s, the combustion rate is independent of the pressure, and when the velocity is very high (i. e. , greater than 100cm/s) , the combustion rate increase with the increase of the pressure.

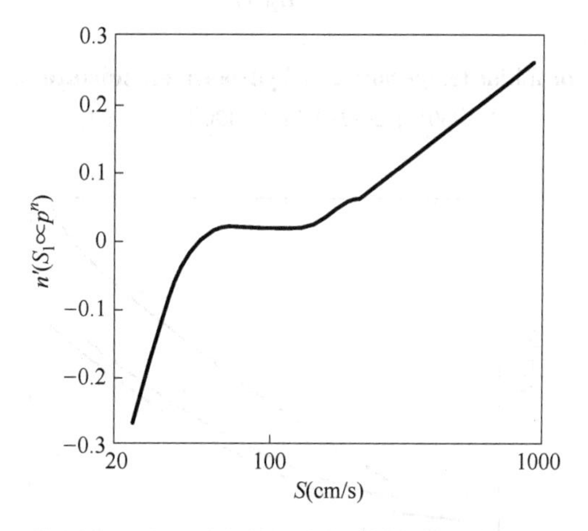

Fig. 4-8 Effect of pressure index on *u*

It will berecognized in the flame propagation mechanism that the above relationship of $u_0(p)$ indicates that the overall combustion reaction order is smaller than 2 when the combustion rate u_0

is <50cm/s; when 50cm/s< u_0 < 100cm/s, the total combustion reaction order is 2; for u_0 > 100cm/s, and the reaction order is greater than 2. Spalding's conclusion from Levine's experiment is that the pressure dependence of the overall order of the combustion reaction when the burning rate of the mixture u_0 = 25cm/s indicates that the overall order of the reaction is 1.4, while the order of reaction for the mixture u_0 = 800cm/s is 2.5.

4.2.4.4 Effect of initial temperature of mixture

Dugger and Heimel's experiment have shown that u_0 is related to T_s. Their tests confirmed Mallard and Lechateler's conjecture that preheating did increase the rate of combustion. Fig. 4-9 shows the effects of the initial temperature of the hydrogen/air mixture on the rate of combustion. Dugger measured the variation of u_0 with T_s for three mixtures, as shown in Fig. 4-10. From these data, $u_0 \propto T_s^m$, and the value of m is 1.5~2.

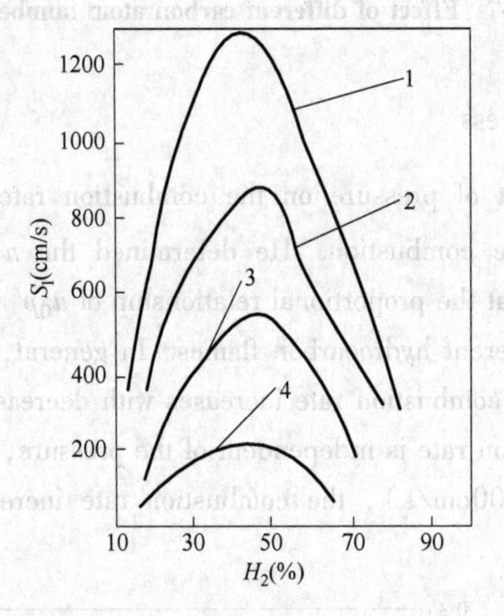

Fig. 4-9 Effect of initial temperature of hydrogen/air mixture on combustion rate

1—430℃; 2—310℃; 3—190℃; 4—20℃

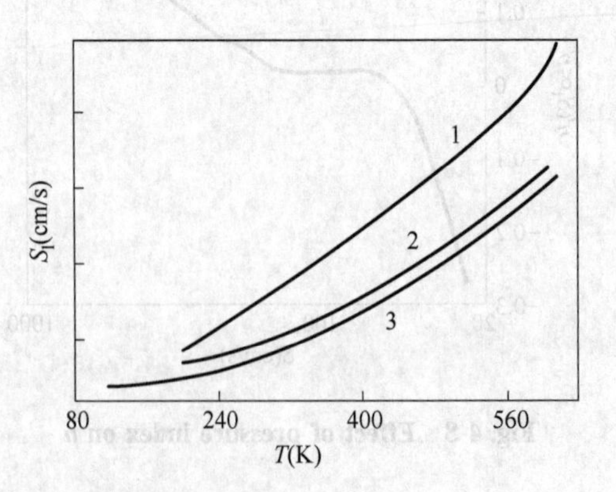

Fig. 4-10 Variation of u_0 with T_s for three mixtures

1—C_2H_4+air; 2—C_3H_8+air; 3—CH_4+air

4. 2. 4. 5 Effect of flame temperature

Fig. 4-11 shows the final combustion temperature, T_f, as a function of u_0 for several mixtures. From the data in this figure, compiled by Bartholorne and Sachsse, it can be seen that the higher the flame temperature, the stronger the effect on u_0. So it can be concluded that as far as the effect of temperature is concerned, it is essentially the flame temperature that affects the combustion rate. Moreover, for most mixtures, the increase in combustion velocity is much more rapidly than the increase in flame temperature. Since the decomposition reaction is apt to occur at a high temperature, the intermediate reaction is accelerated in many circumstance, resulting in this.

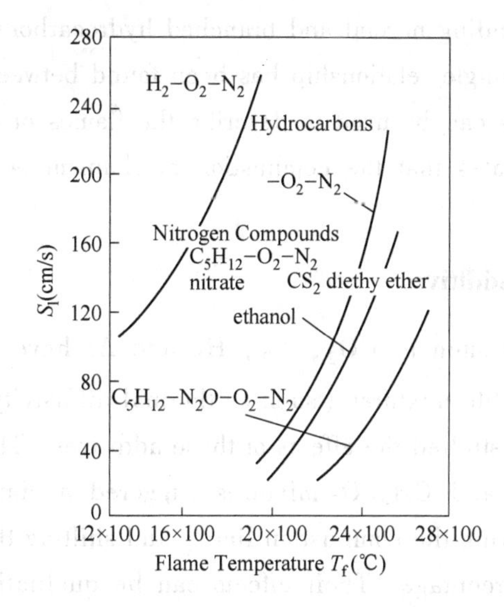

Fig. 4-11 Relationship between final combustion temperature and combustion rate of several mixtures

The decomposition reaction also causes free radicals to appear in the flame, and these free radicals act as chain carriers to promote the reaction and the propagation of the flame. The radicals and atoms produced near the flame temperature are quite easy to diffuse, so atomic H (and to some extent also O and the radical OH) can significantly increase the combustion temperature.

Tanford thought that the action of the H atom in the CO and O_2 flame takes place in the following way (where R stands for CO or H_2 and P stands for CO_2 or H_2O), respectively.

$$\left.\begin{array}{l} H_2 + O_2 + M \longrightarrow HO_2 + M \\ HO_2 + R \longrightarrow P + OH \\ OH + R \longrightarrow P + H \end{array}\right\} \text{Continuous}$$

$$H + H + M \longrightarrow H_2 + M \text{ (chain termination)}$$

The chain termination reaction can occur at the vessel wall (at low pressure) or in the gas phase (at very high pressure).

Hydrogen has a similar catalytic effect in the combustion of coal and some metal chlorides. In many combustible mixtures, the substitution of helium for nitrogen decreases the specific heat capacity of the mixture, thereby increasing its flame temperature, which in turn increases the concentration of activated H atoms and thevelocity of combustion.

Based on these observations, Tanford firmly believes that the diffusion of free radicals plays an extremely important role in certain gas mixtures. The effect of C_H (hydrogen atom concentration) on u_0 measured by Gaydon, Bortholome, and Jahn et al. Simon tested the combustion of 35 hydrocarbons in air in response to Tanford's statement. These fuels include paraffins and olefins (including normal and branched hydrocarbons), acetylenes, benzene, and cyclohexane, etc. A single relationship has been found between the burning rate u_0 and $(6.5C_H+C_O+C_{OH})$. This can be used to describe the flames of all these hydrocarbon and benzene fuels. This indicates that the combustion mechanism is the same for all of these blends.

4.2.4.6 Effect of inert additives

Chemically inert additives such as CO_2, N_2, He and Ar have an effect on the physical characteristics of combustible mixtures (such as thermal diffusivity, specific heat capacity, etc.). Many people have studied the effects of these additives. The addition of CO_2 and N_2 to the H_2/O_2, CO/O_2, and CH_4/O_2 mixtures triggered a similar effect: reducing the combustion speed, narrowing the combustion limit, and shifting the maximum value of u_0 to the side with less fuel percentage. Their effects can be qualitatively represented by Fig. 4-12; Fig. 4-13 is the measured data. The effect of inert gas additives on the combustion rate mainly manifests in the ratio of thermal conductivity to heat capacity. When there is excess fuel or excess oxidant in the fuel mixture, the effect of the excess fuel or oxidant is similar to that of the inert gas.

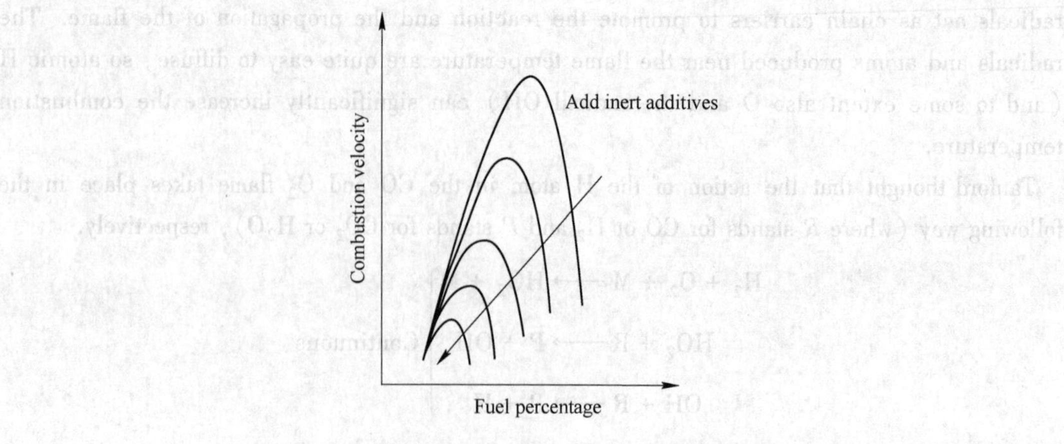

Fig. 4-12 Effect of inert components on combustion rate

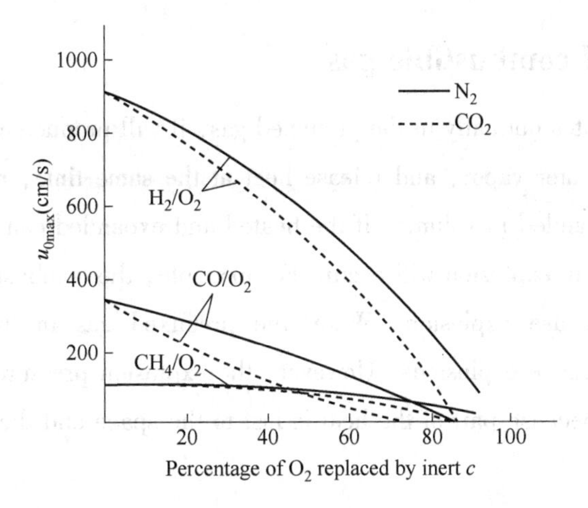

Fig. 4-13 Experimental results of inert components on combustion rate

4.2.4.7 Influcncc of active additives

The addition of active additives in gas mixtures will have certain impacts on the flame propagation speed. However, the addition of active additives in some fuel mixtures has little influence on the combustion rate, and the shift of the combustion rate curve has its own characteristics. For example, when CO is gradually replaced by CH_4in the CO/air mixture, the curve shifts to the left, when only 5% of CO is replaced by CH_4, the increase in $u_0$reaches a maximum, as shown in Fig. 4-14. It is clear that the CO/air's velocity enhancements are due to the H atoms produced by CH_4.

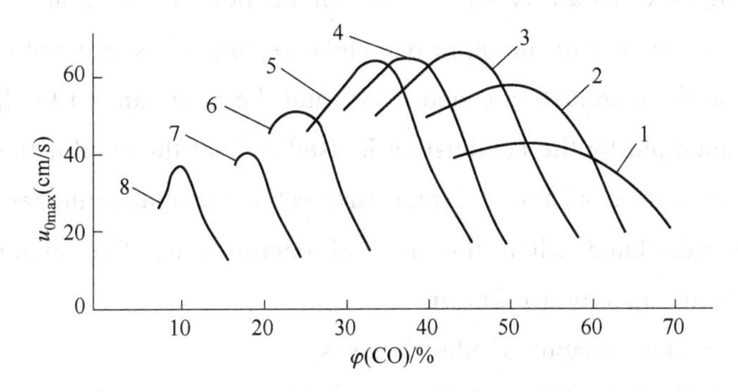

Fig. 4-14 Effect of methane on combustion rate of CO/Air

Fuel percent in air is shown as Table 4-1.

Table 4-1 Fuel percent in air

Curve		1	2	3	4	5	6	7	8
Composition (%)	CO	100	96	95	90	85	70	50	0
	CH_4	0	4	5	10	15	30	50	100

4.3 Explosion of combustible gas

When the flame propagates normally in the premixed gas, it will produce combustion products such as carbon dioxide and water vapor, and release heat at the same time, and making the products heated, heated and expanded in volume. If the heated and expanded combustion products can not be discharged in time, an explosion will occur. For example, the combustion of premixed gas in a closed container will cause explosion. When the premixed gas in the free space is more, combustion can also produce explosions. However, the explosion pressure is generally lower than that in a closed vessel because part of the heat is lost to the space and the products can dilate to a certain extent.

4.3.1 Calculation of explosion temperature of premixed gas

The explosion temperature of a substance is an important parameter to measure the destructive power of an explosion. The explosion temperature of a combustible gas is the maximum temperature at which all of the heat released by the explosion of the substance is used to heat the reaction products. The explosion temperature is essentially the same as the combustion temperature, which can be determined according to the energy conservation relationship of the substances before and after the reaction.

The calculation of the temperature of the premixed gas at the time of explosion is illustrated by taking ether as an example. The combustion of the premixed gas of ether and air in a closed container can be approximately regarded as adiabatic constant-volume combustion because the combustion speed is fast and the heat can not be dissipated in time, and all the heat generated by the combustion is used to heat the combustion products. If the ratio of ether to air is used as the stoichiometric ratio, the maximum temperature of ether explosion can be calculated when the heat of combustion, the amount of combustion products and the heat capacity are known.

The combustion reaction equation of ether in air is:

$$C_4H_{10}O + 6O_2 + 22.6N_2 \longrightarrow 4CO_2 + 5H_2O + 22.6N_2$$

Since the volume ratio of nitrogen to oxygen in air is 79 : 21, and the volume ratio of gases is equal to the ratio of their moles, one kilomole of ether requires six kilomoles of oxygen, which will definitely be accompanied by $6 \times 79/21 = 22.6$ kilomoles of nitrogen.

It can be seen from the combustion reaction equation that the total number of kilomoles of unburned gas mixture before explosion is 29.6, and the total number of kilomoles of burned gas after explosion is 31.6.

Check Table 4-2 for the calculation formula of average molar heat capacity at constant volume of gas.

Table 4-2 Calculation formula of average molar heat capacity at constant volume of gas

Gas	Heat capacity(4186.8 J/(kmol · ℃))
Monatomic gas (Ar, He, metal vapor)	4.93
Diatomic gases (N_2, O_2, H_2, CO, NO, etc.)	$4.80+0.00045t$
CO_2, SO_2	$9.0+0.00058t$
H_2O, H_2S	$4.0+0.00215t$
All tetraatomic gases (NH_3 and others)	$10.00+0.00045t$
All five-atom gases (CH_4 and others)	$12.00+0.00045t$

According to the calculation formula listed in the table, the heat capacity of each component of the combustion product is:

The molar heat capacity at constant volume of N_2 is [$(4.8 + 0.00045t) \times 4186.8$] J/(kmol · ℃).

The molar heat capacity at constant volume of H_2O is [$(4.0 + 0.00215t) \times 4186.8$] J/(kmol · ℃).

The molar heat capacity at constant volume of CO_2 is [$(9.0 + 0.00058t) \times 4186.8$] J/(kmol · ℃).

The total heat capacity of the combustion products is:

22.6[$(4.8+0.00045t) \times 4186.8$]J/(kmol · ℃)+5[$(4.0+0.00215t) \times 4186.8$]J/(kmol · ℃)+4[$(9.0+0.00058t) \times 4186.8$]J/(kmol · ℃) = $(688.4+0.0967t) \times 10^3$J/(kmol · ℃)

The total heat capacity of the combustion products is [$(688.4 + 0.0967t) \times 10^3$] J/(kmol · ℃). The heat capacity here is the heat capacity at constant volume, which corresponds to an explosion in a closed container.

The heat of combustion of diethyl ether is researched to be 2.7×10^6J/mol, i. e. 2.7×10^9J/kmol.

Because the explosion is very fast and almost adiabatic, so the total heat of combustion can be approximately regarded as using to raise the combustion products' temperature, which is equal to the product of the heat capacity of the combustion products and the temperature, that is,

$$2.7 \times 10^6 = (688.4 + 0.0967t) \times 10^3 \cdot t$$

The maximum explosion temperature t obtained by solving the above formula is 2826℃.

In the above calculation, the original temperature is regarded as 0℃. Although there is a difference of several degrees from the normal room temperature, the maximum explosion temperature is high enough to ignore the impacts on the accuracy of the calculation results.

4.3.2 Calculation of explosion pressure of combustible mixture

The pressure produced by the explosion of flammable mixture is related to its initial pressure, initial temperature, concentration, composition, container and other factors. The pressure generated during explosion can be determined according to the law that the pressure is proportional to the temperature and the number of moles. According to this law, the following relationship is obtained:

$$\frac{P_m}{P_0}=\frac{T_m}{T_0}\times\frac{n_m}{n_0}$$

Where P_m, T_m, n_m——maximum pressure, maximum temperature and gas mole number after explosion;

P_0, T_0, n_0—— initial pressure, initial temperature, and number of moles of gas before explosion.

Therefore, the calculation formula of explosion pressure can be obtained as:

$$P_m=\frac{T_m n_m}{T_0 n_0}P_0 \tag{4-22}$$

The explosion pressure of the combustible mixture is the basic cause of the destructive force. The explosion pressure is different when the mixture ratio is different, and the explosion pressure is the largest when the mixture ratio is equivalent. The maximum explosion pressures for some materials are listed in Table 4-3.

Table 4-3 Maximum explosion pressures for certain substances

Substance	Maximum explosion pressure(MPa)	Substance	Maximum explosion pressure(MPa)
Acetaldehyde	0.73	Propane	0.86
Ethylene	0.89	Hydrogen	0.74
Ethanol	0.75	Cyclohexane	0.86
Ether	0.92	Propylene	0.86
Acetylene	1.03	Carbon disulfide	0.78
Acetone	0.89	Hydrogen sulfide	0.50
Benzene	0.90	Vinyl chloride	0.68

See Table 4-4 for the explosion pressure of gasoline at different concentrations.

Table 4-4 Explosion pressure of gasoline at different concentrations

Concentration	Explosion pressure (Pa)	Concentration	Explosion pressure (Pa)
1.36	It doesn't explode	4.02	7.3×10^5
1.58	5.56×10^5	4.28	6.63×10^5
2.08	8.01×10^5	4.44	2.16×10^5
2.58	7.85×10^5	5.04	1.57×10^5
2.78	8.25×10^5	5.84	1.08×10^5
3.14	7.95×10^5	6.08	0.68×10^5
3.4	8.06×10^5	6.48	0.58×10^5
3.98	6.74×10^5		

4.3.3 Speed of pressure rise during explosion

The explosion pressure p_m calculated above is the maximum pressure that the combustible gases' explosion can achieve under this condition. There is a short period of time from the initial pressure rise to the maximum explosion pressure, during which the pressure rises gradually. The instantaneous pressure p at any time in this period (within the range of $p<p_m$) can be calculated using the following equation:

$$p = K_d p_0 \frac{S_1^3 t^3}{V_0} + p_0 \tag{4-23}$$

Among

$$K_d = \frac{4}{3}\pi\left(\frac{n_m T_m}{n_0 T_0}\right)^2 \cdot \left(\frac{n_m T_m}{n_0 T_0} - 1\right) \cdot K \tag{4-24}$$

Where p——instantaneous pressure, Pa;

S_1——flame propagation speed, cm/s;

K_d——coefficient;

K——coefficient, taken as 1.4;

T——time, s.

The average rate of pressure rise is the maximum explosion pressure p_m minus the initial pressure p_0, divided by the time required to reach the maximum pressure:

$$v = \frac{p_m - p_0}{t}$$

Obviously, the maximum explosion pressure and flame propagation speed of different combustible gases are different, and the time required to reach the maximum pressure is also different. Therefore, the rate of pressure rise is different when different combustible gases are exploding. For example, the rate of pressure rise in a hydrogen explosion is faster than in a methane explosion. The rate of pressure rise during a methane explosion is shown in Fig. 4-15. Also that volume of the container are different, the boost speed is also different, the larger the volume, the lower the boost speed, and the smaller the volume, the higher the boost speed. The pressure rise rate of some combustible gases is shown in the Table 4-5.

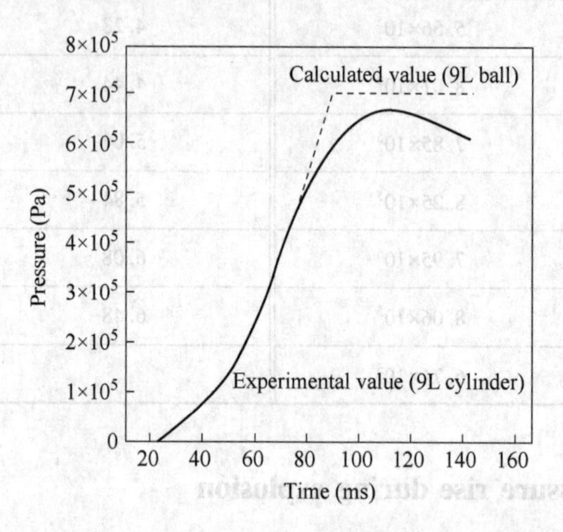

Fig. 4-15 Pressure rise velocity of methane explosion

Table 4-5 Maximum explosion pressure and pressure rise speed of some combustible gases and steam

Substance	Maximum explosion pressure(MPa)	The speed of pressure rise(MPa/s)
Hydrogen	0.74	90
Acetylene	1.03	80
Ethylene	0.89	55
Benzene	0.90	3
Ethanol	0.75	

4.3.4 Explosive power index

The degree of damage to the equipment during explosion is not only related to the maximum explosion pressure, but also to the pressure rise speed.

Explosive power index = maximum explosion pressure × average pressure rise speed. Check Table 4-6 for the explosion power index of some combustible gases.

Table 4-6 Explosive power index of several combustible gases

Substance	Butane	Benzene	Ethane	Hydrogen	Acetylene
Power index	9.30	2.4	12.13	55.80	76.00

4.3.5 Total explosion energy

The total energy of explosion can be calculated by the following formula

$$E = Q_v V \tag{4-25}$$

Where E——total explosion energy of combustible gas, kJ;

Q_v——calorific value of combustible gas, kJ/m^3;

V——volume of combustible gas, m^3.

4.3.6 Determination of explosion parameters

4.3.6.1 Experimental equipment

The test equipment consists of an explosion chamber, a vessel, a nozzle, an igniter, and a pressure measurement system, as shown in Fig. 4-16.

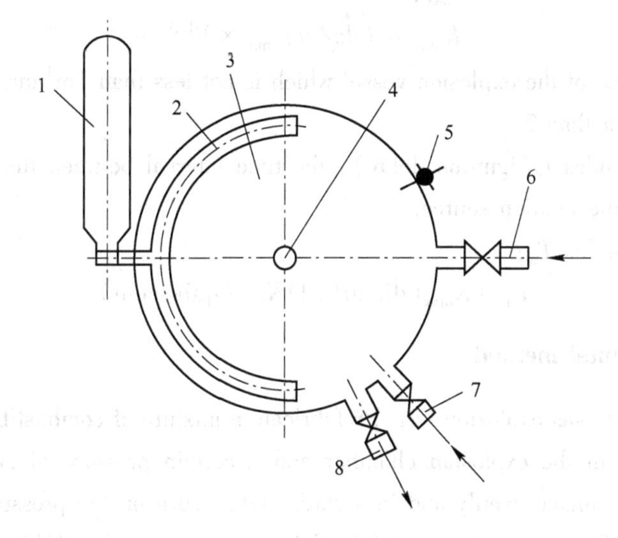

Fig. 4-16 Combustible gas explosion tester

1—Vessel; 2—Semicircular nozzle; 3—Explosion Chamber; 4—Ignition source; 5—Pressure sensor; 6—Combustible gas /air inlet; 7—Purge air; 8—Exhaust pipe

Explosion chamber: cylindrical vessel with volume of 1m^3, height : diameter of 1 : 1.

Vessel: volume is 5L and can be pressurized to 2MPa with air. A 19 mm quick-opening valve is provided adjacent to the vessel to allow the contents to be injected into the chamber within 10ms after opening the valve.

Nozzle: a semicircular tube with an inner diameter of 19mm with holes on the tube whose hole diameter is 4~6mm, and the total area of the holes is about 300mm.

Electronic igniter: the igniter current is about 300W, the output voltage is 15kV, and the spark spacing is 3~5mm, which is located at the geometric center of the experimental device.

4.3.6.2 Test parameters of experimental equipment

(1) Explosion pressure (p_m): the maximum explosion pressure of a combustible gas at a certain concentration.

(2) Maximum explosion pressure (p_{max}): the maximum explosion pressure of a combustible gas in a large concentration range.

(3) Pressure rise rate $(dp/dt)_m$: the pressure rise rate of a certain combustible gas at a certain concentration.

(4) Maximum pressure rise velocity $(dp/dt)_{max}$: the maximum pressure rise velocity of a combustible gas in a large concentration range.

(5) Explosion index K_m:

$$K_m = (dp/dt)_m \times V^{1/3} \tag{4-26}$$

(6) Maximum explosion index K_{max}:

$$K_{max} = (dp/dt)_{max} \times V^{1/3} \tag{4-27}$$

Where V is the volume of the explosion vessel which is not less than $1m^3$ and the ratio of length to diameter is not greater than 2.

(7) Perturbation index t_v (ignition delay): the time interval between the start of air injection and the initiation of the ignition source.

(8) Perturbation index T_u:

$$T_u = K_{max}(\text{disturbed})K_{max}(\text{quiescent})$$

4.3.6.3 Experimental method

(1) Static combustible gas explosion test. Prefabricate a mixture of combustible gas and air with a certain concentration in the explosion chamber and a certain pressure of atmospheric pressure. Ensure that the gas is mixed evenly and in a static state. Turn on the pressure recorder and start the ignition source. The measured p_m and $(dp/dt)_m$ are shown in Fig. 4-17.

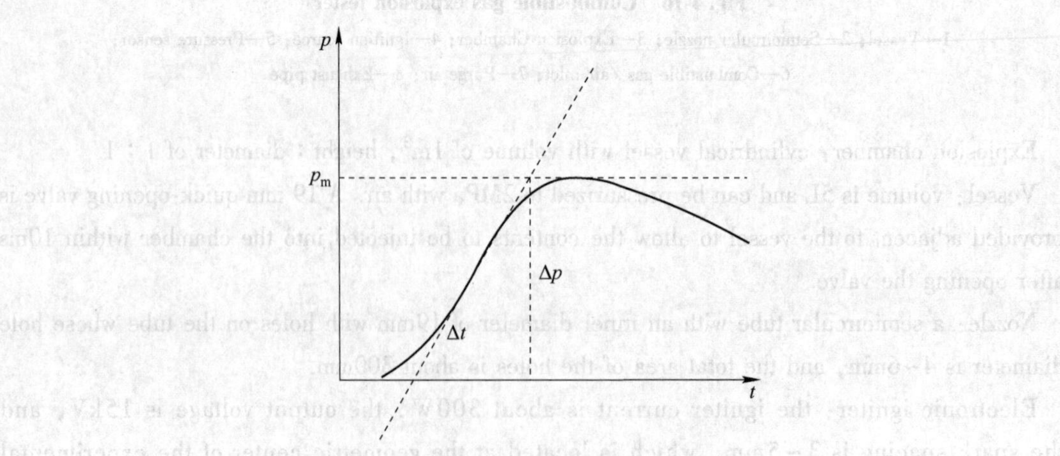

Fig. 4-17 Definition of p_m and $(dp/dt)_m$

Maximum values of p_m and $(dp/dt)_m$ were measured by repeating the test over a wide range of gas concentrations.

(2) Dynamic combustible explosion test. Prefabricate a mixture of combustible gas and air with a certain concentration in the explosion chamber and pressurize a 5L container to 2MPa with air. Open the valve of the container and the pressure recorder. Ignite and disturb the mixture of combustible gas and air under a certain ignition delay condition (the longer the delay time is, the lower the disturbance degree of the mixture is). p_m and $(dp/dt)_m$ can be measured over a wide range of combustible gas concentrations.

4.4 Theory and calculation of explosion limit

4.4.1 Explosion limit theory

Mixtures of flammable gases or vapors with air do not burn or explode in any proportion, and the rate of combustion (flame spread) varies with the proportion of the mixture. It is known from experiments that when the content of combustible gas in the mixture is close to the stoichiometry (which is the content of the substance when it is theoretically completely burned), it burns the fastest or explodes the most violently. If the content decreases or increases, the flame spread rate decreases, and when the concentration is below or above a certain limit, the flame will no longer spread. The lowest concentration that the mixture of flammable gas or vapor and air can make the flame spread is called the lower explosive limit of the gas or vapor; similarly, the highest concentration that can make the flame spread is called the upper explosive limit. Mixtures with concentrations below the lower limit and above the upper limit will not ignite or explode. Yet mixtures above the upper limit are combustible in air.

The explosion limit is generally expressed by the volume percentage of combustible gas or vapor in the mixture, and sometimes expressed by the content of combustible substance in unit volume of gas (g/m^3, mg/L). When the concentration of mixed explosives is below the lower explosive limit, there is excessive air, which prevents the spread of flame due to its cooling effect. Meanwhile, the destruction number of activation centers is greater than the production number. Similarly, if the concentration is above the upper explosive limit, it contains excessive flammable substances, and the air is very insufficient (mainly due to oxygen insufficiency), the flame can not spread. However, if air is supplied in this case immediately, there is also a risk of fire and explosion. Therefore, the mixture above the upper limit can not be considered safe.

There is no difference between combustion and explosion from the point of view of chemical reaction. When the mixture burns, the reaction on the wave surface is as follows:

$$A + B \longrightarrow C + D + Q \tag{4-28}$$

Where A, B——reactants;

C, D——product;

Q——heat of reaction (heat of combustion), J.

A, B, C and D are not necessarily stable molecules. It can also be atoms or radicals. The energy change before and after the reaction is as follows: when the reactant (A + B) is given an activation energy E, it becomes an activated state, and the reaction result becomes a product (C + D). At this time, the energy released is W, and the heat of reaction $Q = W-E$.

If the basic reaction concentration of the combustion wave is taken as n (the number of molecules reacting per unit volume), the energy released per unit volume is nW. If the combustion wave is consecutive, the energy released is taken as the activation energy in a new reaction. If α is taken as that probability of activation ($\alpha \leqslant 1$), then the number of activated basic reactions per unit volume of the second batch is $\alpha nW/E$. The energy released by the second batch is $\alpha nW^2/E$.

The ratio of the energy released during the reaction of the two batches of molecules is β:

$$\beta = \frac{\alpha nW^2/E}{nW} = \alpha\frac{W}{E} = \alpha\left(1 + \frac{Q}{E}\right) \tag{4-29}$$

Now we explore the value of β. When $\beta<1$, it means that after the reaction system is excited by the energy source, the heat release is less and less, the number of molecules causing the reaction is less and less, and finally the reaction stops, and combustion or explosion cannot be formed. When $\beta = 1$, it means that the reaction system can release heat evenly after being excited by the energy source, and a certain number of molecules continue to react. This is the condition that determines the explosion limit (strictly speaking, it can explode only slightly more than that). When $\beta>1$, it means that the heat release is larger and larger, the reaction molecules are more and more, so the explosion is formed.

At the explosion limit, $\beta=1$

Then

$$\alpha\left(1 + \frac{Q}{E}\right) = 1$$

Let the lower explosive limit be L_{low} (% by volume) vs. probability of reaction α proportional, i. e.

$$\alpha = KL_{low}$$

Where K is the constant of proportionality, so that

$$\frac{1}{L_{low}} = K\left(1 + \frac{Q}{E}\right) \tag{4-30}$$

When Q is large compared to E, the above equation can be written approximately as:

$$\frac{1}{L_{low}} = K\frac{Q}{E}$$

The above equation further shows the relationship among the lower explosive limit, the heat of combustion Q and the activation energy E. If the activation energy of each combustible gas does not change much, it can be roughly concluded that:

$$L_{low} * Q = \text{Constant} \tag{4-31}$$

This shows that the lower explosive limit L_{down} is inversely proportional to the combustion heat Q of combustible gas, that is to say, the greater the combustion heat of combustible gas molecules, the lower explosive limit. For example, alkanes' $L_{low} \times Q$ is close to the constant 1091, which proves that the above conclusion is correct.

For other flammable gases, this constant also exists. For alcohols, ethers, ketones and olefins, the constant is close to 1000, while for chlorinated and brominated alkanes, the constant is ralatively higher due to the introduction of halogen atoms, which greatly increases the lower explosion limit. The lower explosive limit of homologues can be calculated by using the constant relationship between the lower explosive limit and the product of combustion heat. However, it cannot be applied to combustible gases such as hydrogen, acetylene, and carbon disulfide.

The lower explosive limit L'_{low} is expressed in mg/L (milligram/liter) express. L_{low} is the lower explosive limit expressed in percent by volume.

Their relationship at 20℃ is:

$$L'_{low} = \frac{L_{low}}{100} \times \frac{1000M}{22.4} \times \frac{273}{273+20} = L_{low} \cdot M/2.4 \tag{4-32}$$

In the formula, M is the molecular weight of the combustible gas.

Substitute $1/L_{low} \approx 2.4K \cdot Q/(ME)$ into the above formula, then

$$1/L'_{low} = 2.4KQ/(ME)$$

Let Q be equivalent to the heat of combustion per gram of combustible gas, and $2.4K = K'$, then the above formula is:

$$1/L'_{low} = K' \cdot q/E \tag{4-33}$$

This formula is exactly the same as the previous one. For example, in the case of hydrocarbons, Q varies with the lengths of the hydrocarbon chain, but the heat of combustion Q per gram of substance is roughly the same. $Q = 42 \sim 46$kJ/g, which is nearly a constant.

Since the activation energy E of hydrocarbons is almost the same, the lower explosion limit nearly in mg/L L'_{low} are substantially equal. $L'_{low} = 40 \sim 45$mg/L$=40 \sim 45$g/m^3. So, $L'_{low} \times q = 1600 \sim 2100$kJ/m^3.

4.4.2 Influencing factors of explosion limit

The explosion limit is not a fixed value and varies with various factors. However, if the influence of the change of external conditions on the explosion limit is grasped, the explosion limit measured under certain conditions still has its general reference value.

The main factors affecting the explosion limit are as follows.

4.4.2.1 Initial temperature

The higher the initial temperature of the explosive mixture is, the larger the explosive limit range is, that is, the lower explosive limit is reduced and the upper explosive limit is increased. As the temperature of the system increases, the intramolecular energy increases, making the original non-combustible mixture a flammable and explosive system, so the temperature increases, also increasing the risk of explosion.

The experimental results of the influence of temperature on the upper and lower explosion limits of methane and hydrogen are shown in Fig. 4-18 and Fig. 4-19. It can be seen from the figure that the explosion range of methane and hydrogen expands with the increase of temperature, and the change is close to a straight line.

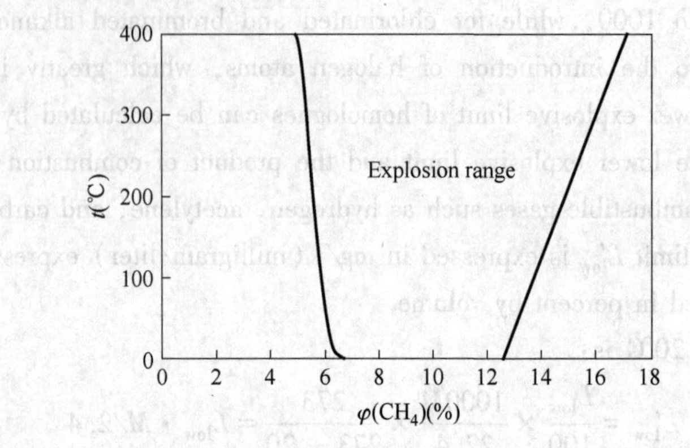

Fig. 4-18 Explosion limits of methane at different temperatures

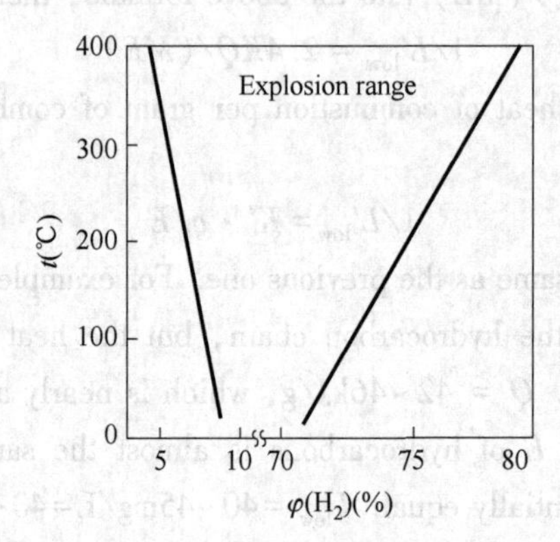

Fig. 4-19 Explosion limits of hydrogen in air at different temperatures

The effect of temperature on the explosion limit can be also seen in Table 4-7.

Table 4-7 Effect of temperature on explosion limit of acetone

Mixture temperature(℃)	Lower explosive limit(%)	Upper explosion limit(%)
0	4.2	8.0
50	4.0	9.8
100	3.2	10.0

4.4.2.2 Initial pressure

The initial pressure of the mixture has a great influence on the explosion limit, and in the case of pressurization, the change of the explosion limit is also complex. Generally, the pressure increases and the explosion limit expands. This is because when the pressure of the system increases, the molecules are closer, and the probability of collision increases, thus making the initial reaction of combustion and the progress of the reaction easier.

When the pressure is reduced, the explosion limit range is reduced. When the pressure drops to a certain value, the lower limit coincides with the upper limit, and the lowest pressure at this time is called the critical pressure of explosion. If the pressure drops below the critical pressure, the system will not explode. Therefore, it is advantageous for safe production to carry out decompression (negative pressure) operation in a closed container. See Fig. 4-20 for explosion limits of methane and hydrogen under different pressures. Generally speaking, the effect of pressure on the upper explosive limit is very significant, while the effect on the lower explosive limit is relatively small.

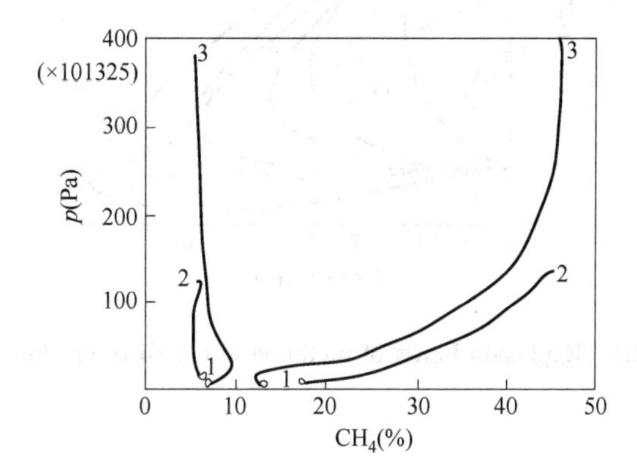

Fig. 4-20 Explosion limit of methane under different pressures

1—Downward propagation of flame the size of the cylindrical container is 37 × 8 cm;
2—End or center point Spherical container; 3—Downward propagation of flame, cylindrical container

4.4.2.3 Inert medium and impurities

If the percentage of inert gas in the mixture is increased, the range of explosion limit is reduced, and the concentration of inert gas is increased to a certain value and the mixture will not explode.

If inert gases (nitrogen, carbon dioxide, water vapor, hydrogen, nitrogen, carbon tetrafluoride, etc.) are added to the mixture of methane, the effects on the upper explosion limit are more significant than that on the lower explosion limit with the increase of the amount of inert gases in the mixture. Because an increase in that concentration of the inert gas indicate a relative decrease in the concentration of oxygen, in the upper limit, the concentration of oxygen is already very small, so a slight increase in the concentration of inert gas will have a great impact, so the upper limit of explosion will be sharply reduced.

For reactions involving gases, impurities also have a large effect. For example, dry chlorine has no oxidizing properties without water, and dry air is completely incapable of oxidizing sodium or phosphorus. A dry mixture of hydrogen and oxygen will not explode at higher temperatures. A small amount of water will sharply accelerate the decomposition of ozone, peroxide and other substances. A small amount of hydrogen sulfide will greatly reduce the ignition point of the water gas and mixture, and thus promote its explosion. It can be seen from Fig. 4-21 that the influence of inert gases on the explosion limit of methane is in the following order: CCl_4>CO_2>water vapor>N_2>He>Ar.

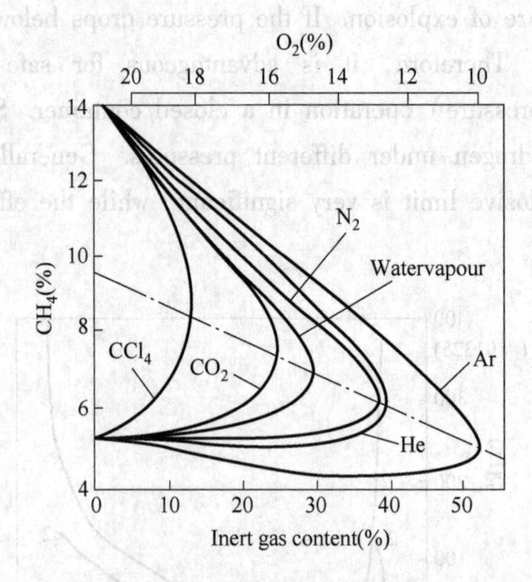

Fig. 4-21 Explosion limits of methane under different inert gases

4.4.2.4 Container

The material and size of the filling container both have their impacts on the explosion limit of the substance. Experiments show that the smaller the diameter of the vessel tube, the narrower the explosion limit range. For the same combustible material, the smaller the pipe diameter, the lower the flame spreading speed. When the pipe diameter (or flame passage) is small enough to a certain extent, the flame cannot pass through. This distance is called the maximum fire extinguishing distance, also known as the critical diameter. When the pipe diameter is less than the maximum fire extinguishing distance, the flame is extinguished because it cannot pass through. The effect of vessel size on explosion limit can also be explained by the wall effect. Combustion is the result of a series of chain reactions produced by free radicals. Combustion can continue only when the new free radicals are greater than the disappeared free radicals. However, with the decrease of pipe diameter (size), the collision probability between free radicals and pipe wall increases correspondingly. When the size is reduced to a certain extent, which also means the destruction of free radicals (collision with the vessel wall) is greater than that of free radicals, then the combustion reaction cannot continue.

With regard to the effects of materials, for example, hydrogen and fluorine mixed in glassware can explode even in the dark at liquid air temperatures. In silver utensils, the reaction takes place only at normal temperatures.

4.4.2.5 Energy for ignition

The energy of the spark, the area of the hot surface, and the contact time between the fire source and the mixture all have an effect on the explosion limit of the general mixture. For example, methane does not explode in any proportion to an electric spark with a voltage of 100V and a current intensity of 1A; the explosion limit is 5.9% ~ 13.6% at 2A; 5.85% ~ 14.8% at 3A. There is therefore a minimum detonation energy (which generally occurs close to the chemical theory) for each explosive mixture. Fig. 4-22 shows the relationship between the explosion limit of methane-air mixture and the energy of the fire source.

In addition to the above factors, light also has an effect on the explosion limit. As we all know, the reaction between hydrogen and chlorine is very slow in the dark, but under sunlight, it will cause a violent reaction, and if the proportion of the two gases is appropriate, it will explode. In addition, surface active substances also have an effect on some media, for example, in a spherical vessel, when the temperature is 530℃, hydrogen and oxygen do not react at all. However, when a quartz, glass, copper or iron rod is inserted into the vessel, an explosion occurs.

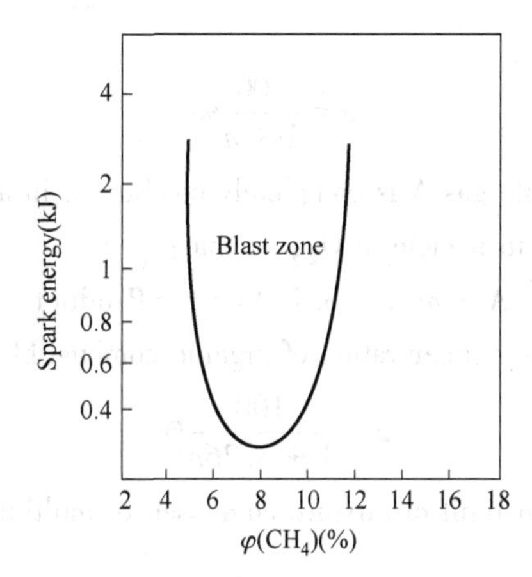

Fig. 4-22 Explosion limits of methane at different spark energies

4.4.3 Calculation of explosion limits of combustible mixture

The explosion limit of combustible mixture can be approximately calculated by empirical formula.

(1) The explosion limit (volume percentage) of organic combustible gas is calculated from the number of moles (N) of oxygen atoms required for one mole of combustible gas in the combustion reaction by the following formula:

$$x_{low} = \frac{100}{4.76(N-1)+1}\% \quad x_{top} = \frac{4 \times 100}{4.76N+4}\% \tag{4-34}$$

Where x_{low}——lower explosion limit of organic combustible gas;

x_{top}——upper explosion limit of organic combustible gas;

N——moles of oxygen atoms required for complete combustion per mole of combustible gas.

(2) Calculate the explosion limit of organic matter by using the stoichiometric concentration x_0 of combustible gas when it is completely burned in air. The calculation formula is as follows:

$$x_{low} = 0.55x_0 \quad x_{top} = 4.8\sqrt{x_0} \tag{4-35}$$

Where x_{low}——lower explosion limit of organic combustible gas, %;

x_{top}——upper explosion limit of organic combustible gas, %;

x_0——stoichiometric concentration of organic combustible gas at complete combustion in air, %.

This formula is suitable for organic combustible gases mainly composed of saturated hydrocarbons, but not for inorganic combustible gases.

Let the stoichiometric concentration of organic combustible gas A for complete combustion in oxygen is $z\%$, and it is assumed that one mole of A requires n moles of oxygen for complete combustion in oxygen. According to the reaction formula:

$$A + nO_2 \longrightarrow \text{Product}$$

There is

$$z = \frac{100}{1+n}\%$$

When the organic combustible gas A is completely combusted in air, it is accompanied by 3.76 n mole of nitrogen in addition to n mole of O_2, so that:

$$A + nO_2 + 3.76N_2 \longrightarrow \text{Product}$$

Therefore, the stoichiometric concentration of organic combustible gas A in air is:

$$z = \frac{100}{1 + 3.76n}\%$$

(3) Calculation of explosion limit of mixture composed of multiple combustible gases.

The calculation formula is:

$$X = \frac{100\%}{\frac{p_1}{N_1} + \frac{p_2}{N_2} + \frac{p_3}{N_3} + \cdots + \frac{p_i}{N_i}} \tag{4-36}$$

Where X——explosion limit of mixed combustible gas;

$p_1, p_2, p_3, \cdots\cdots, p_i$——volume percentage of each component in the gas mixture, %;

$N_1, N_2, N_3, \cdots\cdots, N_i$——explosion limit of each component in the gas mixture, %.

Formula (4-36) is called the Ley-Chartres formula. When the formula is used for calculation, the result calculated by substituting the lower explosion limit of each component of combustible gas into the formula is the lower explosion limit of combustible gas, and the result calculated by substituting the upper explosion limit of each component of combustible gas into the formula is the upper limit of combustible mixed gas.

In applying the Ley-Chartres formula, care should be taken that no chemical reactions occur between the components that make up the gas mixture. For the mixed gases containing hydrogen-acetylene, hydrogen-hydrogen sulfide, hydrogen sulfide-methane and carbon disulfide, the error of the calculated results is relatively large. The lower explosive limit calculated by the Ley-Chartel formula is relatively close to the actual, and the deviation of the upper explosive limit is relatively large.

(4) Calculation method of explosion limit of combustible mixed gas containing inert gas.

If the combustible gas mixture contains inert gases, such as N_2, CO_2, etc., the Ley-Chattel formula is still used to calculate its explosion limit, but each inert gas and a combustible gas need to be grouped together, and this whole group of gases is regarded as one combustible gas component. The volume percentage content of the group in the mixed gas is the sum of the volume percentage contents of the inert gas and the combustible gas in the group. The explosion limit of the group of gases can be calculated by listing the combined ratio of the group of inert gases to the combustible gases, finding it from Fig. 4-23, and then substituting it into the Ley-Chartres formula.

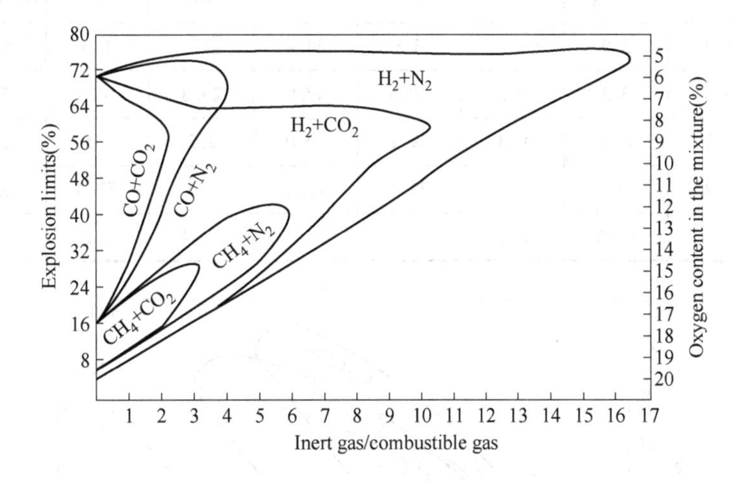

Fig. 4-23 Explosion limit of hydrogen, carbon monoxide, methane, nitrogen and carbon dioxide mixture in air

【Example 4-1】 Solve the explosion limit of gas. The gas composition is: $\varphi(H_2)$ 12.4%; $\varphi(CO)$ 27.3%; $\varphi(CO_2)$ 6.2%; $\varphi(O_2)$ 0%; $\varphi(CH_4)$ 0.7%; $\varphi(N_2)$ 53.4%.

【Solution】 Divide the inert gases and combustible gases in the coal gas into two groups: CO_2 and H_2 in the first group; N_2 and CO are in the second group; CH_4 is in the third group. It is obvious that the volume percentage of CO_2/H_2 group in the total gas mixture should be equal to 6.2% + 12.4% = 18.6%; the volume percentage of N_2 and CO group gases in the total gas mixture is 27.3% + 53.4% = 80.7%; 0.7% for CH_4.

The combination ratio of the inert gas to the combustible gas in each group is:

$$\frac{\varphi(CO_2)}{\varphi(H_2)} = \frac{6.2\%}{12.4\%} = 0.5 \quad \frac{\varphi(N_2)}{\varphi(CO)} = \frac{53.4\%}{27.3\%} = 1.96$$

It can be found from Fig. 4-23:

The explosive limits of H_2/CO_2 group is 6.0%~70%;

The explosive limit of CO/N_2 group is 40%~73%.

The explosion limit of CH_4 is 5%~15%. It can directly adopt the explosion limit of CH_4 in air.

The explosion limit of the gas can be calculated by substituting the above data into the formula (4-36):

$$x_{down} = \frac{100}{\frac{18.6}{6.0} + \frac{80.7}{40} + \frac{0.7}{5.0}}\% = 19\% \quad x_{up} = \frac{100}{\frac{18.6}{70} + \frac{80.7}{73} + \frac{0.7}{15}}\% = 70.93\%$$

See Table 4-8 and Fig. 4-24 for the explosion limits of some mixed gases.

Table 4-8 Explosion limits of some mixed gases

Mixed gas	Gas composition (volume fraction) (%)							Measured value (Volume fraction) (%)	
	CO_2	C_mH_n	O_2	CO	H_2	CH_4	N_2	Lower limit	Upper limit
Coke oven gas	1.9	3.9	0.4	6.3	54.4	31.5	1.6	5.0	28.4
City gas	2.5	3.2	0.5	10.5	47.0	25.8	10.5	5.6	31.7
Water gas	6.2	0.0	0.3	39.2	49.0	2.3	3.5	6.9	69.5
Producer gas	6.2	—	0.0	27.3	12.4	0.7	53.0	20.7	73.7
Natural gas	—	9.16	0.1	—	—	87.4	3.2	4.8	13.46

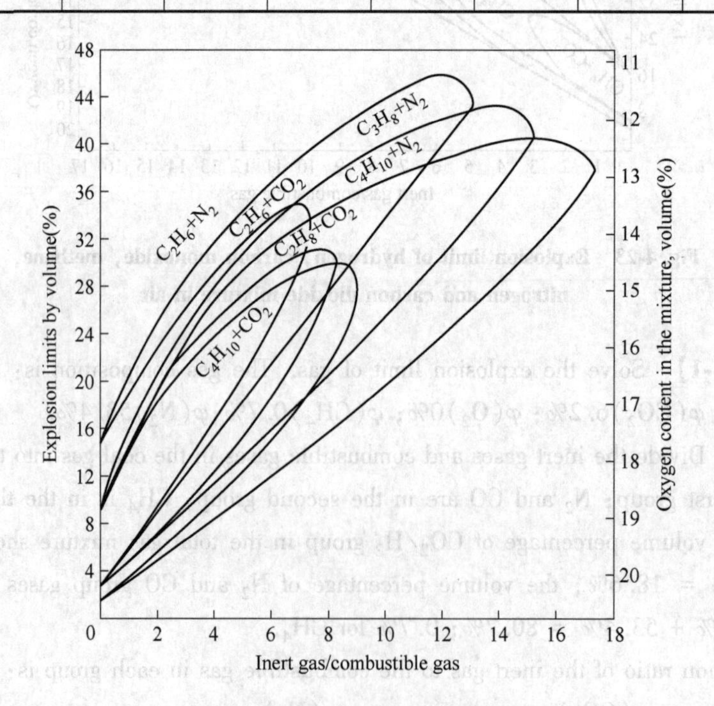

Fig. 4-24 Explosion limits of mixtures of ethane, propane, butane, carbon dioxide and nitrogen (in air)

4.5 Detonation

4.5.1 Shock wave formation

It has been shown that when a gas flows around an object at supersonic speed, a sudden compression wave is formed in front of the object. When the gas flow passes through this compression wave, its pressure, density and temperature suddenly increase by one value, and the flow velocity or Mach number correspondingly decreases by one value. This reflects the gas flow is suddenly compressed. This sudden compression wave is called a shock wave. This is a strong disturbance wave. When the gas flow passes through a strong disturbance wave, it is no longer an isentropic process, but an entropic process. The speed of motion of this wave is greater than the speed of sound of the gas in front of the wave. In other words, if the body is assumed to be motionless, only when the gas blows at supersonic speed can a shock wave be formed in front of the body. When the gas moves in a pipe or flows out of a nozzle at a supersonic speed, shock waves will also form under certain conditions.

Shock waves, just like expansion waves, are another fundamental phenomenon frequently encountered in supersonic flow. It can be said that supersonic flow usually passes through expansion waves when accelerating and through shock waves or isentropic compression waves when decelerating. When aircrafts are in supersonic flight in the gas, or in the supersonic inlet or supersonic gas flows, they almost always come across shock phenomena. Therefore, the study of shock wave is very important to master the law of supersonic flow of gas.

In order to make the discussion of shock wave more general, here we study a typical case, that is, the process of moving shock wave is formed when a piston accelerates in a long pipe and compresses the gas in the pipe.

Imagine a straight pipe with a constant cross section, and the inside of the pipe is initially a static gas with the parameters of P_1, ρ_1 and t_1. There is a piston at the left end of the tube, and the right end is far away so that the conditions of the right-hand side does not affect the state of the gas, as shown in Fig. 4-25.

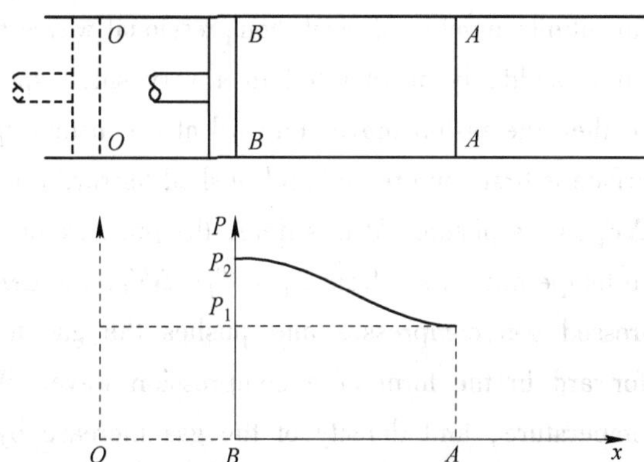

Fig. 4-25 Pressure distribution in the tube at time 0 and time t_1

(Dashed line and solid line represent the conditions at time 0 and time t_1, respectively)

Suppose that from time 0, the piston consecutively accelerates rightward from quiescence to compress the gas in the tube and push the gas in the tube to move rightward. Meanwhile, the pressure of the gas close to the piston surface gradually increases: the pressure increase is a compression disturbance to the gas, and it will propagate forward in the form of a compression wave, as shown by the wave surface *AA*. The pressure, density and temperature of the gas where *AA* passes are slightly increased, and a slight forward speed is obtained. Since the piston is consecutively accelerated, the gas pressure against the piston rises from P_1 to P_2 at time t_1.

In the figure, *OO* is the initial position of the piston, *BB* is the position of the piston at time t_1, and *AA* is the position at which the first weak compression wave formed by compressed gas reaches at time t_1 when the piston just starts to accelerate. Since *AA* is a weak perturbation wave, it advances at a speed equal to the sound speed a_1 of the wavefront gas, so there is:

$$OA = a_1 t_1 \tag{4-37}$$

Between *AA* and *BB*, there are infinitely many weak compression waves formed by the compressed gas due to the consecutive acceleration of the piston during the period from time 0 to time t_1. Each time such a compression wave passes through the gas, the pressure, density, and temperature of the gas increase slightly, and the gas gains a slight increase in forward velocity. The pressure, density and temperature of the gas change a certain amount due to the disturbance of infinitely many weak compression waves between the and *BB*, for example, the pressure increases from P_1 to P_2, and the density and temperature also increase from ρ_1 and T_1 to ρ_2 and T_2 respectively. The velocity of the gas is also increased from 0 to the same velocity as the piston, which is denoted as ΔV. If after t_1, the piston is no longer accelerating, instead it moves forward at a constant speed, then the gas pressure near the piston surface will not increase any more, but will remain unchanged at P_2.

Now, we further analyze the change of the gas state in the tube from time 0 to time t_1.

According to the concept of limit, continuous acceleration can be regarded as consisting of infinite small pulse accelerations. To easily understand, the consecutive acceleration of the piston from time 0 to time t_1 is approximately regarded as consisting of a finite number of very small pulse accelerations. In this way, a finite-track compression wave will be for in that tube, they can approximately represent an infinite number of weak compressional waves between *AA* and *BB*.

Suppose that the piston is suddenly accelerated to a very small speed ΔV_1 in a very short moment from time 0, and then the piston moves forward at a constant speed. At this time, the gas close to the piston surface is first compressed and pushed forward until this part of gas obtains the same moving speed ΔV_1 as the piston. At this time, the pressure of the gas is increased from P_1 to $P_1+\Delta P_1$. When the temperature rises from T_1 to $T_1+\Delta T_1$, the density also rises by a very small value. The compressed gas compresses and pushes the gas to its right, so that the disturbance propagates forward in the form of a compression wave. Where the wave surface passes, the pressure, temperature, and density of the gas increase by ΔP_1, ΔT_1, and $\Delta \rho_1$ respectively, and a very small velocity increment ΔV_1 is obtained. Since the velocity increment ΔV_1 is very small, the corresponding pressure and temperature increments are also very small.

This disturbance is approximately a weak disturbance, and its forward propagation speed is approximately equal to the undisturbed gas's sound speed $a_1(a_1=\sqrt{\gamma gRT_1})$ on the wavefront, as shown for *AA* in Fig. 4-25.

After a small time interval. ΔT has elapsed, the piston is again given a small velocity increment. ΔV_2, and the velocity of the piston movement is increased from. ΔV_1 to $\Delta V_1+\Delta V_2$. It further compresses and pushes the gas to its right, so that the pressure of the gas rises to $P_1+\Delta P_1+\Delta P_2$, and the temperature rises to $T_1+\Delta T_1$. The speed also increases from ΔV_1 to $\Delta V_1+\Delta V_2$. This second disturbance also propagates forward in the form of a compressional wave, and the parameters of the gas change in the same way wherever the wave surface passes. It must be noted here that the second compression wave propagates in the gas which has been compressed by the first compression wave, and for the second compression wave, the temperature of the gas in front of the wave is no longer T_1, but $T_1+\Delta T_1$. The velocity of the second compression wave relative to the gas in front of the wave is not a_1, but approximately $a'_1=\sqrt{\gamma gR(T_1+\Delta T_1)}$, so $a_1'>a_1$. In addition, since the resultant velocity of the forward motion of the latter compressional wave is $a_1+\Delta V_1$, its greater than a_1' than a_1. Therefore, this second compression wave will gradually catch up with the previous first compression wave. Similarly, there is a similar relationship between the compression waves generated by each subsequent small pulse acceleration of the piston, that is, the rear compression wave always moves faster than the front compression wave, so the distance between the rear wave and the front wave will become smaller and smaller with the passage of time. Until the wave behind catches up with the wave in front. For the infinitely many weak compressional waves between *AA* and *BB* shown in Fig. 4-25, the same pattern exists as in the above cases.

Now let us consider the situation after time t_1. Suppose that after t_1, the piston no longer accelerates, but moves forward at a constant speed, that is, the piston does not further compress the gas. At this time, we will see that after time t_1, the distance between *BB* and *AA* gradually decreases, *BB* gradually approaches *AA*, and the latter wave gradually catches up with the former wave. Between the last wave *BB* and the piston face, the pressure will remain at P_2. Let t_2 be a point in time after t_1, when the pressure distribution in the positions of *AA* and *BB* is as shown in Fig. 4-26.

In this way, there will always be a time t_3 when the latter wave catches up with the former wave, and all the compression waves between *BB* and *AA* are superimposed together. At this time, the nature of the waves will change, they will be superimposed from a weak, isentropic compression wave to a strong disturbance wave-shock wave. As shown in Fig. 4-27, *CC*. In front of the shock *CC*, the unperturbed quiescent gas with parameters P_1, ρ_1, T_1, after the shock *CC*, the gas is strongly disturbed, and its parameter values jump to p_2, ρ_2, T_2. The velocity of the gas also suddenly increases from 0 of the wave front to the same velocity ΔV as the piston. Therefore, the shock wave can also be said to be a jump surface of gas parameters, and any gas disturbed by the shock wave has a jump change in parameter values. The speed at which the shock wave *CC* advances V_{sw} is not equal to the speed of the original AA advance a_1, nor is it equal to the speed of the BB advance $a_2+\Delta V$(where $a_2=\sqrt{\gamma gRT^2}$), but in between, that is, $a_2+\Delta V>V_{sw}>a_1$. This indicates that the speed of the shock relative to the gas in front of the wave is supersonic ($V_{sw}>a_1$), which is subsonic with respect to the velocity of the gas behind the wave ($V_{sw}-\Delta V<a_2$).

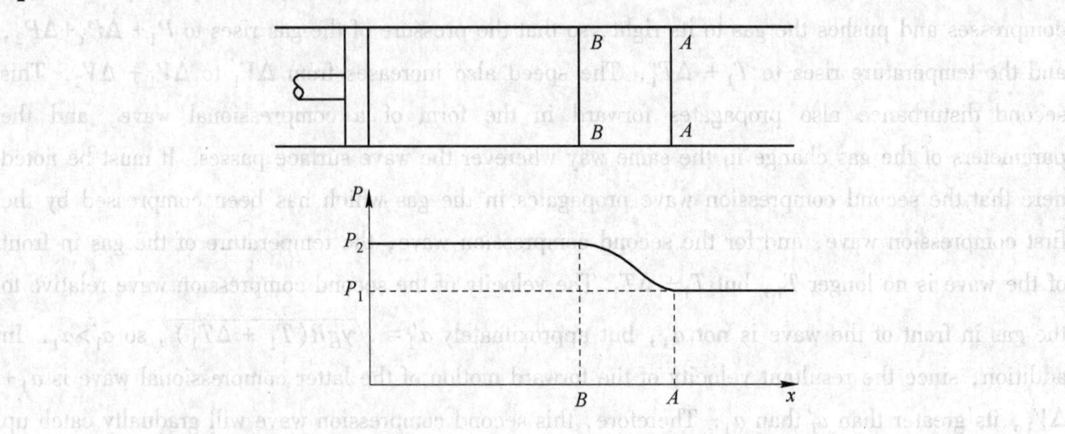

Fig. 4-26 The arrival positions of wave surfaces *AA* and *BB* at time t_2 and the pressure distribution in the hall

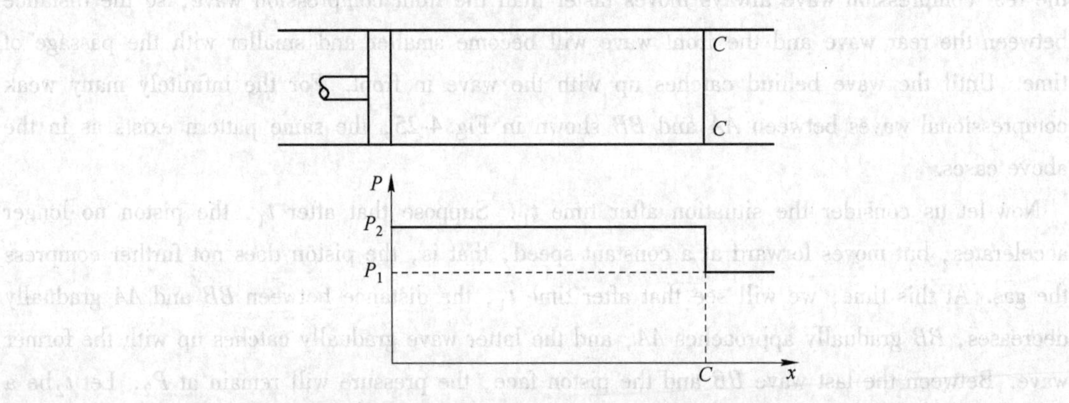

Fig. 4-27 At time t_3, the weak compression wave between *BB* and *AA* surfaces is superimposed as a shock wave

The process of shock wave formation can be observed in experiments, such as the shock tube tester, which is designed according to this principle. Through the shock tube, we can observe the process of shock formation and the motion of the shock, and we can use it to study the properties of the shock and other problems related to the shock.

Finally, we should point out that the weak compression waves can be superimposed to form a strong compression wave, the shock wave, but the expansion waves are no longer superimposed to form a "strong" expansion wave. Taking the movement of a piston in a long tube as an example, if the piston is accelerated to the left from the initial position, the gas in the tube will expand.

Pressure, density, temperature, etc. will decrease accordingly. At the same time, an infinite number of expansion waves will be generated. Like compression waves, they will propagate forward from left to right at the speed of sound of the gas in front of the wave. The latter wave will always move in the gas disturbed by the former wave, but the temperature of the gas disturbed by the expansion wave will decrease. Therefore, the sound speed of the gas in front of the back expansion wave is lower than that in front of the front expansion wave. Therefore, the more backward the expansion wave is, the lower the speed is, and the back expansion wave can never catch up with the front expansion wave. The distance between the expansion waves will become larger and larger, so the expansion waves can not be concentrated or superimposed like the compression waves to form a strong expansion wave.

4.5.2 Properties of shock waves

4.5.2.1 The speed of shock wave motion and the intensity of shock wave

We have pointed out in the previous section that the velocity of the shock wave relative to the gas in front of it is larger than the sound speed of the gas in front of the wave a_1, the speed of the shock relative to the gas behind the wave ($-\Delta V$) is smaller than the sound speed of the gas behind the wave, a_2. Now let's take a closer look at what factors determine the speed of the shock wave.

Referring to Fig. 4-28, suppose that at a certain time t, the shock wave advances to section 2, the parameters of the front wave are P_1, ρ_1, T_1, and the parameters of the back wave are P_2, ρ_2, T_2. The shock advances at a speed of the velocity of the gas behind the wave is ΔV. Assuming that after dt time, the shock wave advances from section 2 to section 1, then the distance between sections 1-2 is:

$$dx = V_t dt \tag{4-38}$$

At the same time, the gas of the 2 section at dt advances to the 2_1 section. The distance between the 2-2_1 sections is

$$dx_{air} = \Delta V dt \tag{4-39}$$

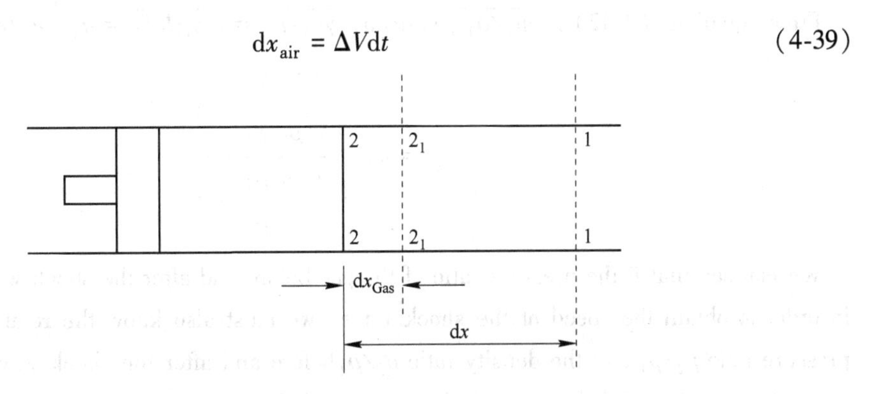

Fig. 4-28 Process of gas compression by shock wave

At this point, the gas at section 1 has just been disturbed and has acquired a velocity ΔV, but has not yet been displaced. During the time dt, the shock wave compresses the gas that originally occupied the area between 1-2 sections to the area between 1-2_1 sections, so that the density of this part of gas increases from ρ_1 to ρ_2. According to the law of conservation of mass, the mass of the gas between 1-2 sections is constant before and after compression.

Suppose the cross-sectional area of the pipe is A, there should be:

$$\rho_1 A \mathrm{d}x = \rho_2 A(\mathrm{d}x - \mathrm{d}x_{\mathrm{gas}})$$

Put dx and dx_{gas} substituting the expressions (4-38) and (4-39), and eliminating A, we obtain

$$V_{\mathrm{T}} = \frac{\rho_2}{\rho_2 - \rho_1}\Delta V \tag{4-40}$$

Let's look at the momentum change of the gas between sections 1-2: before the shock wave passes, the velocity of the gas is 0; after the shock wave passes. the velocity of the gas increases from 0 to ΔV, so the momentum change in dt time between the 1-2 sections is $\rho_1 A\mathrm{d}x\Delta V$. During this time, the impulse acting on this part of the gas is $(p_2 - p_1)A\mathrm{d}t$. According to the law of conservation of momentum, these two should be equal. Eliminating A, we obtain E_{q}.

$$(p_2 - p_1)\mathrm{d}t = \rho_1 \mathrm{d}x\Delta V$$

Substituting the expression (4-38) for dx and eliminating dt gives:

$$p_2 - p_1 = \rho_1 V_{\mathrm{sw}}\Delta V \tag{4-41}$$

Equations (4-40) and (4-41) can be solved:

$$V_{\mathrm{sw}} = \sqrt{\frac{\rho_2}{\rho_1}\frac{p_2 - p_1}{\rho_2 - \rho_1}} \tag{4-42}$$

$$\Delta V = \sqrt{\frac{(\rho_2 - \rho_1)(p_2 - p_1)}{\rho_1\rho_2}} \tag{4-43}$$

From equations (4-42) $\sqrt{\gamma p_1/\rho_1}$, noticed $\sqrt{\gamma p_1/\rho_1} = \sqrt{\gamma g R T_1} = a_1$, so there is:

$$V_{\mathrm{sw}} = a_1\sqrt{\frac{1}{\gamma}\frac{\rho_2}{\rho_1}\frac{\dfrac{p_2}{p_1} - 1}{\dfrac{\rho_2}{\rho_1} - 1}} \tag{4-44}$$

We can see that if the pressure ratio of the gas before and after the shock wave p_2/p_1 is known, in order to obtain the speed of the shock wave, we must also know the relationship between the pressure ratio p_2/p_1 and the density ratio ρ_2/ρ_1 before and after the shock wave. To this end, we study the variation of the gas energy between the 1-2 cross sections.

In dt time, the acting by the external pressure on the gas between 1-2 sections is $P_2 A\Delta V\mathrm{d}t$.

According to the law of conservation of energy, it should be equal to the sum of the increment of kinetic energy and internal energy of this part gas:

$$\rho_1 A dx\left[\frac{\Delta V^2}{2}+\frac{g}{A}(u_2-u_1)\right]$$

Where u is the internal energy per unit weight of the gas and A is the work-heat equivalent. It is known frombasic laws of thermodynamics:

$$u=c_v T=\frac{1}{\gamma-1}ART$$

Where c_v is the specific heat at constant volume of the gas.

Substituting the equation of state of an ideal gas, $T = p/\rho GR$, we obtain:

$$u=\frac{1}{\gamma-1}\frac{A}{g}\frac{p}{\rho}$$

So there should be

$$p_2 A\Delta V dt=\rho_1 A dx\left[\frac{\Delta V^2}{2}+\frac{1}{\gamma-1}\left(\frac{p_2}{\rho_2}-\frac{p_1}{\rho_1}\right)\right]$$

Substituting the expression (4-38) for dx, and eliminating Adt in the both sides of the equation, we get:

$$p_2\Delta V=\rho_1 V_{sw}\left[\frac{\Delta V^2}{2}+\frac{1}{\gamma-1}\left(\frac{p_2}{\rho_2}-\frac{p_1}{\rho_1}\right)\right] \tag{4-45}$$

Will V_{sw} substituting the expressions (4-42) and (4-43) for and (V) into the above equations, we can solve for:

$$\frac{\rho_2}{\rho_1}=\frac{\frac{(\gamma+1)p_2}{(\gamma-1)p_2}+1}{\frac{p_2}{p_1}+\frac{\gamma+1}{\gamma-1}} \tag{4-46}$$

This is the relation between the pressure ratio and the density ratio before and after the shock. Substituting this back into equation (4-44) obtains the formula for the velocity of the shock motion:

$$V_{sw}=a_1\sqrt{\frac{\gamma-1}{2\gamma}\left(1+\frac{\gamma+1}{\gamma-1}\frac{p_2}{p_1}\right)} \tag{4-47}$$

If p_2/p_1 is written in the form of $1+(p_2-p_1)/p_1$, the above equation can also be expressed as:

$$V_{sw}=a_1\sqrt{1+\frac{\gamma+1}{2\gamma}\frac{p_2-p_1}{p_1}} \tag{4-48}$$

Since the pressure p_2 after the wave is always greater than the pressure p_1 before the wave, the value in the square root of the above formula is greater than 1, so $V_{sw} > a_1$. The pressure ratio p_2/p_1 marks the magnitude of the shock strength. Under the condition that a_1 is constant, the larger the shock strength p_2/p_1 is, the faster the shock moves. The larger V_{sw} is, and vice versa, the speed of the shock motion decreases. For example, when $p_2/p_1 \rightarrow 1$, the shock strength weakens to an isentropic compression wave. It can be seen from equation (4-47) that at this time, V_{sw} also approaches the speed of sound a_1.

In a similar way, the velocity of the shock relative to the gas behind the shock can also be derived:

$$V_{sw} - \Delta V = a_2 \sqrt{1 - \frac{\gamma + 1}{2\gamma} \frac{p_2 - p_1}{p_2}} \tag{4-49}$$

Since the value in the square root sign is less than 1, $(V - \Delta V) < a_2$. It shows that the velocity of the shock wave relative to the gas behind the wave is subsonic.

4.5.2.2 Thickness of shock wave and entropy increase

In the previous analysis, we treated the shock as a bump in the gas parameters, i. e., the shock is thickness-less. But this is not actually the case, because if it were, the rate of change of the parameters of the gas as it passes through the shock as $\partial p/\partial x$, $\partial T/\partial x$, $\partial V/\partial x$ and so on, would be infinite, which is impossible. In fact, the shock wave is thick. The velocity gradient in a zone when all the compression waves are close together (i. e., the zone is narrow) $\partial V/\partial x$ and the effect of internal friction caused by gas viscosity and the effect of heat conduction in the wave zone become very strong, so the variation process of gas parameters is no longer isentropic, but entropic. The speed of each compression wave is no longer the sound speed value of the wavefront gas calculated based on the isentropic relation, rather, the entire wave region as a whole is propelled forward at the speed of the shock. But the thickness of this wave zone is really thin. According to the theoretical calculation, when $p_2/p_1 = 2$, the thickness of the shock wave is 4.47×10^{-4} mm (which can be calculated by the formula. $V_{sw}/a_1 = M_{sw} = 1.36$, where $M_{Excitation}$ is the Mach number of the shock motion). For $p_2/p_1 = 10$, the shock thickness is 0.66per 10000millimeters ($M_{Excitation} = 2.95$ in this case). The stronger the shock, the thinner its thickness. In these calculations, although some theoretical assumptions (such as continuous medium assumption) are made, the calculation results are not necessarily very accurate. But at least an approximate concept of an order of magnitude is given. And the fact that the thickness of the shock wave is very thin. Therefore, in the commonly encountered problems, we can still approximately regard the shock wave as a jump surface of gas parameters without thickness, but the entropy of the gas increases when it passes through it.

As the gas passes through the shock wave, it undergoes a sudden compression, and the change of its state parameters is expressed by the previously derived equation (4-46). This is different from the law of state change in the process of isentropic compression of gas. It is known from thermodynamics that in an isentropic process:

$$\frac{\rho_2}{\rho_1} = \left(\frac{p_2}{p_1}\right)^{1/\gamma} \tag{4-50}$$

Increasing the gas pressure from p_1 to p_2, the density ρ_2 achieved by shock compression is lower than the density $\rho_{2\,(\text{isentropic})}$ achieved by isentropic compression,

$$\rho_2 < \rho_{2(\text{isentropic})} \tag{4-51}$$

This is due to the irreversible conversion of a portion of the mechanical energy of the gas into thermal energy in the shock layer. From this relation, it can be shown concretely that the entropy of a gas increases as it passes through a shock wave.

Only when the intensity of the shock wave is not large, the density change caused by shock wave compression is similar to that caused by isentropic compression. This is because when the shock wave is weak, the entropy increment of the gas passing through the shock wave is not large, which is close to an isentropic process.

The increase of gas entropy reflects the loss of mechanical energy. It can be proved that there is the following relationship between the increase of gas entropy and the decrease of total pressure:

$$S_2 - S_1 = -AR\ln(p_{02}/p_{01}) \tag{4-52}$$

Where $S_2 - S_1$ is the entropy increment of the flow passing through the shock, p_{01} is the total pressure of the flow before the shock, and p_{02} is the total pressure of the flow behind the shock.

For the isentropic process, the total pressure of the flow is constant, but when the flow passes through the shock, the total pressure decreases from p_{01} to p_{02}, so $p_{02} < p_{01}$, so $S_2 > S_1$. The smaller the p_{02}/p_{01} is, the more significant the total pressure of the airflow decreases, and the larger the entropy increment $S_2 - S_1$ is. p_{01} is usually already given, p_{02} can be calculated from the shock relations, and therefore the entropy increment $S_2 - S_1$ can also be calculated according to (4-52).

4.5.3 Shock wave moving in space

When an object accelerates in space, it will also form a shock wave, the process and principle of which is the same as that of a piston moving in a tube, i.e. the shock wave is formed by the superposition of weak compression waves. The difference is that when the piston accelerates in the tube and causes a shock wave, due to the existence of the tube wall, no matter how far the piston surface of the shock wave is, only after the piston accelerates. The intensity of the shock wave will not change or weaken if it keeps constant speed, that is, no new disturbance will be given to the gas behind the wave. All the gas compressed by the shock wave will be forced to move at the speed of the piston, and the pressure behind the wave will remain constant at p_2. The motion of the piston can be supersonic or subsonic, but the situation in space is different. At that time, only when the piston or object moves at supersonic speed can a stable shock wave be formed. The analysis is as follows:

Suppose that the piston is moving in space and is surrounded by a pipe wall, as shown by the dotted line in Fig. 4-29. After the formation of the shock wave, it moves forward at the velocity of V_{sw}. The gas pressure behind the wave is p_2 and the velocity is equal to the velocity of the piston motion. Neither the wave front nor the gas outside the tube is compressed, and the pressure is kept at p_1. But in fact, the tube wall does not exist, so the high-pressure gas behind the wave will move to both sides, resulting in a decrease in the pressure of the gas behind the wave, and the farther the shock wave is from the piston surface, the more significant the decrease in the pressure behind the wave. If the piston or body is moving at subsonic speed, the distance between the shock wave and the body surface will become larger and larger, and the shock wave strength p_2/p_1 will become smaller and smaller, because the speed of the shock wave is always greater than the speed of sound, until the shock is a weak compression wave at infinity. Therefore, the piston or object moving at subsonic speed in space will not form a stable shock wave. But this will not be the case when the piston or object is moving in space at supersonic speeds. It can be seen from the shock wave velocity formulas (4-47) and (4-48) that V_{sw} decreases with the weakening of the shock strength. Therefore, when the speed of the shock wave in front of the object is reduced to the speed of the object (they are all supersonic), the distance between the shock wave and the object surface will not increase, the strength of the shock wave will not be further weakened, and the speed will be constant, and it will move forward with the object at the same speed. At this point, there will be a stable shock wave in front of the object.

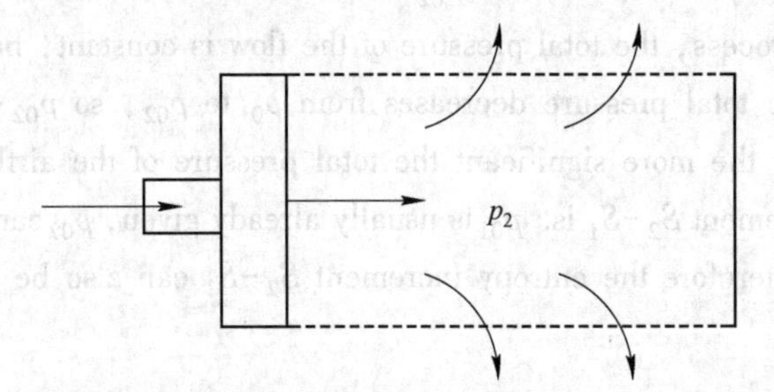

Fig. 4-29 Motion diagram of shock wave in space

4.5.4 Occurrence of detonation

There is a long tube filled with combustible premixed gas. One end of the tube is closed, and the mixed gas is ignited at the closed end forming a combustion wave. The initial combustion wave is normal flame propagation, and the burned gas generated by normal flame propagation will expand in volume due to the increase in temperature. The volume-expanded burned gas is equivalent to a piston-gas piston, which compresses the unburned gas mixture. A series of compression waves are generated, and these waves propagate toward the unburnt mixture, causing a small increment in P, ρ, T of the unburnt mixture in front of the wave respectively, and causing the unburned mixture to acquire a small forward velocity. So the velocity of the compression wave in the rear is greater than that in the front. When the tube is long enough, the following compression waves may catch up with one another and eventually overlap forming a shock wave. It follows that the shock wave must be generated in front of the normal flame that is beginning to form. Once the shock wave is formed, the unburned mixture ignites due to the very high pressure behind the shock wave. After a period of time, that normal flame propagation merges with the shock-induced combustion. So wherever the shock wave travels, the gas mixture catches fire. The flame propagation speed is the same as the shock speed. The burned gas behind the shock wave continuously transmits a series of compression waves forward, and continuously provides energy to prevent the attenuation of the shock wave intensity, thus obtaining a stable detonation wave. The formation process of detonation wave is shown in Fig. 4-30.

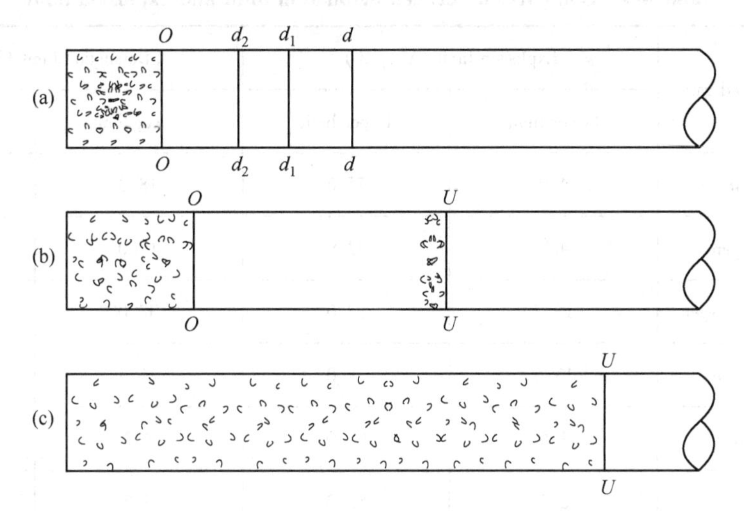

Fig. 4-30 Detonation formation process

(a) Normal flame propagation $O—O$ is preceded by a series of compression waves $d — d$, $d_1— d_1$, $d_2—d_2$

(b) Detonation wave $U—U$ has been formed in front of normal flame propagation $O—O$, and the unburned mixture is ignited;

(c) The normal flame propagation is combined with the combustion caused by the detonation wave

4.5.5 Detonation formation conditions

(1) The initial normal flame propagation can form compressive disturbances. The detonation wave is essentially a shock wave formed by the compression disturbance produced by combustion. Whether the initial normal flame propagation can form a compressive disturbance is the key to the generation of detonation waves. Because only the compression wave has the characteristic that the wave speed of the back is faster than that of the front.

(2) The pipe shall be long enough or the volume of premixed gas in the free space shall be large enough. There is a process in the formation of a shock wave by the superposition of a series of successive compression waves, which requires a certain distance. If the tube is not long enough or the volume of premixed gas in the free space is not large enough, the initial normal flame propagation can not form a shock wave. The detonation is formed in front of the normal flame

front, the distance between the normal flame front and the detonation formation position is called the pre-detonation spacing. Other things being equal. The pre-detonation spacing is closely related to the tube diameter, so it can be expressed by the multiple of the tube diameter. For a smooth tube, the pre-detonation spacing is tens of times the tube diameter. For the tube with rough surface, the pre-detonation spacing is 2~4 times of the tube diameter.

(3) Combustible gas concentration shall be within the detonation limit. Like explosion, detonation also has the problem of limit, but the range of detonation limit is generally narrower than that of explosion limit. Table 4-9 shows the detonation and explosion limits for several combustible mixtures.

Table 4-9 Comparison between detonation limit and explosion limit

Combustible mixed gas	Explosion limit (Vol, %)		Detonation limit (Vol, %)	
	Lower limit	Upper limit	Lower limit	Upper limit
Hydrogen-air	4.0	75.6	18.3	59.0
Hydrogen-oxygen	4.7	93.9	15.0	90.0
Carbon monoxide-oxygen	15.7	94.0	38.0	90.0
Ammonia-oxygen	13.5	79.0	25.4	75.0
Acetylene-air	1.5	82.0	4.2	50.0
Propane-oxygen	2.3	55.0	3.2	37.0
Ether-air	1.7	36.0	2.8	4.5
Ether-oxygen	2.1	82.0	2.6	24.0

(4) The diameter of the tube is larger than the critical diameter of detonation. The smaller the tube diameter is, the greater the heat loss of the flame is, the more chances the free radicals in the flame will be destroyed by touching the tube wall which means the slower the flame propagation is. When the tube diameter is small to a certain extent, the flame can not propagate and detonation can not be formed. The minimum diameter of the tube that can form detonation is called the critical diameter of detonation, which is about 12~15mm.

4.5.6 Detonation wave velocity and pressure

It can be seen from the formation process of detonation wave that the detonation wave is supersonic relative to the gas in front of the wave, and the propagation speed of detonation wave is much higher than that of normal flame. Table 4-10 shows the measurement results of propagation speed (U_0) when some combustible mixed gas forms detonation wave.

Table 4-10 Measured results of detonation wave velocity of some combustible mixture

Mixture	U_0(m/s)
CH_4+2O_2	2146
$2CO+O_2$	1264
$2H_2+O_2$	2821
$C_2H_6+3.5O_2$	2363
$C_2H_4+3O_2$	2209
$C_2H_4+2O_2+8N_2$	1734
$C_3H_8+3O_2$	2600
$C_3H_8+6O_2$	2280
$C_6H_6+22.5O_2$	1658

The detonation wave velocity can not only be measured accurately, but also be obtained by calculation, and the calculated value is in good agreement with the measured value. For example, under the conditions of initial temperature T_∞ = 291K, pressure P_∞ = 1.01325 × 10^5 Pa, the calculated detonation wave velocity U_1 of stoichiometric hydrogen-oxygen mixture is 2806 (m/s), and the experimental value is 2819 (m/s), the deviation is less than 1%. The measured and calculated detonation wave velocities are listed in Table 4-11. It can be seen that most of the calculated values are higher than the measured values. This is because the product dissociates at a higher temperature, and if dissociation is taken into account, the calculated value will be lower, which is closer to the measured value.

Table 4-11 Detonation wave velocities of stoichiometric Hydrogen-Oxygen mixtures

Mixture	P_m(Pa)	T_m(K)	U_1(m/s)	
			Calculated value	Experimental value
$2H_2+O_2$	18.5×10^5	3583	2806	2819
$(2H_2+O_2)+5O_2$	14.13×10^5	2620	1732	1700
$(2H_2+O_2)+5N_2$	14.39×10^5	2685	1850	1822
$(2H_2+O_2)+5H_2$	15.97×10^5	2975	3627	3527
$(2H_2+O_2)+5He$	16.32×10^5	2097	3617	3160
$(2H_2+O_2)+5Ar$	16.32×10^5	3097	1762	1700

4.6 Prevention of gas explosion

Combustible gas is mixed with air or oxygen to form premixed combustible gas. Once it meets the ignition source, it will explode and cause major accidents. Gas explosion is one of the most common explosions. It is very important to take effective measures to prevent gas explosion.

Gas explosion must have three conditions, gas, air (the ratio of gas to air must be within a certain range) and ignition source.

Therefore, the methods to prevent combustible gas explosion as follows: (1) strictly control the fire source; (2) prevent the production of premixed combustible gas; (3) Use inert gas to prevent gas explosion; (4) Use fire arresters to prevent explosion from spreading; (5) Suppression of detonation with detonation suppressor; (6) Protect the equipment with pressure relief device prevent the expansion of explosion disaster and reduce the loss; (7) Use explosion suppression device to suppress the explosion.

In the places with flammable and explosive gases, the use of electrical equipment and the burial of pipes are particularly important for the prevention of explosion. This section focuses on them.

4.6.1 Strict control of fire sources

There are many types of fire sources, such as open fire sources generated by electric welding and gas welding; electric sparks caused by electrical equipment startup, shutdown and short circuit; sparks caused by electrostatic discharge; sparks produced when objects collide or rub against each other. The generation of various ignition sources shall be strictly controlled.

The temperature of electrical equipment or power lines rises due to the short circuit of equipment and circuit during operation, excessive contact resistance, overload or poor ventilation and heat dissipation, resulting in sparks and arcs, which become a major ignition source of combustible gas explosion.

Electric sparks can be divided into working spark and accident spark. The former is the sparks generated by the normal operation of electrical equipment (such as DC welding machine), and the latter is the sparks generated by the fault or wrong operation of electrical equipment and circuit. Electric sparks generally have a high temperature, especially the temperature of electric arc can reach 5000~6000K, which can not only cause the combustion of combustible materials, but also make the metal melt and splash. These can constitute dangerous sources of ignition.

Explosion-proof electrical equipment shall be used in workshops and mines with explosion hazards according to their degrees of danger. According to the different characteristics of explosion-proof structure and explosion-proof performance, explosion-proof electrical equipment can be divided into increased safety type, flameproof type, oil-filled type, sand-filled type, ventilated and inflated type, intrinsic safety type, non-spark type and special type. See Table 4-12 for signs of various types of explosion-proof electrical equipment.

Table 4-12 Types and Marking of Explosion-proof Electrical Equipment

Type		Logo		
		Factory use		Used in coal mines
Old	New	Old	New	
Explosion-proof safety type	Increased safety type	A	c	KA
Flameproof type	Flameproof type	B	d	KB
Explosion-proof oil-filled type	Oil filled type	C	o	KC
—	Sand washing mold	—	s	—
Explosion-proof ventilation and inflation	Ventilated inflatable type	F	p	KF
Safety spark type	Intrinsically safe	H	i	KH
—	Non-sparking type	—	n	—
Explosion-proof special type	Special type	T	s	KT

Increased-safety type (formerly known as explosion-proof-safety type) refers to electrical equipment which do not produce electric sparks, arcs and dangerous temperatures during normal operation, such as explosion-proof safety high-pressure mercury fluorescent lamps.

Flameproof type refers to the electrical equipment, such as flameproof motor, whose shells can withstand the pressure generated by the explosion of explosive mixture in the shell, and can prevent the explosion flame from spreading around the shells, so as not to cause the explosion of external explosive mixture.

Oil-filled type (formerly known as explosion-proof oil-filled type) refers to the electrical equipment that immerses the electrical equipment that may produce sparks, arcs or live parts at dangerous temperatures in insulating oil, so as not to cause the explosion of explosive mixtures on the oil surface.

Ventilated inflatable type (formerly known as explosion-proof ventilated inflatable type or positive pressure type) refers to the electrical equipment in which fresh air or inert gas is introduced into the equipment and positive pressure is maintained to prevent the external explosive mixture from entering the interior and causing explosion.

Intrinsically safe type (formerly known as safety spark type) refers to the electrical equipment that will not cause explosion because the current value of the electric spark generated under normal or fault conditions is less than the minimum detonation current of the explosive mixture in the place.

Special type (formerly known as explosion-proof special type) refers to the explosion-proof electrical equipment that do not belong to the above types in structure, such as the explosion-proof electrical equipment poured with epoxy resin and filled with quartz sand.

Electrical equipment shall be selected according to the levels of explosion hazard area. The classification of explosion hazardous areas is shown in Table 4-13.

Table 4-13 Classification of explosion and fire hazard locations

Category	Level	Features
Places where have the a risk of explosion of flammable gas or flammable liquid vapor	Q-1	Explosive mixture can be formed under normal conditions (such as startup, operation, shutdown, etc.)
	Q-2	Explosive mixture can not be formed under normal conditions, but can form under abnormal conditions (such as equipment damage, misoperation, maintenance, etc.)
	Q-3	Although explosive mixtures can be formed under abnormal conditions, the possibility or range is small. For example, the quantity of explosive hazardous substances is small, the lower explosive limit is small, and the density of the explosive mixture formed is very small and difficult to accumulate
Places with explosion risk of combustible dust and combustible fiber	G-1	Under normal conditions, it can form explosive mixture
	G-2	Can not be formed under normal conditions, but can form explosive mixtures under abnormal conditions
Places with fire risk	H-1	In the production process, the production, using, storage and delivery of flammable liquids with a flash point higher than the ambient temperature of the site can cause fire hazards in terms of quantity and configuration
	H-2	Places where suspended or accumulative combustible dust or combustible fibers of explosive mixture cannot be formed during production, but can cause fire hazard in terms of quantity and configuration
	H-3	Locations with solid combustible substances that can cause fire hazards in terms of quantity and configuration

The type selection of electrical equipment in explosion hazardous area are listed in Table 4-14.

Table 4-14 Selection of electrical equipment in explosion hazardous area

Site level	Q-1	Q-2	Q-3	G-1	G-2
Electric machinery	Flameproof type, ventilated and inflated type	Any type of explosion protection	Type H43	Any level of explosion-proof type, ventilation and inflation type	Type H44

Continuted Table 4-14

Site level		Q-1	Q-2	Q-3	G-1	G-2
Electrical appliances and instruments	Fixed mounting type	Flameproof, oil-filled, ventilated, intrinsically safe	Type H45	Type H45	Any level of flameproof type, ventilated and inflated type, and oil-filled type	Type H45
	Mobile	Flameproof, inflatable, intrinsically safe	Flameproof, inflatable, intrinsically safe	Any explosion-proof type other than oil-filled type or even H57	Any level of explosion-proof type, ventilation and inflation type	
	Portable	Flameproof type, intrinsically safe type	Flameproof type, intrinsically safe type	Flameproof, increased safety, H57	Any level of flameproof type	
Lighting fixtures	Fixed and mobile	Flameproof type, ventilated and inflated type	Increased safety type	Type H45	Any level of flameproof type	Type H45
	Portable	Flameproof type	Flameproof type	Flameproof, increased safety, H57	Any level of flameproof type	Any level of flameproof type
Transformer		Flameproof type, ventilated and inflated type	Increased safety type, oil-filled type	Type H45	Any level of flameproof type, oil-filled type, ventilated and inflated type	Type H45
Communication appliance		Flameproof, oil-filled, ventilated, intrinsically safe	Increased safety type	Type H57	Any level of flameproof type, oil-filled type, ventilated and inflated type	Type H45
Power distribution unit		Flameproof type, ventilated and inflated type	Any type of explosion protection	Type H57	Any level of explosion-proof type, ventilation and inflation type	Type H45

The explosion-proof type has better explosion-proof performance and should be preferentially used in Class I explosion hazardous areas. The explosion-proof performance of the increased safety type is relatively poor, and it is suitable for places with low degree of danger. According to the different conditions of using, the equipment can be divided into fixed-installation type, mobile type, portable type and some other kinds. The oil-filled type cannot be used in a mobile and portable type because frequent movement can easily cause the fluctuation of oil level or oil leakage of equipment, it is bound to produce sparks or expose the oil level of high temperature parts, thus losing explosion-proof performance.

Electrical equipment in fire hazard areas shall be selected according to the types listed in Table 4-15 according to the different levels of the places.

Table 4-15 Selection of electrical equipment in fire hazard area

Electrical equipment and service conditions		Site level		
		Class H-1	Class H-2	Class H-3
Electric machinery	Fixed type	Splash-proof	Closed type	Drip-proof
	Mobile or portable	Closed type	Closed type	Closed type
Electrical appliances and instruments	Fixed type	Oil-filled, fireproof dust-proof type, protective type	Dust-proof type	Open type
	Mobile or portable	Waterproof and dustproof	Dust-proof type	Protective type
Lighting fixtures	Fixed type	Protective type	Dust-proof type	Open type
	Mobile or portable	Dust-proof type	Dust-proof type	Protective type
Power distribution unit		Dust-proof type		Protective type
Junction box		Dust-proof type		Protective type

4.6.2 Prevent the generation of premixed combustible gas

Equipment and pipelines for production, storage and transportation of combustible gas should be strictly sealed to prevent combustible gas from leaking into the atmosphere and forming explosive mixed gas with air. Monitors should be installed in important explosion-proof locations so as to monitor the leakage of combustible gas at any time.

At the condition which it is impossible to protect the equipment and make it absolutely sealed, the plant and workshop should be kept in good ventilation conditions, so that a small amount of leaked combustible gas can be discharged at any time without forming explosive gas mixture. The relative density of the combustible gas should be considered in the design of the ventilation and exhaust system. Sometimes combustible gas is lighter than air (such as hydrogen), and when it leaks out, it often accumulates on the roof which forms explosive mixed gas with the roof air, so the roof should have exhaust passage, such as skylights. Some combustible gases are heavier than air, which may accumulate in low-lying areas such as trenches and form explosive mixed gases with air, and measures should be taken to discharge them. The blower blades of the explosion-proof ventilation and exhaust system should be made of materials that will not produce sparks under collide.

4.6.3 Using inert gas to prevent gas explosions

Inert gases (nitrogen, carbon dioxide, etc.) can be used for dilution when the plant or equipment is full of explosive mixed gases which are not easy to be discharged, or when the combustible gas is inevitably in contact with air (or oxygen) in some production processes (such as the production of nitric acid from ammonia and oxygen, the production of formaldehyde from methanol and oxygen, and the mixing of oil vapor and air on the liquid level of gasoline tanks and tanks). The mixed gas formed is not within the explosion limit and is not explosive. This method is called inert gas protection. Inert gas may also be used for protection during crushing, grinding, screening, mixing of flammable solids and conveying of powdery materials.

When inert gas is added, only the oxygen content in the mixed gas is below the critical value, the mixed gas will not explode with encountering fire. The critical oxygen concentration of methane is 12% (under the conditions of temperature 26℃, standard atmospheric pressure). See Table 4-16 for the critical oxygen concentration of various combustible gases at normal temperature and pressure.

Table 4-16 Critical Oxygen content of combustible gases (Under normal temperature and pressure)

Combustible substance	Critical oxygen concentration(%)		Combustible substance	Critical oxygen concentration(%)	
	CO_2 Diluent	N_2 Diluent		CO_2 Diluent	N_2 Diluent
Methane	14.6	12.1	Ethylene	11.7	10.6
Ethane	13.4	11.0	Propylene	14.1	11.5
Propane	14.3	11.4	Butadiene	13.9	10.4
Butane	14.5	12.1	Hydrogen	5.9	5.0
Pentane	14.4	12.1	Carbon monoxide	5.9	5.6
Hexane	14.5	11.9	Acetone	15	13.5

4.6.4 Prevent explosion propagation with fire arrestor

When the flammable gas explodes, it is necessary to set up a fire retardant device to prevent the spread of flame. The equipment is installed with fire retardant device includes: opening of petroleum tank, input pipeline of combustible gas, solvent recovery pipeline, gas chimney, exhaust pipe of dryer, gas welding equipment and pipeline, etc. Its function is to prevent flame from entering equipment, containers and pipelines, or to prevent the spread of flames in equipment and piping. Its working principle is to set a fire-retardant medium between the two sides of the combustible gas inlet and outlet, so that when either side is on fire, the spread of the flame is prevented and will not burn to the other side. The commonly used fire retarding devices are safety liquid seal, flame arrester and one-way valve.

In some explosive gas mixtures, the flame propagation speed increasing with the increase of propagation distance and finally the flame becomes detonation. Once it does become a detonation, it is necessary to install a detonation suppressor to prevent its propagation.

4.6.4.1 Safety liquid seal

This type of fire retarding device uses liquid as the fire retarding medium. Safety water seals are the most widely used at present. They use water as a fire-retardant medium and are generally installed between the gas pipeline and the production equipment. For example, various gas generators or gas holders often use safety water seals for fire resistance. Combustible gas from the gas generator or gas holder enters the production equipment through the safe water seal. Either side of the safety water seal is on fire, the flame is prevented from spreading to the other side of the safety water seal by the action of water when the flame is transmitted to the water seal. There are two types of safety water seals in common use: open type and closed type.

(1) Open type safety water seals: their structures and working principles are shown in Fig. 4-31. They mainly comprises a tank body, an air inlet pipe, an air outlet pipe and a safety pipe, wherein the air inlet pipe is deeply inserted into the liquid level, while the safety pipe is shallowly inserted into the liquid level. During normal operation, the combustible gas enters the tank through the gas inlet pipe and then flows out from the gas outlet. When flame flashback occurs, the gas pressure in the tank rises and presses the water surface. When the liquid level drops, the safety pipe leaves the water surface first and relieves the pressure of the tank. Because the safety pipe is inserted into the liquid level shallowly, the flame is stopped by the water from entering the other side. Open safety water seals are suitable for the gas system with low pressure.

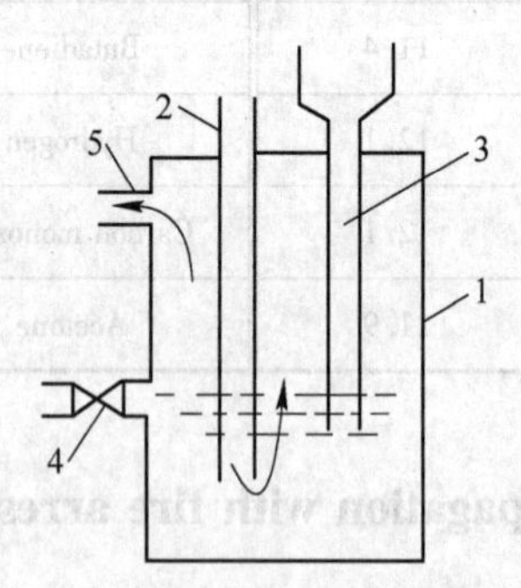

Fig. 4-31 Schematic diagram of open safety water seal

1—Shell; 2—Air inlet pipe; 3—Safety pipe; 4—Water test plug; 5—Gas outlet

(2) Closed safety water seals: their structures and working principles are shown in Fig. 4-32. During normal operation, the combustible gas flows in from the air inlet pipe and flows out from the air outlet through the one-way valve. When the flame back flaming occurs, the pressure in the

tank increases to press the water surface, and the one-way valve is closed instantaneously through the water layer, thus effectively preventing the flame from entering the other side. If the gas pressure is high, the bursting disc on the top of the tank breaks, which just right protects the tank body. The closed safety water seal is suitable for the gas system with higher pressure.

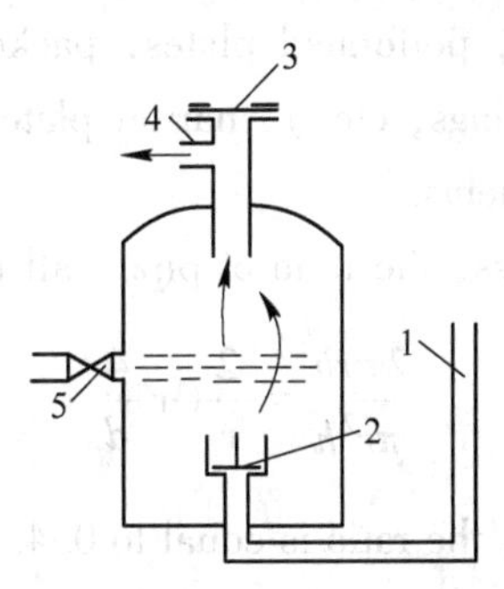

Fig. 4-32 Schematic diagram of closed safety water seal

1—Gas inlet; 2—Check valve; 3—Explosion-proof film; 4—Gas outlet; 5—Water test column

When safety water seal is used, the water level should not be lower than the position marked by the water level valve. However, the water level should not be too high, otherwise it will affect the flow of combustible gas, and water may enter the outlet pipe with the combustible gas. The water level should be checked and replenished in time after each flame flashback. The safety water seal should be kept in a vertical position. When the safety water seal is used in winter, water should be prevented from freezing. If freezing is found, only hot water or steam can be used to heat and thaw, and it is strictly prohibited to bake with open fire. In order to prevent freezing, small amounts of salt can be added to the water to lower the condensation point of the water.

4.6.4.2 Flame arrestor

Flame arrester is a kind of dry safety device which uses gap flame suppression to prevent flame propagation. The safety device has the advantages of simple structure, low cost, convenient installation and maintenance, and wide application. The flame arrestor used in the high temperature equipment, combustion chamber, high temperature oxidation furnace, high temperature reactor, etc. are easy to cause explosion and the pipelines for conveying combustible gas and flammable liquid vapor. As well as on the exhaust pipes of containers, pipelines and equipment of flammable liquids and gases, flame arresters are often used for fire resistance.

Gap flame extinction means that the flame passing through the metal mesh lose their part of the active groups (free radicals), due to contact with the mesh surface, so that the chain free radical reaction is stopped. This kind of phenomena is called gap flameout phenomenon, which is how the flame arresters work.

Flame extinction diameter is an important parameter for designing flame arrester. The so-called flame extinction diameter refers to the critical diameter of the pipeline that does not propagate flame when the mixed gas is ignited.

The flame arrester element is a collection of many gaps. It is the most important component of the flame arrester. Whether it is properly selected or not has a decisive impact on the ability of the device. Porous materials with incombustibility and air permeability are generally used as flame suppression elements, and they should have certain strength. Metal mesh is the most commonly used, and corrugated metal sheets, perforated plates, packed layers of fine particles (such as sand, glass balls, iron or copper filings, etc.), narrow plates, metal thin tube bundles, etc, are also used as flame suppression elements.

From the point of view of heat loss, the ratio of pipe wall heating area to mixture volume is:

$$\frac{2\pi rh}{\pi r^2 h}=\frac{2}{r}\text{Or}\frac{4}{d} \tag{4-53}$$

When the pipe diameter is 10cm, the ratio is equal to 0. 4. When the pipe diameter is 2cm, the ratio is equal to 0. 2. So it can be seen that whith the diameter of the tube decreases, the heat loss gradually increases, and the combustion temperature and flame propagation speed decrease accordingly. When the pipe diameter is small to a certain limit value, the heat loss of the pipe wall will be greater than the reaction heat, then the flame is extinguished. The factors affecting the performance of the flame arrester are the thickness of the flame arrester layer, the diameter of its gap and the size of its channels.

The metal mesh flame arrester is shown in Fig. 4-33, and its flame suppression element is composed of several metal meshes with a certain aperture. The number of layers of metal mesh is usually 10~12, the wire is 0. 4mm diameter steel wire or copper wire, and the mesh density is 210~250 holes/cm^2. However, organic solvents can usually prevent flame propagation by using four layers of metal mesh.

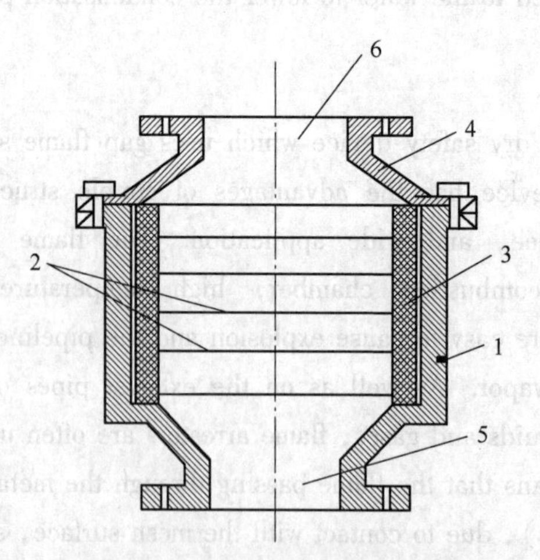

Fig. 4-33 Metal mesh flame arrester

1—Valve body; 2—Metal mesh; 3—Gasket; 4—Upper cover; 5—Inlet; 6—Exit

In addition to metal mesh flame arresters, gravel flame arresters are often used. This kind of flame arresters use sand grains, pebbles, glass, iron or copper filings as filling materials, the fire-retardant media separate the space in the flame arresters into a plurality of nonlinear small pores, and the nonlinear small pores can effectively prevent the flame from spreading when the combustible gas is backfired, whose fire retarding effect is better than that of metal mesh flame arrester.

4.6.4.3 Check valve

One-way valve is also called check valve. Its function is to allow only the combustible gas or liquid to flow in one direction, and it will automatically close in case of backflow, so as to avoid the backflow of fluid in the gas or fuel oil system, or the explosion of the vessel pipeline caused by high pressure entering low pressure, or the backflow and spread of flame in case of backfire. In industrial production, a check valve is usually placed between the inlet and outlet of the fluid, on the auxiliary line connecting the gas or oil piping and equipment, on the low pressure system between the high pressure and low pressure systems, or on the outlet line of the compressor and oil pump.

4.7 Turbulent combustion and diffusion combustion

4.7.1 Turbulent combustion

We know that real fluids are always viscous. There are two distinct flowstates in the motion of this real viscous fluid, i. e. laminar flow and turbulent flow.

When the Reynolds number Re is greater than or equal to a critical value, the steady laminar flow will change into the unsteady turbulent flow. In the state of turbulence, the motion parameters (the magnitude and direction of velocity) and dynamic parameters (the magnitude of pressure) of fluid particles will change continuously and irregularly with time. In the turbulent flow field, numerous irregular and transient eddies of different scales are intermingled and distributed in the whole flow space, and the eddies themselves undergo the process of occurrence, development and disappearance. The phenomenon that the motion parameters and dynamic parameters of fluid particles or micro-clusters change instantaneously with time is called pulsation, which generally behaves as a nonlinear random motion. It can be found through experimental observation that, the velocity and pressure in the turbulent state fluctuate above and below an average value, which has a certain regularity.

Turbulent flow has the following macroscopic characteristics:

(1) The turbulent flow field is a fluid motion field in which many eddies of different scales and shapes are mixed with each other. A single fluid micelle has a completely irregular and instantaneously changing pulsation characteristic. Fluctuation is the main feature that distinguishes turbulent flow from laminar flow.

(2) All the physical quantities in the turbulent flow field are random quantities varying with time and space, and they all have some regular statistical characteristics to a certain extent. Therefore, any instantaneous physical quantity at a space point can be expressed by the sum of its average value and pulsation value. Its average value can be regarded as not changing with time or

changing slowly with time according to a constant law. This kind of turbulent flow field has quasi stationarity, which is called quasi-steady turbulence.

(3) The physical quantities of any two adjacent space points in the turbulent flow field have a certain degree of correlation with each other, such as the correlation between the velocity of two points, between pressure and velocity, and between density and velocity, etc. Different correlation properties are manifested in the presence of various correlation terms in the turbulence equations, which depend on different turbulent structures and boundary conditions to occur. They cause various changes in the turbulent motion.

(4) Turbulence is composed of numerous irregular eddies, which mix with the fluid around them and show turbulent transport characteristics. The gradual deformation and splitting of the eddies form a process of turbulent energy transfer, a larger scale into those of a smaller scale, and then into those of a smaller scale, and finally up to a certain limit value. It can be considered that the vortex cluster no longer exists. At this time, the energy of turbulent fluctuation is dissipated into the heat energy of molecular turbulence. If there is no external energy source to make the turbulent motion occur continuously, the turbulent motion will gradually decay and eventually disappear.

4.7.1.2 Characteristics of turbulent combustion

Turbulent flame is different from laminar flame in some obvious characteristics, as shown in Fig. 4-34, its flame length is short, its thickness is thick, its luminous area is blurred, and its noise is obvious, whose basic characteristics are combustion intensification and reaction rate increase. Turbulent flame may be caused by one or a combination of the following three factors:

(1) Turbulence may make the flame surface curved and wrinkled, increasing the reaction area, but the laminar flame speed is still maintained in the normal direction of the curved flame surface;

(2) Turbulence may increase the transport rate of heat and reactive species, thus increasing the combustion velocity perpendicular to the flame plane;

(3) Turbulence can quickly mix the burned gas and unburned fresh gas, making the flame become a homogeneous reactant in essence, thus shortening the mixing time and increasing the combustion velocity.

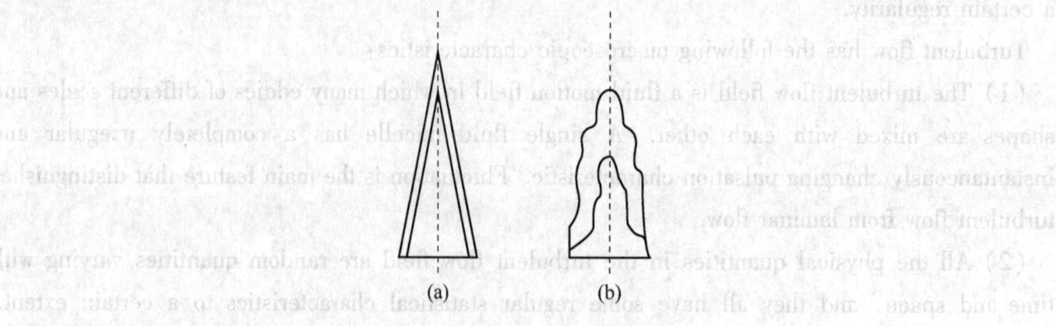

Fig. 4-34 Laminar and turbulent flame profile

(a) Laminar Flame; (b) Turbulent Flame

The homogeneous reaction velocity mainly depends on the proportion of burnt and unburnt gases produced during the mixing process. It can also be seen that turbulent combustion is affected by both the flow properties of turbulence and the chemical reaction kinetics, in which the flow plays a more important role. In particular, in laminar combustion, the transport coefficient is a property of the burning substance, while in turbulent combustion, all the transport coefficients are closely related to the flow characteristics, and the transport flow plays a more important role. Therefore, the problem of turbulent combustion is much more complex than that in laminar combustion, but it is not affected by the further increase of Reynolds number, which makes it possible to simplify it in some aspects.

4.7.1.3 Dunkler — Serkin wrinkled flamelet model

Early work on turbulent combustion was initiated by Dunklerfrom Germany and Serkin from the Soviet Union. They used the concept of laminar flame propagation to explain the mechanism of turbulent combustion, and used the turbulent flame speed to explain the turbulent combustion process. The incoming flow was assumed to be turbulent, which distorts the flame but does not destroy the flame front. The curved and wrinkled flame surface is still a laminar flame. In this way, the surface area of the flame is greatly increased, thereby increasing the space heating rate. As shown in Fig. 4-35, it was assumed that the turbulent flame is one-dimensional, the flow field is uniform and isotropic, and the turbulent flame propagation speed S_t is related to the incoming flow speed u_∞ as follows:

$$S_t = u_\infty \cos\psi$$

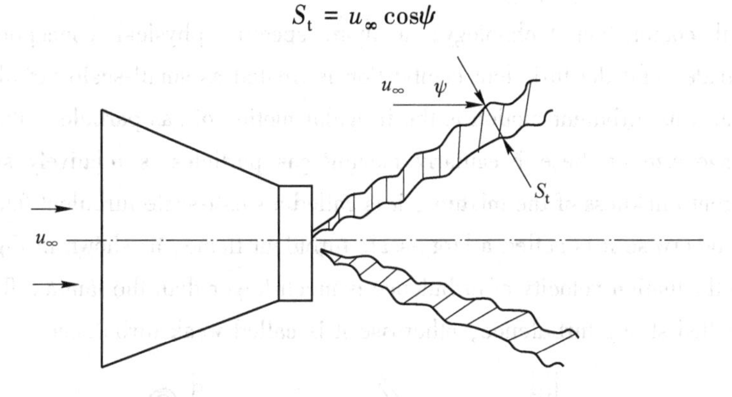

Fig. 4-35 Turbulent flame propagation speed diagram

Mimicking the one-dimensional laminar flame propagation problem, the one-dimensional quasi-steady turbulent flame energy balance equation can be written as:

$$\rho_\infty c_p S_t \frac{dT}{dt} = \frac{d}{dx}\left[(\lambda + \lambda_t)\frac{dT}{dx}\right] + w_s Q_s \tag{4-54}$$

The molecular thermal conductivity λ and the turbulent thermal conductivity λ_t are constants.

$$\lambda_1 = \rho_\infty c_p \sqrt{v_x'^2} L_h$$

Where, L_h is turbulent micelles' scale.

Take dimensionless quantities.

Dimensionless temperature: $\theta = \dfrac{T_m - T}{T_m - T_\infty}$

dimensionless velocity: $\overline{S}_t = \dfrac{S_t}{u_\infty}$

dimensionless coordinates: $\varepsilon = \dfrac{x}{L}$

Where L is the characteristic dimension.

Substituting equation (4-54), the dimensionless form of this equation is:

$$\overline{S}_t \frac{d\theta}{d\varepsilon} = \frac{\alpha_\infty + \sqrt{\overline{v_x'^2}} L_h}{u_\infty L} \cdot \frac{d^2\theta}{d\varepsilon^2} - \frac{L Q_s W_s}{\rho_\infty c_p u_\infty (T_m - T_\infty)} \tag{4-55}$$

With further simplification, it is possible to obtain,

$$S_t = A\left(\frac{\sqrt{\overline{v_x'^2}}}{u_\infty}\right)^{\alpha}\left(\frac{S_l}{u_\infty}\right)^{\beta} \tag{4-56}$$

$$S_t = A\left(\sqrt{\overline{v_x'^2}}\right)^{\alpha}(S_l)^{\beta} \tag{4-57}$$

Where S_l——laminar flame propagation speed;

$$\alpha + \beta = 1.$$

This means that the turbulent flame propagation speed depends on the turbulent fluctuation speed and the laminar flame propagation speed.

In the actual combustion technology, a more specific physical conception of turbulent combustion is made, and the turbulent combustion is treated as small-scale turbulence and large-scale turbulence. Gas turbulent motion is the irregular motion of gas particles with different sizes. When the average size of these irregularly moving gas particles is relatively smaller than the laminar flame front thickness of the mixture, it is called a small-scale turbulent flame, as shown in Fig. 4-36 (a), otherwise it is called a large-scale turbulent flame, as shown in Fig. 4-36 (b) and (c). When the fluctuation velocity of turbulence is much larger than the laminar flame propagation velocity, it is called strong turbulence, otherwise it is called weak turbulence.

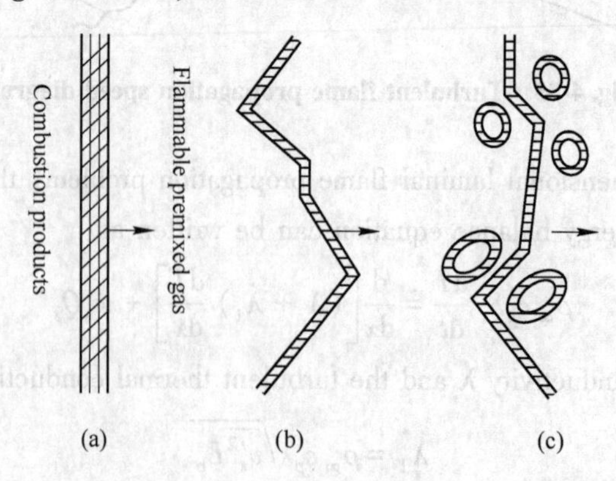

Fig. 4-36 Three turbulent flame models

(a) Small-scale eddy current; (b) Large scale weak eddy current; (c) Large scale strong eddy current

A Small-scale turbulence

Turbulence is small-scale in the range $2300 < Re < 6000$. The vortex size and mixing length are much smaller than the thickness of the flame front. The effect of small-scale eddies is mainly to increase the intensity of the transport process in the flame front. Under this condition, the transport of heat and mass (species) is proportional to the turbulent diffusivity μ_t but not to the molecular diffusivity D_i(or $\lambda/\rho c_p$). The laminar flame propagation speed, S_l, is in direct proportion to $\sqrt{D_i}$ or $\sqrt{\lambda/\rho c_p}$. Therefore, it is reasonable to infer that the small-scale turbulent flame propagation speed S_t is proportional to $\sqrt{\mu_t}$ in direct proportion. Therefore

$$\frac{S_t}{S_l} = \left(\frac{\mu_t}{\lambda/\rho c_p}\right)^{1/2} \approx \left(\frac{\mu_t}{D_i}\right)^{1/2} \approx \left(\frac{\mu_t}{v}\right)^{1/2} \tag{4-58}$$

For flow in a tube, from $\mu_t/v \approx 0.01Re$, there is an approximate relation:

$$\frac{S_t}{S_l} \approx 0.1Re^{1/2}$$

The analytical results are in good agreement with the experimental results.

B Large-scale turbulence

When $Re > 6000$, the size of the turbulent eddies is quite large and exceeds the thickness of the laminar flame. At this time, the turbulent fluctuation velocity is generally small, but enough to distort the flame surface and produce wrinkled flames. In Fig. 4-37, some points on the flame front are due to the forward pulsation velocity v'_x running faster than the overall flame surface advancing at the average velocity v_x, which forms a convex cone. But behind the average flame surface, several other places due to the backward pulsation velocity. v'_x, forms a concave conical surface. These two make the flame front uneven. On this uneven and wrinkled flame surface, the flame is pushed forward to the unburned side along the normal direction of the flame surface at this point at the laminar flame propagation speed S_l. So the amount of combustible mixture burned per unit time is the product of the laminar flame propagation speed S_l and the surface area A_l. If this amount of burned mixture is calculated as the forward advance of the average position of the entire flame surface, it should be the product of the turbulent flame propagation speed S_t and the average position of the flame surface of the plane area A_t.

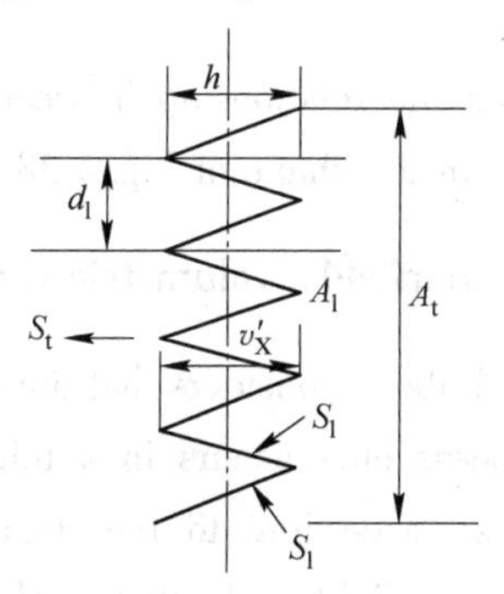

Fig. 4-37 Simplified model of wrinkled flame caused by large-scale turbulence

$$S_t A_t = S_l A_l$$

Shelkin regarded the corrugated flame surface as a conical surface, the diameter of whose base is equivalent to the average vortex diameter d_e, and the height of which is equal to the root-mean-square value of the fluctuating velocity $\sqrt{v_x'^2}$ multiplied by the time t. The pulsation time t can be considered as approximately equal to d_e/S_l, i. e. $h = \sqrt{v_x'^2}\, d_e/S_l$. So according to the relationship of geometry, we get

$$\frac{A_l}{A_t} = \frac{\frac{\pi}{2} d_e \sqrt{\left(\frac{d_e}{2}\right)^2 + h^2}}{\frac{\pi}{4} d_e^2} = \sqrt{1 + \frac{4h^2}{d_e^2}} = \sqrt{1 + \left(\frac{2\sqrt{v_x'^2}}{S_l}\right)^2} \tag{4-59}$$

For large-scale weak turbulence, $\sqrt{v_x'^2} \ll S_l$, then the above equation can be expanded into a Taylor series and the higher order terms are omitted to obtain

$$\frac{S_t}{S_l} = \frac{A_l}{A_t} \approx 1 + 2\left(\frac{\sqrt{v_x'^2}}{S_l}\right)^2 \tag{4-60}$$

For strong turbulence on large scales, $\sqrt{v_x'^2} \gg S_l$, then 1 in the radical sign can be omitted in the formula (4-62) to obtain

$$\frac{S_t}{S_l} \approx \sqrt{\left(\frac{2\sqrt{v_x'^2}}{S_l}\right)^2} \approx \frac{2\sqrt{v_x'^2}}{S_l}\left(2\sqrt{v_x'^2} \gg S_l\right) \approx \frac{\sqrt{v_x'^2}}{S_l} \tag{4-61}$$

At this time, the flame combustion model can be assumed that the unburned combustible mixture breaks through the flame front and enters the surrounding of high temperature combustion products, and the high temperature combustion products also break through the flame front and enter the unburned premixed gas to form island-like closed small blocks. These small blocks all keep their own independence. At the same time, flame propagation is carried out in the surrounding air mixture. It can be said that where these small pieces, which depend on the pulsation speed, move, the flame propagates. Therefore, the flame propagation speed is equal to the pulsation speed.

The experimental results show that the relationship between turbulent flame propagation speed and Reynolds number can be drawn in the shape of Fig. 4-38.

4.7.1.4 Summerfield (M. Summerfield) volumetric combustion model

The wrinkled flame model mentioned above considers that the combustion chemical reaction rate is very high, and the combustion process only occurs in a thin flame surface, so it is a surface combustion model. In recent years, according to the data of concentration and temperature distribution in turbulent flame, summerfield and others believed that the chemical reaction of combustion occurs at different speeds everywhere in the flame. Turbulent transport causes mixing

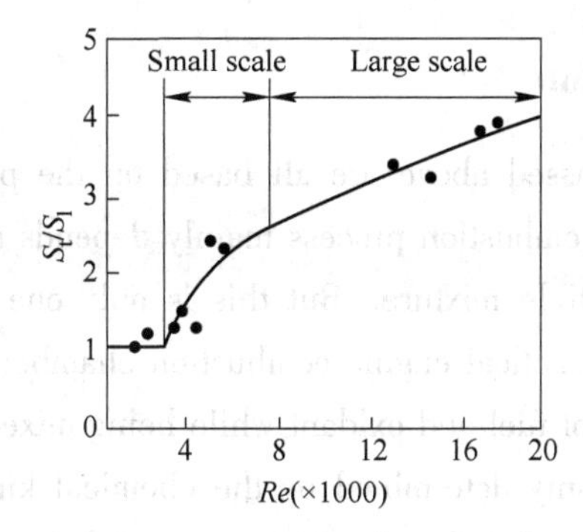

Fig. 4-38 Effect of *Re* number on flame speed figure

of gases of different composition in the flame zone simultaneously with combustion. The combination of combustion and mixing results in flame propagation, which is the volumetric combustion model.

The differences between the volumetric combustion model and the pleated flame model can be seen in Fig. 4-39 (a), (b).

Fig. 4-39 (a) is a schematic representation of a wrinkled flame model. In this surface combustion model, both the fresh air and the combustion products are in the form of clusters, and the reaction zone is located at the interface between them. The reduction in the volume of fresh gas is on the one hand due to the laminar combustion taking place on the surface of the air mass and, on the other hand, to the rupture of the air masses due to pulsations.

Fig. 4-39 (b) is a volumetric combustion model. In the volumetric combustion models, the combustion rate is affected not only by the mixing rate, but also by the chemical reaction rate, and the turbulent mixing is the main factor. The combustion process is no longer limited to laminar combustion at the surface, and takes place inside the volume. During the time of existence of the reaction zone, it can be considered that the concentration and temperature in the reaction zone are locally balanced. However, different micells have different concentrations, temperatures, and degrees of reactivity. This model can be used to calculate the turbulent combustion characteristics by simple numerical calculation.

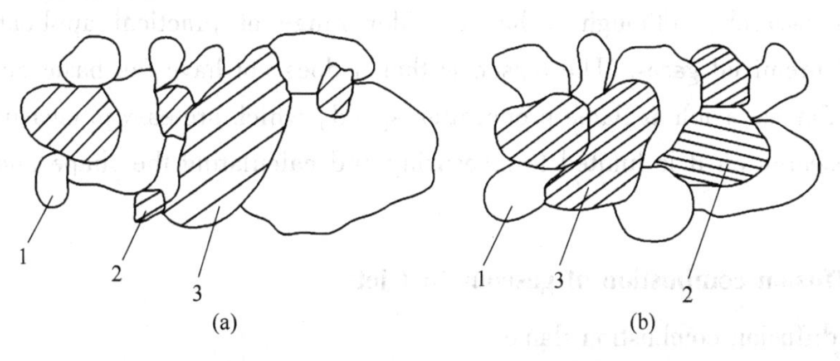

Fig. 4-39 Two models of turbulent flames

(a) Wrinkled flame model; (b) Volumetric combustion model

1—Fresh air; 2—Combustion products; 3—Reaction zone

4. 7. 2 Diffusion combustion

The combustion problems discussed above are all based on the pre-mixed combustible mixture, and the progress of the whole combustion process mainly depends on the chemical kinetic process of the oxidation of the combustible mixture. But this is only one way of burning fuel. There is another way of combustion in practical engine combustion chambers, boilers, industrial furnaces, and fires, such as combustion of fuel and oxidant while being mixed. At this time, the progress of the combustion process is not only determined by the chemical kinetic process of fuel oxidation, but also by the diffusion process of fuel and oxidant (usually air) mixing.

According to the different conditions of the combustion process, the combustion process can be generally divided into two types: chemical dynamic combustion and diffusion combustion.

If the progress of the process is mainly determined by the chemical kinetic process of fuel oxidation, that is, when the mixing rate of fuel and air is greater than the combustion rate, such as the combustion of homogeneous combustible mixture mentioned above, this combustion process is called as chemically powered combustion.

If the progress of the process is mainly determined by the diffusion mixing process of fuel and air, that is, the chemical reaction rate is greater than the mixing rate, this combustion process is called as diffusion combustion.

Diffusion combustion is one of the earliest forms of combustion in which humans used fire. To this day, the diffusion flame is the most common type of flame we have. Flames such as dancing fires and torches used in camping, candles and kerosene lamps used in homes, combustion in coal stoves, and droplet combustion in various engines and industrial kilns are all diffuse flames. All kinds of destructive fires that threaten and destroy human civilization and life and property are also constituted by diffuse flames.

The diffusion combustion can be single-phase or polyphase. The combustion of petroleum and coal in air belongs to polyphase diffusion combustion, while the jet combustion of gas fuel belongs to single phase diffusion combustion.

In the field of combustion, the diffusion combustion of gaseous fuels has received little attention and research, although it has a wider range of practical applications than the combustion of premixed gases. The reason is that it does not have the basic characteristics of premixed gas flames, such as flame propagation speed, which are easy to measure. Therefore, the current research on it is limited to measuring and calculating the shape and length of the diffusion flame.

4. 7. 2. 1 Diffusion combustion of gaseous fuel jets

A Types of diffusion combustion flame

Gaseous diffusion combustion is combustion in which gaseous fuel is fed separately from air and simultaneously into a combustion chamber.

In diffusion combustion, the oxygen required for combustion is obtained by air diffusion, so that a diffusion flame is apparently generated at the interface between the fuel and the oxidant. The fuel and oxidant diffuse from both sides of the flame to the interface, while the combustion products diffuse to both sides of the flame. Therefore, for a diffusion flame, there is no propagation of flame.

There are three types of diffusion flames in which fuel and air are supplied separately:

(1) Free jet diffusion flame. It is generated at the interface of the fuel jet formed after the gas fuel is ejected from the burner into the still air in a large space, as shown in Fig. 4-40 (a).

(2) Coaxial flow diffusion flame. Produced at the interface of a fuel jet of gaseous fuel exiting the nozzle on the same axis as the air stream, as shown in Fig. 4-40 (b).

The coaxial-flow diffusion flame, like the free-jet diffusion flame, is also a jet flame. The difference is that in the coaxial-flow diffusion flame, the fuel jet is injected into the combustion chamber with a limited space, so it will be affected by the walls of the combustion chamber. Therefore, such a jet flame is also called a confined jet diffusion flame.

(3) Reversed jet diffusion flame. It is generated at the interface of the fuel jet ejected against the air flow, as shown in Fig. 4-40 (c).

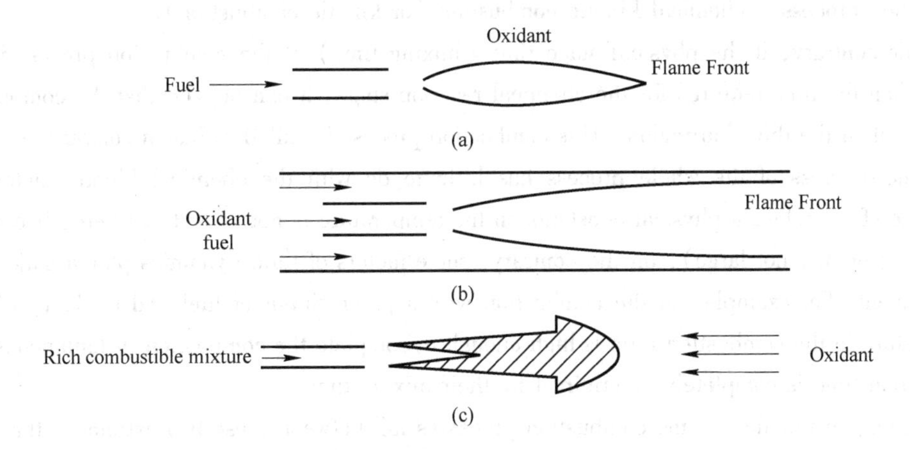

Fig. 4-40 Types of diffusion flames

(a) Free jet diffusion flame; (b) Coaxial flow diffusion flame; (c) Reverse jet diffusion flame

The flame produced around the oil droplet is also a gaseous diffusion flame essentially. It is a kind of diffusion flame produced by the diffusion and mixing of the fuel vapor produced by the evaporation of the fuel droplet surface and the surrounding air at the interface of the two.

Jet diffusion flame can also be divided into laminar jet diffusion flame and turbulent jet diffusion flame according to the state of jet flow. It is obvious that the diffusion mixing of turbulent jet is better than that of laminar jet, so the length of turbulent jet flame is much shorter than that of laminar jet. Because the diffusion flame does not have a backfire phenomenon and has good stability, and it does not need to mix fuel and oxidant in advance before combustion, it is widely used in industry. In addition, in order to obtain high space heating rate in industrial combustion equipment, turbulent jet diffusion flame is generally used.

B Diffusion combustion and power combustion

In general, the total time required for the fuel to burn is made up of two parts: the time required for the gaseous fuel to mix with air and the chemical reaction time for the fuel to oxidize. If the overlap of these two processes is not taken into account, the overall combustion process time should be the sum of the above two times. The mixture of fuel and air can have molecular diffusion or turbulent diffusion. If the mixing and diffusion time is so small as to be negligible compared with the time of the oxidation reaction, the total combustion time can be approximately equal to the time required for the oxidation reaction.

The combustion process is carried out in the kinetic region of the chemical reaction. This is an example of the combustion of a homogeneous combustible mixture. At this point, the progress of the combustion process (or the rate of combustion) will be strongly controlled by chemical kinetic factors, such as the nature of the combustible mixture, the temperature, the pressure in the combustion space and the concentration of the reactants, will strongly affect the combustion rate, while the diffusion factors of fluid dynamics, such as the velocity of the gas flow, the shape and size of the object through which the gas flow flows, are independent of the combustion rate. This combustion process is chemical kinetic combustion (or kinetic combustion).

On the contrary, if the physical stage time (mixing time) of the combustion process is much longer than the time required for the chemical reaction stage, it can be said that the combustion is carried out in the diffusion region. This combustion process is called diffusion combustion. At this time, the progress of the whole process has little to do with the chemical kinetic factors (the influence of the relevant physical constants on the temperature is not considered here, because this effect is generally not large), on the contrary, some factors of fluid dynamics play a major role at this moment. For example, in the combustion with separate input of fuel and oxidant, when the temperature in the combustion zone is high enough to complete the combustion instantaneously, the combustion time is completely determined by their mixing time.

However, in practice, some combustion processes fall between these two extremes, that is, the mixing time required for combustion is approximately equal to the chemical kinetic time of oxidation, which is the most complex case because it depends on both chemical and hydrodynamic factors.

4.7.2.2 Laminar diffusion combustion of a gas-fuel jet

The most common laminar diffusion flames in daily life are candle flames, or unpremixed Bunsen flames. A non-premixed Bunsen burner flame can be achieved by closing the primary air holes in the bottom of the ordinary Bunsen burner. Laminar diffusion flames were first studied by Burke and Schumann (1928), who used two concentric tubes as shown in Fig. 4-41, with gaseous fuel in the inner tube. The outer tube is filled with air, which flows in the tube at the same speed. Two types of diffusion flame shapes can be observed. One is when the amount of air supplied in the outer tube is more than that required for complete combustion of the fuel in the inner tube, or when the fuel jet is injected into the static air in a large space (that is, the ratio of d'/d is pretty big at this time), the diffusion flame is a closed convergent conical flame (called air excess diffusion flame). The other is that the amount of air provided in the outer tube is not enough to supply the fuel jet in the inner tube for complete combustion, and the flame shape is a diffuse inverted trumpet flame (called insufficient air diffusion flame). It can be seen that the shape of a laminar diffusion flame depends on the mixture concentration of fuel and air. The characteristics of a diffusion flame are usually described by the surface on which the chemical reaction occurs instantaneously, and this surface is generally assumed to coincide with the luminous combustion surface, which is exactly the shape of the above diffusion flame.

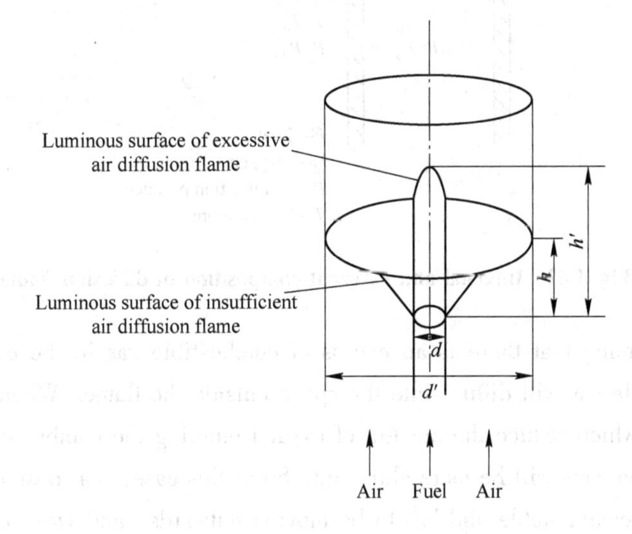

Fig. 4-41 Profile of laminar diffusion flame

We know that in laminar flow, the oxygen needed for fuel jet combustion is obtained from the surrounding air by molecular diffusion. If the shape of the burner is circular and in the case of an excess of air supply, the shape of the combustion flame is conical. This is because along the flow direction, the fuel gas flow is continuously consumed by combustion, so the combustion zone gradually moves closer to the center of the gas flow, and finally converges on the center line of the gas flow to become the vertex of cone.

It is clear that there is only fuel and no oxygen (air) on the inner side of the flame front (i. e., the combustion zone) and only oxygen and no fuel on the outer side (see Fig. 4-42). Fuel and oxygen are transported to each other by molecular diffusion, and a stable combustion zone (i. e., a flame front) is formed at each position where the ratio of fuel to oxygen reaches a stoichiometric ratio, in which combustion proceeds rapidly. It can be considered that the chemical reaction rate is many times greater than the diffusion rate of combustible matter, and the speed of the whole combustion process depends on the molecular diffusion rate between fuel and oxygen entirely.

Now let's ask ourselves this question: why is the composition of the mixture on the surface of the stable combustion zone, or flame front, exactly stoichiometric? This is because there can be no excess oxygen or fuel in the combustion zone, otherwise the position of the combustion zone will

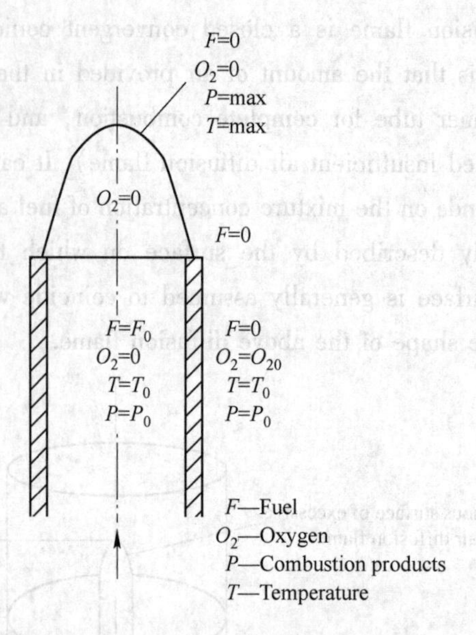

Fig. 4-42 Internal and external composition of diffusion flame

not be stable. Assuming that there is an excess of combustible gas in the combustion zone, the unburned combustible gas will diffuse into the space outside the flame. When it meets oxygen, it ignites and burns, which reduce the amount of oxygen entering the combustion. The combustible gas in the combustion zone will be more abundant. So in this case, that position of the combustion zone is unlikely to remain stable and has to be moving outwards, and vice versa. Then it can be seen that the diffusion flame can be stable only on the surface where the composition ratio of combustible gas and oxygen is in accordance with the stoichiometric ratio.

The combustible mixture formed by the combustible gas (fuel) and oxygen entering the combustion zone ignites and burns due to the heat transmitted from the flame front, and the generated combustion products diffuse to both sides of the flame, diluting and heating the combustible gas and oxygen. As a result, that flame front divide the combustion space into two regions: the outside of the flame has only oxygen and combustion products but no combustible gases, it is an oxidation zone. while the inner side of the flame has only combustible gas and combustion products but no oxygen, which is the reduction zone.

Because of the very high chemical reaction rate in the combustion zone, the combustible mixture that reaches the combustion zone actually burns out instantaneously, so that its concentration value in the combustion zone is zero, while the concentration and temperature of the combustion products reach their maxima. In addition, the thickness of the combustion zone (i. e., the width of the flame front) will become very thin due to the large chemical reaction rate. Therefore, in an ideal diffusion flame, it can be regarded as a geometric surface with zero surface thickness, which is impermeable to both oxygen and fuel, with oxygen on one side and fuel on the other. Therefore, the shape of the flame front of the laminar diffusion flame depends only on the conditions of molecular diffusion and has nothing to do with the chemical kinetics. It can be found by mathematical analysis as a geometric surface. The ratio of the rate of outward diffusion of combustible gas to the rate of inward diffusion of oxygen on this surface should be equal to the stoichiometric ratio at complete combustion.

Fig. 4-43 shows the radial distribution of the concentration of each species in the diffusion flame at a certain height from the fuel jet orifice. It can be seen from the figure that the concentration of fuel and oxidant is the smallest (equal to zero) at the flame front, while the concentration of combustion products is the largest there and penetrates to both sides of the flame by diffusion. This concentration profile is also suitable for a fuel jet directed into the surrounding static atmosphere.

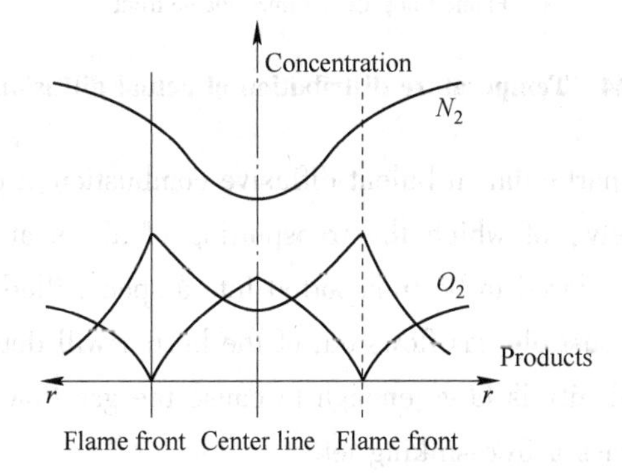

Fig. 4-43 Distribution of transverse material concentration of diffusion flame at a certain height from the fuel nozzle

As a practical matter, that reaction zone in a diffusion flame is not infinitely thin as theoretically described above. As shown in Fig. 4-44, the experiment shows that the combustion temperature reaches the maximum in the main reaction zone, in which the various gas compositions are in a state of thermodynamic equilibrium. On both sides of the main reaction zone are preheating zones, which are characterized by a relatively steep temperature gradient. The fuel and oxidant undergo chemical reaction in that preheat zone. Since little oxygen passes through the primary reaction zone into the fuel jet, the fuel is heat in the preheat zone by heat conduction and diffusion of the high temperature combustion products. The chemical reaction that occurs is mainly thermal decomposition. At this time, the hydrocarbons in the combustible gas will decompose into carbon particles. The higher the temperature, the more violent the decomposition will be. At the same time, it may increase the content of heavy hydrocarbons that are complex and difficult to burn. These carbon particles and heavy hydrocarbons are often carried away by the combustion products in the form of soot before they can be burned, resulting in chemical incomplete combustion losses. Therefore, a significant feature of diffusion combustion is that it will produce incomplete combustion losses, which is not the case with premixed flames.

4.7.2.3 Turbulent diffusion combustion in a gaseous fuel jet

The most widely used diffusion combustion in industrial applications and under fire conditions is generally turbulent diffusion combustion.

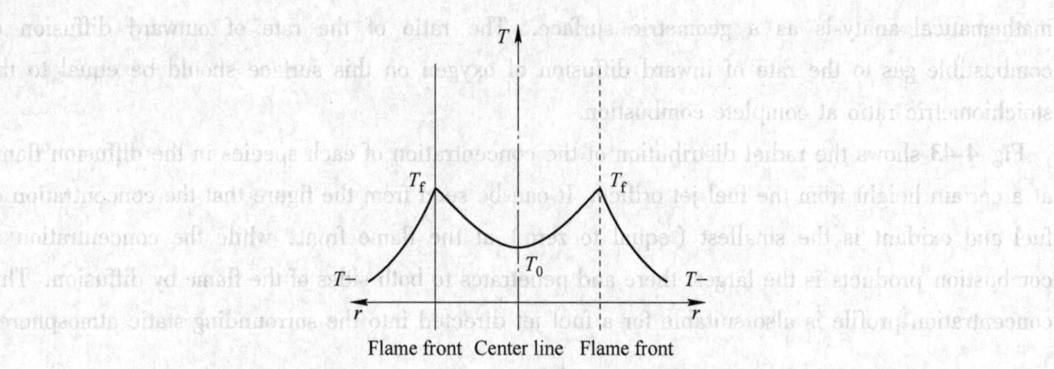

Fig. 4-44 Temperature distribution of actual diffusion flame

Now let us study such a particular turbulent diffusive combustion: a combustible gas (fuel) and air are transported separately, of which the transporting of air is at a very small rate and the combustible gas can be considered to be transported into a space filled with still air. In this way, the speed at which the combustible gas flows out of the burner will determine the flow state of the gas flow. If the gas flow velocity is large enough to cause the gas flow to be in a turbulent state, then the turbulent jet becomes a free-sinking jet.

As shown in Fig. 4-45, after the outlet of the burner, the jet absorbs air from the surrounding space in the process of turbulent diffusion, so that continuously the mass of the airflow increases and the width of the jet expands, while the velocity of the airflow decreases continuously and becomes uniform gradually. At the same time, mixtures of different concentrations are formed on the width of the jet.

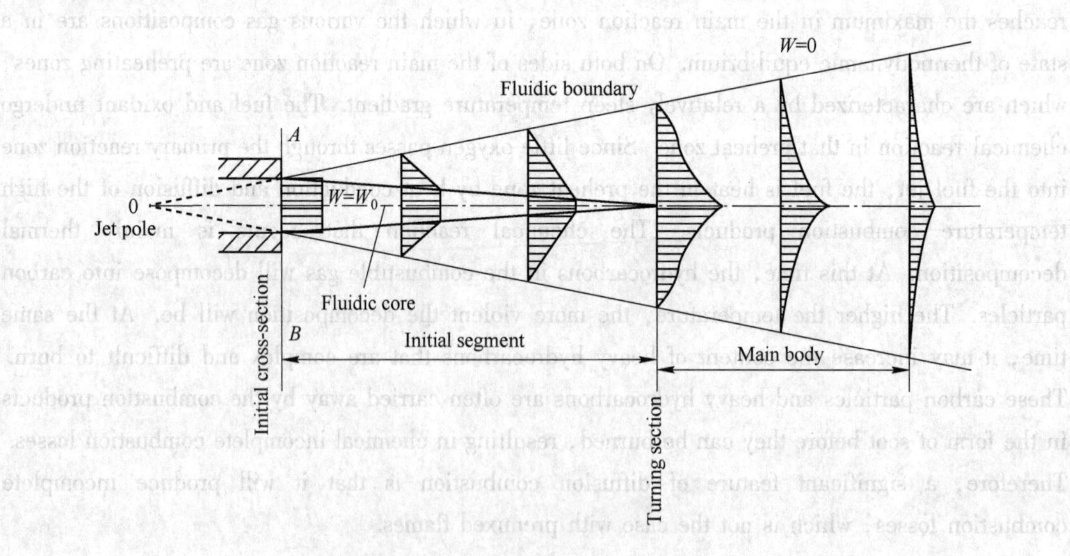

Fig. 4-45 Free submersible jet

There is only combustible gas in the constant velocity core region of the initial section of the jet, and the mixture of combustible gas and air exists only in the turbulent boundary layer. In the main section of the jet, the concentration distribution of the combustible gas on any section is shown in Fig. 4-46. The concentration of the combustible gas is maximum on the jet axis and gradually decreases near the jet boundary. On the other hand, the gas concentration is zero on the boundary, and the concentration of combustible gas becomes smaller and smaller as it goes away from the burner. The air concentration is the smallest on the jet axis, and the closer to the jet boundary and the farther away from the burner, the larger the air concentration is.

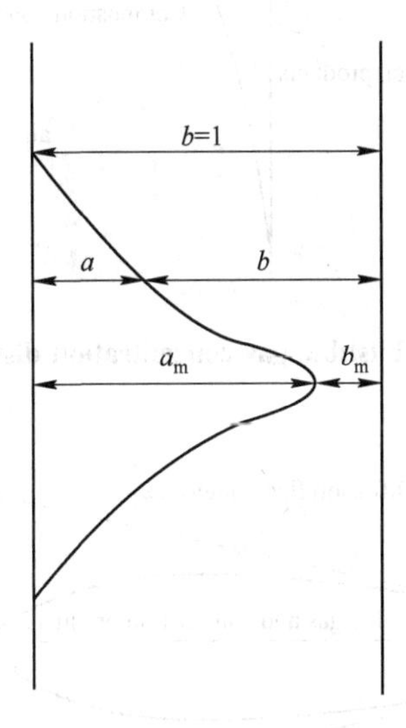

Fig. 4-46 Formation of turbulent diffusion flame on any section of jet body section

Thus, the combustible mixture formed on the jet boundary layer is obviously different in its composition ratio (i. e. , a) at different locations. A similar analysis was done for a laminar diffusion flame to obtain: the stable combustion zone (i. e. , the flame front) in the gas stream at the time of ignition is located at the surface where the composition ratio of the mixture corresponds to the stoichiometric ratio at the time of theoretical complete combustion. It can be seen that the position of the combustion zone is completely determined by the condition of turbulent diffusion, and the combustion speed is determined by its diffusion speed.

Suppose that the concentration distribution of combustible gas and air on a certain section is as shown in Fig. 4-47, the mixture of chemical equivalence ratio is formed at point *A* at a certain distance from the axis of the jet, and the combustion zone (flame front) is formed by the circle formed by these points on the same section. An elongated conical diffusion flame front is formed at each section through these corresponding circles (see Fig. 4-48). The oxygen diffused into the combustion zone reacts with the combustible gas to release the corresponding heat, and the combustion products generated by combustion diffuse to both sides of the combustion zone (flame). Therefore, there is a mixture of combustible gas and combustion products inside the flame while there is no oxygen. On the outside of the flame is a mixture of combustion products and oxygen (air) while there is no combustible gas.

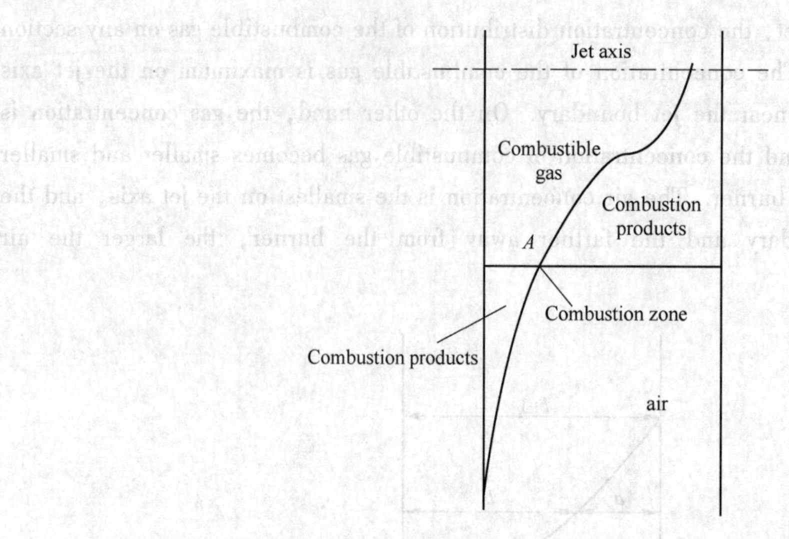

Fig. 4-47 Combustible gas concentration distribution curves

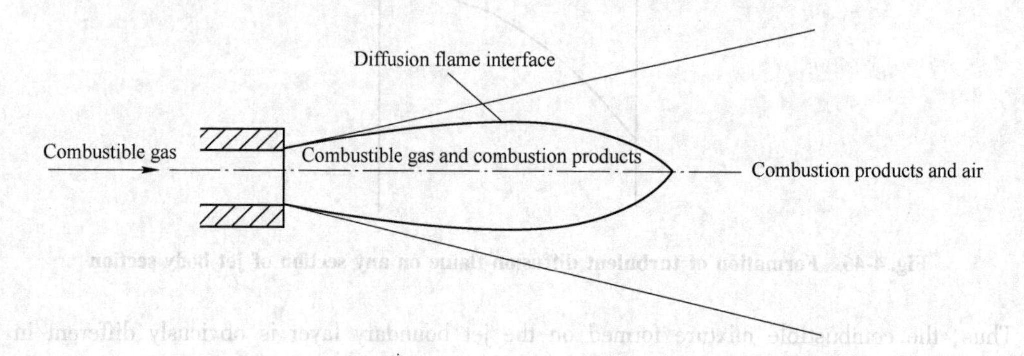

Fig. 4-48 Flame front of turbulent diffusion combustion

Fig. 4-49 shows the experimental results of the flame shape and height as the jet velocity increases. It can be seen from Fig. 4-49 that when the flow velocity is relatively low, that is, in the laminar flow state, the flame height is approximately proportional to the increase in flow velocity, while when the flow velocity is relatively high, that is, in the turbulent flow state, the flame height is almost independent of the flow velocity.

Also shown in Fig. 4-49 is the evolution of the diffusion flame from laminar to turbulent flow. It can be seen from the figure that the flame front of the laminar diffusion flame has a smooth edge, a sharp contour, and a stable shape. With the increase of the flow rate (or *Re* number), the height of the flame front increases almost linearly until it reaches a maximum value. After that, an increase in flow velocity will destabilize the tip of the flame front, and make it begin to vibrate. With the further increase of flow velocity, the instability will gradually develop into a turbulent brush flame with noise, which starts from a certain point at the top of the flame and breaks up into a turbulent jet. Due to the turbulent diffusion, the combustion is accelerated, and the height of the flame is rapidly shortened, meanwhile the breaking point from the laminar flame breaking to the turbulent flame is moved toward the burner. When the jet velocity reaches the condition that the breakup point is very close to the nozzle, that is, when the fully developed turbulent flame is reached, if the velocity is further increased, the height of the flame and the length S of the breakup point do not change and remain constant, but the noise of the flame will continue to increase. The brightness of the flame will also continue to decrease. Finally, at a certain speed, which depends on the type of combustible gas and the size of the burner, the flame will be blown away from under the nozzle.

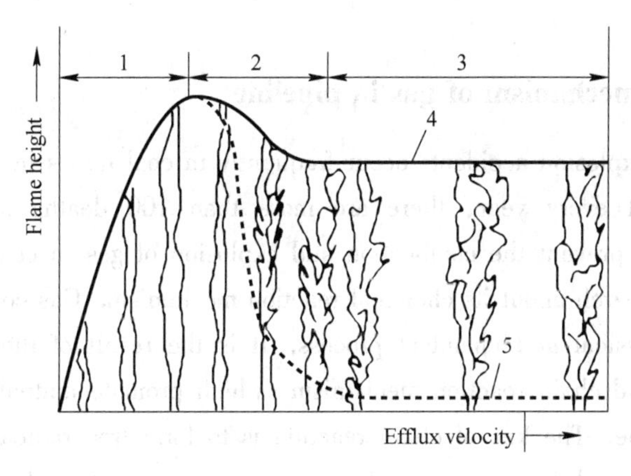

Fig. 4-49 Variation of diffusion flame height with the flow rate of combustible gas

1—Laminar flame zone; 2—Transition flame zone; 3—Fully developed turbulent flame region; 4—Flame height envelope; 5—Breakpoint envelope

The transition of the diffusion flame from laminar flow to turbulent flow generally occurs in the critical range of Reynolds number from 2000 to 10000. The reason for such a wide transition range is that the viscosity of the gas is strongly temperature dependent, and a flame with a relatively high adiabatic temperature can be expected to enter turbulence at a relatively high number of *Re*. Conversely, a flame with a relatively low adiabatic temperature should enter turbulence at a relatively low Reynolds number.

It is also experimentally found that the height of the diffusion laminar flame is related to the stoichiometric ratio of oxygen to combustible gas. The more the number of moles of oxygen required for one mole of combustible gas, the higher the height of the diffusion flame is, and vice versa, the lower the height of the diffusion flame. As the oxygen content in the environment decreases, the flame height increases.

4.8 Explosion ignition and suppression of pipeline gas

When studying the combustion and explosion of combustible gas, pipeline gas is an important part. The combustion of pipeline gas will cause explosion, causing serious damage to facilities and equipment, and endanger personal safety. Therefore, it is necessary to study the combustion and explosion law of pipeline gas and put forward the corresponding explosion suppression technology. China has also done a lot of research on the gas explosion flame in the pipeline. Lin Baiquan and others had adopted a special gas explosion experimental chamber (80mm × 80mm, 24m long) to study the propagation of gas explosion flame in the pipeline. In the underground production process, gas has always been an important factor affecting the underground safety production. Especially in the transportation process, if the gas concentration is lower than 30%, the gas concentration may reach the explosion concentration range during the transportation process, therefore an explosion risk exists.

This section takes gas as an example to introduce the combustion and explosion law of pipeline gas.

4. 8. 1 Explosion mechanism of gas in pipeline

In recent years, gas explosion accidents occur frequently in coal mines in China, and there is an upward trend. Almost every year, there are more than 100 deaths in major gas explosion accidents. In order to prevent the combustion and explosion of gas in coal mine, many scholars have done a lot of research about its chemical reaction mechanism. Gas combustion and explosion is a very complex physical and chemical process. It is the result of the interaction of thermal reaction mechanism and chain reaction mechanism, which promote mutually and make the chain reaction of gas continue. The key of chain reaction is to form free radicals with strong activity, which can be regenerated by means of their own reaction heat under certain environmental conditions. The process of chain reaction in gas explosion includes three stages: chain initiation, chain continuation and chain scission.

According to the above reaction mechanism, three basic conditions for gas explosion accident are determined: first, under normal temperature and pressure, the gas concentration should be within the explosion limit range (5% ~ 16%); second, the minimum concentration of oxygen is 12%; third, the presence of an ignition source that is greater than the minimum ignition energy (0. 28mJ) of the pilot gas.

4. 8. 2 Influence factors of explosion propagation of pipeline gas

Gas deflagration in the pipeline is actually a kind of combustion with pressure wave. When there is a constraint or obstacle at the rear boundary of the combustion front, the combustion products can build up a certain pressure, and a pressure difference is formed on both sides of the wave front, which leads to a series of waves propagating forward at the local speed of sound, which is what we call pressure waves. Because this pressure wave propagates much faster than the flame front, it travels in front of the combustion front, so it is also called the precursor shock wave. It can be seen that the gas deflagration is composed of the preceding shock wave and the following flame front. If the boundary constraint is strengthened after deflagration, the flame will accelerate until the flame front catches up with the precursor shock wave front, then the flame front and the shock front combine, forming a shock wave with a chemical reaction zone, which is the detonation wave. It can be seen that the propagation of gas explosion is actually a coupling process of pressure wave and gas combustion. According to the characteristics of shock wave propagation, gas explosion propagation has obvious entrainment effect: in the process of propagation, the shock wave will carry the gas passing through the site together, which makes the combustion area of gas explosion much larger than the original gas distribution area, which has been verified by relevant experiments.

There are many factors affecting the propagation of gas explosion, mainly involving obstacles in the roadway or pipeline, wall roughness, bends and slopes, bifurcations, gas concentration, volume, distribution, ignition energy, ignition location, etc. The following is a brief introduction to several factors:

(1) Obstacles. When the explosion shock wave passes through the obstacle, the pressure near the obstacle changes obviously and rises significantly. Obstacle excitation effect exists in both combustion zone and non-combustion zone, but the degree of excitation depends on the state of gas explosion and its peak pressure.

The obstacles set in the pipe can accelerate the gas flame, and the acceleration mechanism can be understood as the positive feedback of the turbulent flow zone induced by the obstacles to the gas combustion process. In the process of flame propagation in the duct, the turbulent effect is the main factor to produce pressure wave, and the flame propagation speed directly affects the generation and strengthening of explosion shock wave. Due to the formation of a high-concentration viscous boundary layer near the obstacle, the turbulence accelerates the pressure wave and flame, and the accelerated pressure wave and flame enhance the turbulence. This positive feedback effect accelerates the pressure waves and flame repeatedly. In the course of this action, due to the effect of the high concentration viscous boundary layer formed by the flame near the obstacle is greater than that of the high concentration viscous boundary layer formed by the pressure wave near the obstacle, the acceleration effect of the obstacle on the flame is greater than that of the pressure wave. The existence of the obstacle leads to the increase of the fold degree of the flame front, and increases the turbulence intensity of the unburned gas in front of the flame and the flow field inside the flame, thereby enhancing the acceleration of the flame.

(2) Pipe bifurcation and section. If there is a bifurcation in the pipeline, the bifurcation part of the pipeline is a disturbance source, and induces additional turbulence to increase the turbulence intensity of the air flow, so that the flame propagation speed of gas explosion is rapidly improved, the flame propagation speed in the branch pipe of the bifurcated pipeline is increased at the front end, and then rapidly reduced, while the reflection generated by the closure of the straight pipe end of the bifurcated pipeline has little effect on the flame transmission speed in the straight pipe section. The flame propagates at an accelerated rate within the straight sections of the bifurcated pipeline. The sudden change of the cross-sectional area of the pipeline also has an important influence on the propagation of gas explosion. The flame propagation speed is much higher when the cross-sectional area of the pipe is suddenly enlarged than when it is suddenly reduced, and the maximum flame propagation speed is not at the point where the cross-sectional area of the pipe is suddenly reduced, but at the point where $L/d = 70$. Turbulence is most intense when the flame enters the region of sudden expansion of the cross section; when it enters the region of sudden

contraction of the cross section, the maximum turbulence intensity is not actually at the sudden reduction of the cross section, but at a certain cross section.

(3) Pipe wall roughness and thermal effect. The pipe wall roughness has a great influence on the process of gas explosion. Compared with smooth pipe, the physical parameters such as flame propagation speed and peak pressure of gas explosion in rough pipe are greatly improved. The thermal effect of pipeline wall has a great influence on the propagation characteristics of gas explosion.

After the insulation material is pasted on the inner wall of the pipe, the heat dissipation of the wall is greatly reduced (about 1/3 of the original), part of the reduced heat is transferred to the unburned gas through heat conduction and diffusion, and the other part of the reduced heat does work through expansion to increase the pressure wave intensity, both of which increase the flame propagation speed and the pressure wave intensity, and can induce the generation of shock waves. When the pressure wave encounters a solid wall (especially a tunnel or pipe with a closed end), it will produce a reflected wave, which will accelerate the propagation of the flame.

(4) Ignition energy. The minimum ignition energy is one of the basic conditions to characterize the flame propagation and safety of gas explosion combustion, which refers to the minimum energy value from a very small capacitance spark that can ignite the gas mixture and make the flame propagate from the ignition source to the surrounding. If the energy produced by the capacitor spark is less than the minimum ignition energy required for the gas to burn, only a small amount of the gas mixture in the vicinity of the spark is caused to burn, but the energy imparted to the gas by the capacitive spark is insufficient to produce a flame wave needed to support the propagation of the spark.

The ignition energy required for gas explosion is relatively low. Under standard conditions, the minimum ignition energy of gas is 0. 28 mJ, which can be easily achieved. There are many kinds of ignition sources of gas explosion in coal mine, such as blasting, coal spontaneous combustion and other chemical ignition sources; objects colliding or rubbing with each other, impacts high-temperature fire source generated in that approximate adiabatic compression processes of a vacuum pump, an air compressor and the like; electric spark, high voltage arc and static electricity produce electric fire source; the hot surface and thermal radiation of high temperature objects can also cause underground gas explosion.

Combustible gas (or steam) and air (or oxygen) must be uniformly mixed within a certain concentration range to form a premixed gas, which will explode when it meets the ignition source. This concentration range is called the gas explosion limit (or the gas explosion concentration limit). Usually, the explosion limit of combustible gas is affected by many factors, such as temperature, pressure, oxygen content and energy. With the increase of ignition energy, the greater the energy transmitted by the ignition source to the nearby gas mixture laminar flow is, the wider the spontaneous propagation velocity range of combustion is, and the range of gas explosion limit is increased, as shown in Table 4-17.

Table 4-17 Influence of ignition energy on gas explosion limit

Energy Output(J)	Lower explosive limit, V(%)	Upper explosive limit, V(%)	Explosive concentration (%)
1	4.9	13.8	8.9
10	4.6	14.2	9.6
100	4.25	15.1	10.8
1000	3.6	17.5	13.9

It can be seen that the limits of mine gas explosion are not fixed, but also affected by many factors such as temperature, pressure, coal dust, other flammable gases and inertgases. Meanwhile the influence of gas concentration on ignition energy is also significant, and the general law is shown in Fig. 4-50: combustible has a lower explosion limit (assuming that the concentration at this time is C_1). When the concentration is lower than C_1, even if the ignition energy is large, it is impossible to make the combustible material explode; the concentration begins to be higher than C_1, the required energy decreases gradually, and the ignition energy is the lowest near the equivalent concentration C_2, and then the ignition energy begins to increase gradually. In general, the ignition temperature of gas ranges from 650℃ to 750℃. However, in practice, it will change due to the influence of gas concentration, fire source nature and gas pressure. When the gas concentration is 7%~8%, the gas is most easily ignited, and the ignition temperature of the gas decreases with the increase of the pressure of the mixed gas; when the ignition temperature is the same, the larger the fire source area and the longer the ignition time, the easier the gas is to be ignited. Fire source energy and ignition location have a significant impact on the propagation of explosion pressure wave. If the fire source energy is large, the time of pressure peak will be shortened, but the pressure peak will not be changed. The ignition location will not only affect the time of pressure peak, but will also change the magnitude of the pressure peak. For example, under strong ignition conditions such as detonators and explosives, the explosion of combustible gas will be directly converted into detonation.

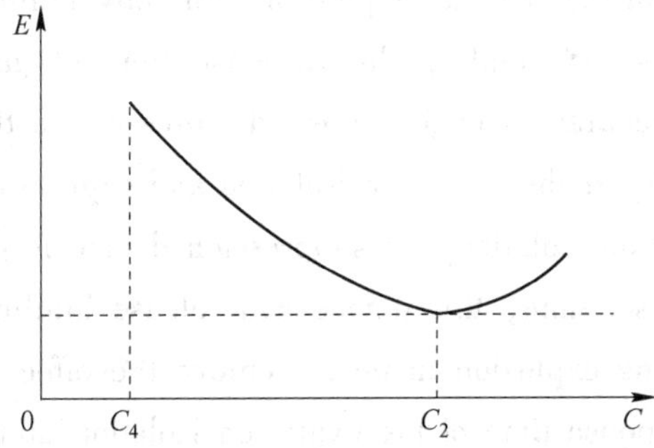

Fig. 4-50 Effect of concentration on ignition energy

It can be seen from the above that in the gas explosion accidents that have occurred, most of the coal mine gas explosions are ignited by low energy fire source. According to statistics, the categories of fire sources causing mine gas explosion accidents in China during the 10 years from 1970 to 1979 are as follows: electric spark is 56.2%, explosion spark is 27.2%, impact fire source is 6.9%, and open fire is 9.2%. It is not difficult to see that the vast majority of mine gas explosions in China are ignited by ignition sources with very small energy, and there are very few gas explosions caused by strong ignition sources.

4.8.3 Explosion suppression technology of pipeline gas

At present, many countries in the world, such as the United States, Russia, Poland, Germany and other major coal-producing countries, have built large-scale underground explosion experimental tunnels of different scales, and have successfully developed a variety of automatic explosion-proof devices and passive explosion-proof sheds with sensitive action and good atomization effect. In recent years, great progress has also been made in the research of gas explosion suppression in China. China mainly uses the passive explosion isolation and suppression device, whose principle is to use the precursor shock wave generated by gas explosion to trigger the explosion isolation and suppression device and make it release the inhibitor, cool, dilute and extinguish the explosion flame, so as to achieve the purpose of explosion prevention and suppression.

These devices of course play their certain role in the control of gas explosion accidents in coal mines. However, in the practical application and explosion suppression test, the device fails to achieve the effect of explosion suppression from time to time. The main reason may be that the propagation law of gas explosion flame and pressure wave has not been fully understood. The pressure value set by the gas explosion isolation and suppression device is not very appropriate. If the action pressure of the gas explosion isolation and suppression device is set too low, the device may cause malfunction when there is no gas explosion accident; on the contrary, the action pressure is set too high. It may be unable to operate due to the small pressure value generated during the gas explosion. Thereby failing to achieve the effect of inhibiting the gas explosion. Secondly, the response time of gas explosion isolation and suppression device is not accurate enough. When the pressure of the gas explosion precursor shock wave is enough to trigger the explosion isolation and suppression device, and the action response time of the explosion isolation and suppression device is just matched with the time for the flame to reach the set time, this device will release inhibitor, cooling, diluting and extinguishing the propagating explosion flame to achieve the effect of explosion isolation and suppression; but if the response time of the explosion isolation and suppression device is not accurate, the inhibitor will be released either too early or too late, then it will fail to achieve the desired purpose of explosion suppression.

Exercises

1. What are the two modes of combustion wave propagation in premixed combustible gas, What are the characteristics of each?
2. What is flame front? What are the characteristics of flame fronts, what is the mechanism of flame propagation in premixed combustible gases?
3. How is laminar premixed flame propagation velocity S_l derived, what are the main factors affecting the laminar premixed flame propagation speed?
4. What is the lower explosive limit? What is the upper limit of explosion? Briefly describe the theory of explosion limit.
5. What are the main factors affecting the explosion limit, What is the relationship between lower explosive limit and heat of combustion if the activation energy of various combustible gases does not change much?
6. What are the methods for calculating the explosion limits of combustible mixture? Calculate the upper and lower explosive limits of ethane while making use of the number of moles of oxygen atoms required for the combustion reaction of 1 mol of combustible gas.
7. Briefly describe the process of detonation, and explain what its essence is.
8. What are the conditions for detonation formation, what are the characteristics of detonation damage to equipment?
9. Briefly describe the basic methods to prevent and control the explosion of combustible gas.
10. Briefly describe the differences between laminar and turbulent flames.
11. Briefly describe the types and characteristics of diffusion combustion flames.

Chapter 5 Combustion and Explosion of Combustible Liquid

扫码获得数字资源

5.1 Combustion characteristics of liquid fuel

At present, the main body of liquid fuel is petroleum products, so the discussion of liquid fuel combustion mainly involves the combustion of fuel oil. The boiling point of liquid fuel is lower than its ignition point, so the combustion of liquid fuel is to evaporate first, produce fuel vapor, and then mix with air, and then burn. Unlike gaseous fuels, liquid fuel is vaporized before it is mixed with air. For heavy liquid fuels, there is also a thermal decomposition process, that is, the fuel is cracked into light hydrocarbons and carbon black due to heat. Light hydrocarbons are burned in a gaseous form, while soot is burned in a solid phase.

According to the characteristics of evaporation and vaporization of liquid fuel, its combustion forms can be divided into four types: liquid surface combustion, wick combustion, evaporation combustion and atomization combustion.

Liquid level combustion is combustion that occurs directly on the surface of the liquid fuel. If there is a heat source or fire source near the liquid fuel container, the liquid surface is heated under the influence of radiation and convection, resulting in accelerated evaporation and increased fuel vapor above the liquid surface. When it forms a certain concentration of combustible mixture with the surrounding air and reaches the ignition temperature, combustion can occur. In the process of liquid surface combustion, if the mixing condition of fuel vapor and air is not good, it will lead to serious thermal decomposition of fuel, in which the heavy components usually do not undergo combustion reaction, thus emitting a large amount of black smoke and causing serious pollution. It is often a form of disaster burning, such as oil tank fire, oil spill fire on the sea surface, etc. This combustion mode is not suitable for engineering combustion.

Wick combustion is the combustion that occurs after the fuel is sucked up from the container by adsorption and vapor is generated on the surface of the wick. This kind of combustion mode has small power and is generally only used in household life or other small-scale burners, such as kerosene stoves, oil lamps, etc.

Evaporation combustion is to make the liquid fuel pass through a certain evaporation pipe, use part of the heat (such as high temperature flue gas) released during combustion to heat the fuel in the pipe, make it evaporate, and then burn like gas fuel. Evaporative combustion is suitable for light liquid fuels with low viscosity and low boiling point, and has certain application in engineering combustion.

Atomization combustion is the use of various forms of atomizers to break up liquid fuel into many small droplets with diameters ranging from several microns to hundreds of microns, which are suspended in the air while evaporating and burning. Because the evaporation surface area of the fuel is increased by thousands of times, the rapid combustion of the liquid fuel is facilitated. Atomization combustion is the main combustion mode of liquid combustion engineering.

For different liquid fuels, different atomization methods should be adopted according to the difficulty of evaporation. Atomization of readily vaporizable liquid fuels, such as gasoline, is often accomplished with a "carburetor". For liquid fuels that are more difficult to vaporize, some kind of nozzle is usually used to achieve atomization.

5.2 Evaporation of liquid

5.2.1 Evaporation process

When the liquid is placed in a closed vacuum container, the molecules with high energy on the surface of the liquid will overcome the attraction of the molecules adjacent to the liquid surface and leave the liquid surface and enter the space above the liquid surface to become vapor molecules. Due to thermal motion, some of the molecules entering the space may hit the surface of the liquid and be attracted by the liquid surface and condense. At that begin, since the space above the liquid surface is free of vapor molecules. The evaporation rate is maximum and the condensation rate is zero. As the evaporation process continues, the concentration of vapor molecules increases, the rate of condensation also increases, and finally the rate of condensation is equal to that of evaporation, and the liquid (liquid phase) and its vapor (gas phase) are in equilibrium. However, this equilibrium is a dynamic equilibrium, that is, the surface molecules are still evaporating and the vapor molecules are still condensing. It's just that the rate of evaporation is equal to the rate of condensation.

5.2.2 Vapor pressure

At a certain temperature, the liquid and its vapor are in equilibrium, and the pressure of the vapor is called saturated vapor pressure, or vapor pressure for short. The vapor pressure of a liquid is an important property of the liquid, which is only related to the nature and temperature of the liquid, and has nothing to do with the amount of liquid and the size of the space above the liquid surface.

At the same temperature, the stronger the attraction between liquid molecules is, the more difficult it is for liquid molecules to overcome the attraction and run into space, and the vapor pressure is lower. Vice versa, the vapor pressure is high. The attraction between molecules is called intermolecular force, also known as van der Waals force. The most important inter molecular force is the dispersion force. The dispersion force is due to the fact that molecules in motion, the attraction between molecules due to the instantaneous relative motion between the electron cloud and the atomic nucleus, resulting in an instantaneous dipole. The larger the molecular weight, the more deformable the molecule and the greater the dispersion force. Therefore, in the same kind of substances, the larger the molecular weight, the more difficult the evaporation and the lower the vapor pressure. However, in water molecules (H_2O), hydrogen fluoride (HF), ammonia (NH_3) molecules, and many organic compounds, due to the existence of hydrogen bonds, the intermolecular force will be greatly enhanced, the evaporation is not easy, and the vapor pressure is low. For the same liquid, when the temperature rises, the number of molecules with high energy in the liquid increases, and the number of molecules that can overcome the surface attraction of the liquid and run into the air increases, so the vapor pressure is high; On the contrary, when the temperature is low, the vapor pressure is low.

The relationship between the vapor pressure of the liquid (P^0) and the temperature (T) obeys the Clausius-Clapeyron equation:

$$\ln P^0 = -\frac{L_V}{RT} + C \tag{5-1}$$

$$\lg P^0 = -\frac{L_V}{2.303RT} + C' \tag{5-2}$$

When $R = 8.314\text{J/(K}\cdot\text{mol)}$, the formula (5-2) becomes:

$$\lg P^0 = \left(-0.052\frac{L_V}{T}\right) + C' \tag{5-3}$$

Where P^0——equilibrium pressure, Pa;

T——temperature, K;

L_V——heat of vaporization, kJ;

C, C'——constant.

Table 5-1 gives the L_V and C' values for several common organic compounds.

Table 5-1 L_V and C' values of common organic compounds

Compound	Molecular formula	L_V evaporation (J/mol)	C'	Temperature range(℃)
N-pentane	n-C_5H_{12}	27567	9.6116	−77~191
Toluene	C_6H_5OH	35866	9.8443	−28~31
N-decane	n-$C_{10}H_{22}$	45612	10.3730	17~173
Methanol	CH_3OH	37531	10.7647	−44~224
Ethanol	C_2H_5OH	40436	10.9523	−31~242
Benzene	n-C_6H_6	34052	9.9586	−37~290

The Clausius-Clapeyron equation is applicable only to pure liquids of a single component. For a dilute solution, the vapor pressure of the solvent P_A is equal to the vapor pressure of the pure solvent P_A^0 multiplied by the mole fraction of solvent in the solution X_A, this is Raoult's law:

$$P_A = P_A^0 \cdot x_A \tag{5-4}$$

A solution in which any component obeys Raoult's law over the entire range of concentrations is called an ideal solution. For non-ideal solutions, Raoult's law should be modified as

$$P_i = P_i^0 \cdot a_i \quad (5\text{-}5)$$

$$a_i = r_i \cdot x_i \quad (5\text{-}6)$$

Where P_i——vapor pressure of component i in the solution;

P_i^0——vapor pressure of pure i component;

a_i——activity of i component;

r_i——activity coefficients of icomponents.

For an ideal solution, $r_i = 1$, $a_i = x_i$.

[Example 5-1] A mixture containing 3% by volume of cyclohexane and 97% by volume of decane can be approximately regarded as an ideal solution. Try to calculate the $P_{Cyclohexane}$ and P_{Decane} of the liquid surface at 28℃ and 60℃. It is known that $P_{Cyclohexane}$ = 660kg/m^3 and P_{Decane} = 730kg/m^3.

[Solution] (1)

$$P_{cyclohexane} = \frac{3 \times 660/86}{\frac{3 \times 660}{86} + \frac{97 \times 730}{142}} = 0.045$$

$$P_{decane} = 1 - P_{cyclohexane} = 0.955$$

(2) Substitute relevant values in Table 5-1 into formula (5-3) to obtain:

$$\lg P^0_{cyclohexane} = -\frac{0.2185 \times 7830.9}{T} + 9.7870$$

$$\lg P^0_{decane} = -\frac{0.2185 \times 10912.0}{T} + 10.373$$

Substituting T=301K (28℃) and T=333K (60℃):

$$(P^0_{cyclohexane})_{301K} = 12660(Pa)$$

$$(P^0_{decane})_{301K} = 283(Pa)$$

$$(P^0_{cyclohexane})_{333K} = 44543(Pa)$$

$$(P^0_{decane})_{333K} = 1633(Pa)$$

(3) Calculate the vapor pressure on the liquid surface according to Raoult's law:

$$(P_{cyclohexane})_{301K} = 12660 \times 0.045 = 570(Pa)$$

$$(P_{decane})_{301K} = 283 \times 0.955 = 270(Pa)$$

$$(P_{cyclohexane})_{333K} = 44543 \times 0.045 = 2004(Pa)$$

$$(P_{decane})_{333K} = 1633 \times 0.955 = 1560(Pa)$$

5.2.3 Heat of evaporation

In the process of liquid evaporation, high-energy molecules leave the liquid surface and enter the space, which makes the internal energy of the remaining liquid lower and lower, and the temperature of the liquid lower and lower. To keep the liquid at its original temperature, heat must be absorbed from the outside world. This means that in order for a liquid to evaporate at constant temperature and pressure, heat must be absorbed from the surrounding environment. It is generally defined that at a certain temperature and pressure. The heat absorbed by the complete evaporation of a unit mass of liquid is the heat of evaporation of the liquid.

The heat of evaporation is mainly to increase the kinetic energy of liquid molecules to overcome the intermolecular attraction and escape from the liquid surface. Therefore, the greater the intermolecular attraction, the higher the heat of evaporation. In addition, the heat of vaporization is also consumed by the work done by the volume expansion during gasification.

5.2.4 Boiling point of liquid

When the vapor pressure of the liquid is equal to the ambient pressure, the evaporation takes place throughout the liquid, which is called liquid boiling; when the vapor pressure is lower than the ambient pressure, evaporation is limited to the liquid surface. The boiling point of a liquid refers to the temperature of the liquid when the saturated evaporation pressure of the liquid is equal to the external pressure. Obviously, the boiling point of a liquid is closely related to the outside air pressure. Table 5-2 shows the boiling points of some liquids.

Table 5-2 Boiling points of common liquids

Name of substance	Molecular formula	Boiling point (℃)
Methane	CH_4	-161
Ethane	C_2H_6	-89
Propane	C_3H_8	-30
Butane	C_4H_{10}	0
Hexane	C_6H_{14}	68
Octane	C_8H_{18}	125
Decane	$C_{10}H_{22}$	160
Hydrogen fluoride	HF	17
Hydrogen chloride	HCl	-84
Hydrogen Bromide	HBr	-70
Hydrogen iodide	HI	-37
Water	H_2O	100
Hydrogen sulfide	H_2S	-61
Ammonia	NH_3	-33
Phosphine	PH_3	-88
Silane	SiH_4	-112

5.3 Flashover and explosion temperature limit

5.3.1 Flashover and flash point

When the liquid temperature is low, due to the slow evaporation rate, the vapor concentration on the liquid surface is less than the lower explosion limit, and the mixed gas of vapor and air cannot ignite when encountering the fire source. As the liquid temperature increases, the vapor molecular concentration increases. When the vapor molecular concentration increases to the lower explosion limit, the mixed gas of vapor and air will flash sparks when it meets the fire source, but it will be extinguished immediately. This kind of instantaneous combustion phenomenon, which occurs when the mixture of steam and air above the combustible liquid meets the fire source, is called flashover. Under the specified experimental conditions, the lowest temperature at which the liquid surface can produce flash ignition is called flash point.

Liquid flashover occurs because its surface temperature is not high, the evaporation rate is less than the combustion rate, and the steam has no time to supplement the burned steam, but can only maintain a moment of combustion.

Generally, the flash point of liquid should be measured by special open cup or closed cup flash point tester. When the open cup flash point tester is used, because the gas phase space cannot produce saturated vapor air mixture like the closed cup flash point tester, the measured flash point is greater than that measured by the latter. The open cup flash point tester is generally applicable to the liquid with flash point higher than 100℃, while the latter is applicable to the liquid with flash point lower than 100℃.

5. 3. 2 Variation law of flash point of similar liquids

In general, flammable liquids are mostly organic compounds. Organic compounds are divided into several classes according to their molecular structure. Similar organic compounds are similar in structure but differ in one or more series in composition. A series of compounds that differ in composition by one or more series and are structurally similar are called homologous. The compounds in the same series are referred to as homologs.

Although the structures of homologues are similar, their molecular weights are different. The molecular structure with large molecular weight has large deformation, large intermolecular force, difficult evaporation, low vapor concentration and high flash point; otherwise, the flash point is low. Therefore, the flash points of homologues have the following rules:

(1) The flash point of homologue increases with the increasing of molecular weight, as shown in Table 5-3;

(2) The flash point of homologue rises with the rise of boiling point, as shown in Table 5-3;

(3) The flash point of homologues increases with the increase of specific gravity, as shown in Table 5-3;

(4) The flash point of homologues increases with the decrease of vapor pressure, as shown in Table 5-3;

(5) In the homologue, the normal structure has a higher flash point than the isomer, as shown in Table 5-4.

In the isomers with the same number of carbon atoms, the number of branched chains increases, resulting in the increase of spatial barriers and the distance between molecules becomes farther, thus reducing the intermolecular force and the flash point.

Table 5-3 Physical properties of some alcohols and aromatics

Matter		Molecular Formula	Molecular weight	Specific gravity 20℃/4℃	Boiling point (Temperature)	Vapor pressure at 20℃ (kPa)	Flash point (℃)
Alcohol class	Methanol	CH_3OH	32	0.792	64.56	11.82	7
	Ethanol	C_2H_6OH	46	0.789	78.4	5.87	9
	N-propyl alcohol	C_3H_7OH	60	0.804	97.2	1.93	22.5
	N-butanol	C_4H_9OH	74	0.810	117.8	0.63	34
	N-pentanol	$C_5H_{11}OH$	88	0.817	137.8	0.37	46
Fang hydrocarbon class	Benzene	C_6H_6	78	0.873	80.36	9.97	−12
	Toluene	$C_6H_5CH_3$	92	0.866	110.36	2.97	5
	Xylene	$C_6H_4(CH_3)_2$	106	0.879	146.0	2.18	23

Table 5-4 Comparison of flash points of normal and isomer

Name of substance	Boiling point(℃)	Flash point (℃)	Name of substance	Boiling point (℃)	Flash point (℃)
N-pentane	36	−40	Hexanone	127.5	35
Isopentane	28	−52	Isohexanone	119	17
N-octane	125.6	16.5	N-propane	91	−11.5
Isooctane	99	−12.5	Isopropane	69	−13
Butyl chloride	79	−11.5	Amyl formate	132	33
Chloroisobutane	70	−24	Isoamyl formate	123.5	25.5

5.3.3 Flash point of mixed liquid

5.3.3.1 Flash point of a mixture of two perfectly miscible flammable liquids

The flash point of such mixed liquids is generally lower than the arithmetic mean of the flash points of the components and is close to the flash point of the component with a large content. For example, the flash point of pure methanol is 7℃, and the flash point of pure amyl acetate is 28℃. When 60% methanol is mixed with 40% amyl acetate, its flash point is not equal to $7 \times 60\% + 28 \times 40\% = 15.4$℃, but equal to 10℃, as shown in Fig. 5-1. The solid line in the figure is the actual flash point change curve of the mixed liquid; the dotted line is the arithmetic average flash point of the mixed liquid. For a 1 : 1 mixture of methanol and butanol (flash point 36℃), the flash point is equal to 13℃ instead of $\frac{1}{2}(7+36)=21.5$℃, see Fig. 5-2. When 1% gasoline is added to kerosene, the flash point of kerosene will be reduced by more than 10℃.

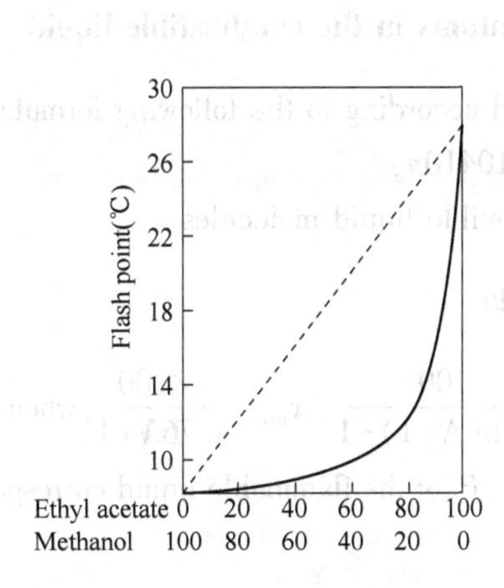

Fig. 5-1 Flash point of methanol and amyl acetate mixture

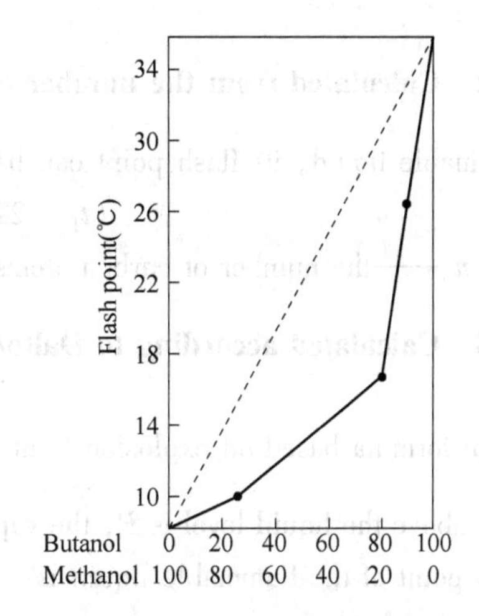

Fig. 5-2 Flash point of mixed solution of methanol and butanol

5.3.2.2 Flash point of mixed liquid of combustible liquid and noncombustible liquid

When the combustible liquid is mixed with the mutually soluble non-combustible liquid, the flash point of the mixed liquid increases with the increase of the content of the non-combustible liquid, and when the content of the non-combustible component reaches a certain value, the mixed liquid no longer flashover. Table 5-5 lists the flash points of alcohol-water solutions.

Table 5-5 Flash point of alcohol-water solution

Alcohol content in solution (%)	Flash point (℃)	
	Methanol	Ethanol
100	7	11
75	18	22
55	22	23
40	30	25
10	60	50
5	None	60
3	None	None

5.3.4 Flash point calculation

5.3.4.1 Calculation of flash point of hydrocarbons based on the formula of wave channel

For hydrocarbon flammable liquid, its flash point shall comply with the formula of wave channel:

$$t_f = 0.6946t_b - 73.7 \tag{5-7}$$

Where t_f——flash point, ℃;

t_b——boiling point, ℃.

5.3.4.2 Calculated from the number of carbon atoms in the combustible liquid

For flammable liquid, its flash point can be calculated according to the following formula:

$$(t_f + 277.3)^2 = 10410 n_c \tag{5-8}$$

Where n_c——the number of carbon atoms in combustible liquid molecules.

5.3.4.3 Calculated according to Dalton's formula

Empirical formula based on explosion limit $x_{down} = \dfrac{100}{4.76(N-1)+1}$, $x_{up} = \dfrac{4 \times 100}{4.76N+4}$, when the total pressure above the liquid level is P, the vapor pressure P_f of the flammable liquid corresponding to the flash point of the flammable liquid is:

$$P_f = \frac{P}{1 + 4.76(N - 1)} \tag{5-9}$$

This is the Dalton formula, where N is the number of moles of oxygen atoms required to burn one mole of flammable liquid.

Table 5-6 gives the vapor pressures of common flammable and combustible liquids. Based on this table and equation (5-9), the flash point of the liquid can be calculated by interpolation.

Table 5-6 Saturated vapor pressure of common flammable and combustible liquids (Pa)

Name of the liquid	Temperature								
	-20	-10	0	+10	+20	+30	+40	+50	+60
Acetone	—	5159.56	8443.28	14708.08	24531.25	37330.16	55901.91	81167.77	115510.18
Benzene	990.58	1950.50	3546.37	5966.16	9972.49	15785.32	24197.94	35823.62	52328.89
Butyl acetate		479.96	933.25	1853.18	3333.05	5826.17	9452.53		
Aviation gasoline			11732.34	15198.17	20531.59	27997.62	37730.13	50262.39	
Motor gasoline			5332.88	6666.1	9332.54	13065.56	18131.79	23997.96	
Methanol	835.93	1795.85	3575.70	6690.10	11821.66	19998.3	32453.91	50889.01	83326.25
Carbon disulfide	6463.45	10799.08	17959.84	27064.37	40236.58	58261.71	82259.67	114216.95	154060.05
Turpentine			275.98	391.97	593.28	915.92	1439.88	2263.81	
Toluene	231.98	455.96	889.26	1693.19	2973.08	4959.58	7905.99	12398.95	18531.74
Ethanol	333.31	746.60	1626.53	3137.06	5866.17	10412.45	17785.15	29304.18	46862.08
Ether	8932.57	14972.0	24583.24	38236.75	57688.43	84632.81	120923.05	168625.66	216408.27
Ethyl acetate	866.59	1719.85	3226.39	5839.50	9705.84	15825.32	24491.25	37636.8	55368.63
Methyl acetate	-2533.12	4686.27	8279.29	13972.15	22638.08	35330.33			
Propanol			435.96	951.92	1933.17	3706.35	6772.76	11798.99	19598.33
Butanol				270.64	627.95	1226.56	2386.46	4412.96	7892.66
Amyl alcohol			79.99	177.32	369.30	738.60	1409.21	2581.11	4545.28
Propyl acetate			933.25	2173.25	3413.04	6432.79	9452.53	16185.29	22918.05

[Example 5-2] It is known that the atmospheric pressure is 1.01325×10^5Pa, find the flash point of benzene.

[Solution] Write the combustion reaction equation of benzene:

$$C_6H_6 + 7.5O_2 \longrightarrow 6CO_2 + 3H_2O$$

From the reaction equation: $N = 15$. Substitute the known data into equation (5-9):

$$P_f = \frac{1.01325 \times 10^5}{1 + 4.76(15 - 1)} = 1498.0(\text{Pa})$$

According to Table 5-6, the vapor pressure of benzene at −20℃ and −10℃ is 990.58Pa and 1950.5Pa respectively. According to the interpolation method, the flash point of benzene is:

$$P_f = -20 + \frac{1498.0 - 990.58}{1950.5 - 990.58} \times 10 = -14.7(℃)$$

5.3.4.4 Calculated according to Brinov formula

The calculation formula is:

$$P_f = \frac{AP}{D_0 \beta} \tag{5-10}$$

Where P_f——saturated vapor pressure of flammable liquid at flash point, Pa;

P——the total pressure of the mixture of flammable liquid vapor and air, usually equal to 1.01325×10^5Pa;

A——instrument constant;

D_0——diffusion coefficient of flammable liquid vapor in air under standard state, see Table 5-7;

β——the number of moles of oxygen molecules required to burn one mole of flammable liquid.

Table 5-7 Diffusion coefficient of common liquid vapor in air (D_0)

Name of the liquid	The diffusion coefficient under standard condition	Name of the liquid	The diffusion coefficient under standard condition
Methanol	0.1325	Ethyl acetate	0.0715
Ethanol	0.102	Butyl acetate	0.058
Propanol	0.085	Carbon disulfide	0.0892
Benzene	0.077	Butanol	0.0703
Toluene	0.0709	Amyl alcohol	0.0589
Ether	0.0778	Acetone	0.086
Acetic acid	0.1064		

[Example 5-3] Given that the flash point of toluene is 5.5℃ and the atmospheric pressure is

1.01325×10^{5}Pa, find the flash point of benzene.

[Solution] First, calculate the instrument constant A according to the flash point of toluene. Since the flash point of toluene is 5.5℃, its saturated vapor pressure range is 889.26 ~ 1693.19Pa from Table 5-6, so the saturated vapor pressure P_f of toluene at the flash point is:

$$P_f = 889 + \frac{1693.19 - 889.26}{10} \times 5.5 = 1333(\text{Pa})$$

From Table 5-7, toluene $D_0 = 0.079$, $\beta = 9$, then

$$A = \frac{P_f D_0 \beta}{P} = \frac{1333 \times 0.0709 \times 9}{1.01325 \times 10^5} = 0.0084$$

According to Table 5-7, the diffusion coefficient of benzene is $D_0 = 0.077$ and $\beta = 7.5$. Then the saturated vapor pressure of benzene at the flash point is obtained by using the formula (5-10):

$$P_f = \frac{AP}{D_0\beta} = \frac{0.0084 \times 1.01325 \times 10^5}{0.077 \times 7.5} = 1473(\text{Pa})$$

According to Table 5-6, when the saturated vapor pressure of benzene is 1473Pa, the corresponding flash point shall be −20~ −10℃, and the flash point of benzene can be calculated by interpolation method as follows:

$$t_f = -20 + \frac{(-10) - (-20)}{1951 - 991} \times (1951 - 1473) = -15(℃)$$

5.3.4.5 Calculated by the lower limit of combustible liquid explosion

The vapor concentration of a liquid at the flash point temperature is the lower explosive limit of the liquid vapor. The relationship between the saturated vapor concentration and the vapor pressure of a liquid is:

$$P_f = \frac{LP}{100}$$

Where L——lower limit of vapor explosion (volume percentage concentration), %;

P——the total pressure of the mixed gas of steam and air, which is generally 1.01325×10^{5}Pa.

[Example 5-4] It is known that the lower explosive limit of ethanol is 3.3%, and the total atmospheric pressure is 1.01325×10^{5}Pa, find the flash point of ethanol.

[Solution] First, P_f at the flash point is

$$P_f = \frac{3.3 \times 1.01325 \times 10^5}{100} = 3344(\text{Pa})$$

According to Table 5-6, when the saturated vapor pressure of ethanol is 3344Pa and the corresponding temperature is between 10℃ and 20℃, the flash point is calculated by interpolation method:

$$t_f = 10 + \frac{20 - 10}{5866 - 3173} \times (3344 - 3173) = 10.6(℃)$$

5.3.4.6 According to Clausius -Clapeyron equation calculation

The vapor concentration corresponding to the flash point is the lower explosive limit. When the lower explosive limit and total pressure of the vapor are known, the vapor pressure P^0 corresponding to the flash point can be calculated, and the flash point t_f can be calculated according to equation(5-2).

【Example 5-5】 The lower explosive limit of decane is known to be 0.75%, and the ambient pressure is 1.01325×10^5 Pa, find the flash point.

The vapor pressure corresponding to the flash point is:

$$P_f = 0.75\% \times 1.01325 \times 10^5 = 760(\text{Pa})$$

According to Table 5-5, decane has $L = 45612$J/mol and $C' = 10.3730$.

Substitute the known value into the formula (5-2), and the flash point is:

$$t_f = \frac{L_V}{2.303 \times R \times (C' - \lg P_f)} = \frac{45612}{2.303 \times 8.314 \times (10.3730 - \lg 760)} = 318(\text{K})$$

Then $t_f = 318 - 273 = 45$(℃).

5.3.5 Explosion temperature limit

5.3.5.1 Explosion temperature limit

When the concentration of flammable liquid vapor in the mixture of saturated vapor and air in the space above the liquid level reaches the explosive concentration limit, the mixture will explode when it meets the fire source. According to the theory of vapor pressure, the saturated vapor pressure (or the corresponding vapor concentration) is related to the temperature for a specific flammable liquid. The liquid temperature corresponding to the upper and lower limits of vapor explosion concentration is called the upper and lower limits of explosion temperature of flammable liquid, which are expressed by t_{up} and t_{down} respectively. Table 5-8 lists the explosion concentration limits and explosion temperature limits for several flammable liquids.

Table 5-8 Comparison of explosive concentration limit and explosive temperature limit of combustible liquid

Explosive concentration limit (%)		Name of the liquid	Explosion temperature limit (℃)	
Lower limit	Upper limit		Lower limit	Upper limit
3.3	18.0	Alcohol	+11	+40
1.5	7.0	Toluene	+5.5	+31
0.8	62.0	Turpentine	+33.5	+53
1.7	7.2	Motor gasoline	−38	−8
1.4	7.5	Kerosene for lamps	+40	+86
1.85	40	Ether	−45	+13
1.5	9.5	Benzene	−14	+19

Obviously, when the liquid temperature is within the explosion temperature limit, the mixture of vapor and air above the liquid level will explode when it meets the fire source. It can be seen that it is more convenient to judge the risk of vapor explosion of flammable liquids by using the explosion temperature limit than the explosion concentration limit.

Assuming that the liquid temperature is equal to the room temperature, there are several relationships between the liquid temperature and the explosion temperature limit in the following examples (assuming that the room temperature is 0 ~ 28℃):

(1) Benzene: lower limit of explosion temperature t_{down} = −14℃, t_{up} = +19℃, and the relationship with room temperature is:

t_{down}(−14℃) t_{up}(+19℃)
0℃ 28℃ t(℃)

It is obvious that benzene vapor is explosive in the range of 0 ~ 19℃.

(2) Alcohol: t_{down} = +11℃, t_{up} = +40℃, the relationship with room temperature is:

t_{down}(+11℃) t_{up}(+40℃)
0℃ 28℃ t(℃)

It is obvious that in the temperature range of room temperature between 11℃ and 28℃, alcohol vapor is just within the explosive concentration limit and can explode.

(3) Kerosene: t_{down} = +40℃, t_{up} = +86℃, the relationship with room temperature is:

t_{down}(+40℃) t_{up}(+86℃)
0℃ 28℃ t(℃)

In the room temperature range, the vapor concentration of kerosene does not reach the lower explosive limit, and the kerosene vapor will not explode.

(4) Gasoline: t_{down} = −38℃, t_{up} = −8℃, the relationship with room temperature is:

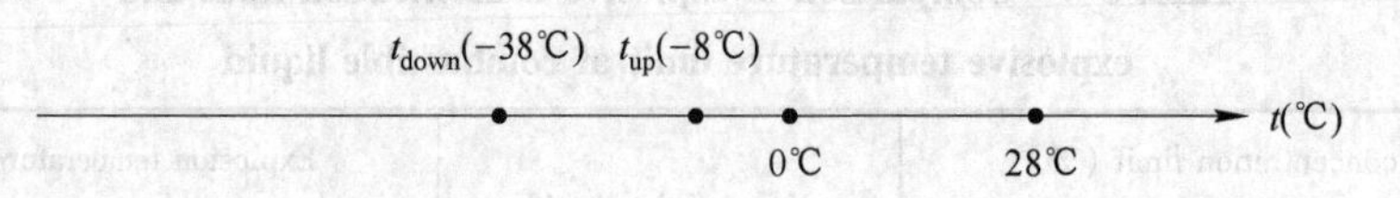

From the coordinates, it can be seen that the saturated vapor concentration of gasoline has exceeded the upper explosion limit in the room temperature range, and the mixed gas of gasoline and air will not explode when it meets the fire source. However, under the actual storage conditions of the warehouse, due to the ventilation of the warehouse, the gasoline vapor is often in an unsaturated state rather than a saturated state. The mixture of steam and air will explode when it meets the fire source.

Through the above analysis, the following conclusions can be drawn:

(1) For the flammable liquid with the lower limit of explosion temperature (t_{down}) less than the

maximum room temperature, the mixture of vapor and air can explode when exposed to fire source.

(2) For the flammable liquid with lower explosion temperature limit (t_{down}) higher than the maximum room temperature, the mixture of vapor and air shall not explode when exposed to fire source.

(3) For the flammable liquid with the upper limit of explosion temperature (t_{up}) less than the minimum room temperature, the mixture of saturated vapor and air will not explode when it meets the fire source, while the mixture of unsaturated vapor and air may explode when it meets the fire source.

5.3.5.2 Calculation of explosion temperature limit

The lower limit of explosion temperature is the flash point of the liquid, and its calculation is the same as the flash point calculation. For the calculation of the upper limit of explosion temperature, the corresponding saturated vapor pressure can be calculated according to the known upper limit of explosive concentration, and then the temperature corresponding to the saturated vapor pressurc is calculated by Clausius-Clapeyron equation and other methods, which is the upper limit of explosion temperature.

[Example 5-6] Toluene is known to have an explosive concentration limit of 1.27% ~ 6.75%, find the explosion temperature limit at 1.01325×10^5Pa atmospheric pressure.

[Solution] (1) Find the saturated vapor pressure corresponding to the explosive concentration limit:

$$P_{\text{lower saturation limit}} = 101325 \times 1.27\% = 1287(\text{Pa})$$

$$P_{\text{upper saturation limit}} = 101325 \times 6.75\% = 6839(\text{Pa})$$

(2) Calculation of explosion temperature limit by Clausius-Clapeyron equation.

When the saturated vapor pressure is 1287Pa and 6839Pa, the temperature range of toluene is 0 ~ 10℃ and 30 ~ 40℃, respectively.

Lower limit:

$$\begin{aligned} t_{\text{lower}} &= \frac{L_V}{2.303 \times R \times (C' - \lg P_f)} \\ &= \frac{35866}{2.303 \times 8.314 \times (9.8443 - \lg 1287)} \\ &= 278.1\text{K} \\ &= 5.0℃ \end{aligned}$$

Upper limit:

$$\begin{aligned} t_{\text{upper}} &= \frac{L_V}{2.303 \times R \times (C' - \lg P_f)} \\ &= \frac{35866}{2.303 \times 8.314 \times (9.8443 - \lg 6839)} \\ &= 311.7\text{K} \\ &= 38.6℃ \end{aligned}$$

5.3.5.3 Influence factors of explosion temperature limit

(1) Properties of flammable liquids. The lower the vapor explosion concentration limit of liquid is, the lower the corresponding explosion temperature limit of liquid is, and the easier the liquid evaporates, the lower the explosion temperature limit.

(2) Pressure. The higher the pressure, the higher the upper and lower limits of the explosion temperature, and vice versa. This is mainly because when the total pressure increases, the vapor pressure needs to be increased accordingly in order to make the vapor concentration reach the explosive concentration limit. Table 5-9 shows the results of the effect of pressure on the flash point of toluene, from which it can be seen that the flash point increases with the increase of pressure, that is, the lower limit of explosion temperature increases.

Table 5-9 Effect of pressure on flash point of toluene

Total pressure (Pa)	Saturated vapor pressure of toluene (Pa)	Closed cup flash point of toluene (℃)
74078	889	0.1
100000	1200	4.9
197368	2368	16.3

When the aircraft takes off, the pressure in the fuel tank changes greatly, so the explosion temperature limit of the fuel also changes greatly. When the fuel temperature is within the explosion temperature limit, the vapor/air mixture above the fuel surface will become flammable, which is very dangerous when encountering lightning and other discharge accidents. Fig. 5-3 (a) and (b) show the variation of fuel flammability regions during flight for aircraft fueled with

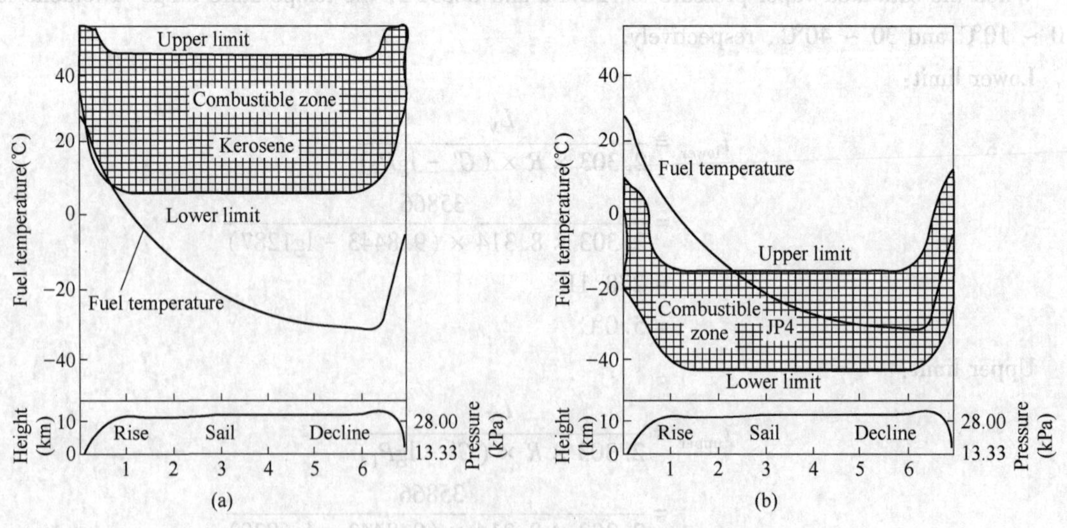

Fig. 5-3 Fuel flammability area diagram during aircraft flight

(a) Aviation kerosene as fuel; (b) JP-4 as fuel

aviation kerosene and JP-4, respectively. In the figure, the fuel explosion temperature limit and the fuel temperature change with the flight process.

(3) Moisture or other substance content. Because water vapor acts as an inert gas in the flammable vapor-air mixture on the liquid surface, adding water to the flammable liquid will increase its explosive temperature limit. If a flammable liquid with a low flash point is added to a flammable liquid with a high flash point, the explosion temperature limit of the mixed liquid is lower than that of the former, but higher than that of the latter. It was found that even a small amount of low flash point liquid could make the flash point of the mixed liquid much lower than that of the high flash point liquid. For example, if 1% of gasoline is added to kerosene, the flash point of kerosene will be reduced by more than 10℃.

(4) Ignition source intensity and ignition time. Generally speaking, when other conditions are the same, the higher the intensity of the ignition source on the liquid surface, or the longer the ignition time, the lower the lower explosive temperature limit (or flash point) of the liquid. This is because at this time, the liquid receives a lot of heat, and the amount of steam evaporated from the liquid surface increases. For example, when the welding arc acts on the liquid surface, due to the high energy of the arc, the liquid will also flash when the initial temperature is lower than the flash point under normal experimental conditions; A large mechanical part stays on the oil surface for a period of time before entering the quenching oil, which may cause flash ignition or ignition of the quenching oil at a low initial temperature.

5.4 Fire spread of liquid combustible materials

5.4.1 Oil pool fire

In the pool fire, the descending speed of the oil level is generally used to represent the burning speed of the pool fire (fuel consumption per unit time and per unit area), and the law shown in Fig. 5-4 is obtained. Why is there such a rule? Of course, this is related to the characteristics of

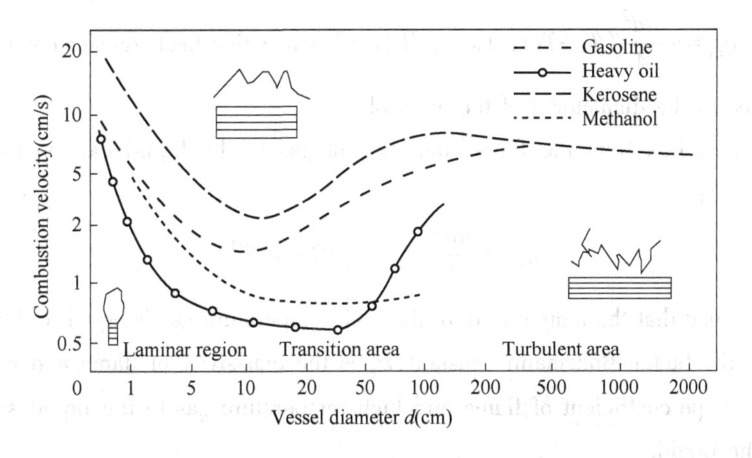

Fig. 5-4 Relationship between descending speed of oil pool fire level and oil pool diameter

the combustion flame, which forms a diffusion flame of fuel vapor, and this must be noted.

When the oil pool diameter is small, a laminar diffusion flame is formed. The flame length becomes shorter as the pool diameter increases. Therefore, the descending speed of the liquid level decreases with the increase of the diameter of the oil pool. When the pool diameter increases to a certain range, which is related to the nature of the liquid fuel, the flame transits from a laminar diffusion flame to a turbulent diffusion flame. In the transition region, the descending speed of the liquid surface varies slowly with the diameter of the oil pool, and sometimes it is even irrelevant. At this time, there is a lot of black smoke in the flame, the flame gradually changes to turbulent diffusion flame, and the height of the flame is difficult to judge. After that, the descending speed of the liquid level increases with the increase of the diameter of the oil pool, and finally tends to a fixed value. The whole process reflects the characteristics of the transition from laminar diffusion flame to turbulent diffusion flame.

It is obvious that the descending speed of the liquid level in the oil pool should be equal to the descending speed of the liquid level caused by the evaporation of liquid caused by the heat transmitted from the flame to the liquid. The heat transferred from the flame to the liquid includes: (1) the heat transfer from the wall of the container to the liquid; (2) the convection heat transfer from the high-temperature gas above the liquid surface to the liquid; (3) the radiation heat transfer of the flame and the high-temperature gas from the liquid.

Since the wall of the vessel is very close to the base of the flame, the wall temperature may be taken as the temperature of the liquid (T). Thus, the temperature difference in the gas near the wall can be taken as $T_F - T_1$, where T_F is the flame temperature. The heat flux from the wall to the liquid can be expressed by the following equation:

$$q_{cd} = k\pi d(T_F - T_1) \tag{5-11}$$

Where d is the pool diameter and k is the heat transfer coefficient.

The heat flux from the high temperature gas above the liquid level to the liquid can be represented by $q_{cv} = h\frac{\pi d^2}{4}(T_F - T_1)$. Here, H is the convective heat transfer coefficient, which is generally related to the diameter d of the oil pool.

The radiant heat flux from the flame and the hot gas to the liquid can be expressed by the following equation:

$$q_{ra} = \frac{\pi d^2}{4}\sigma(\varepsilon_F \varphi_F T_F^4 - \varepsilon_1 T_1^4) \tag{5-12}$$

Where it is assumed that the temperature of the high-temperature gas is equal to the temperature of the flame. σ is the Stefan-Boltzmann constant, ε_F is the emissivity of flame and high temperature gas, φ_F is the shape coefficient of flame and high temperature gas to the liquid surface, ε_1 is the emissivity of the liquid.

Obviously, the sum of these heat flows should be equal to the sum of the heat required for the evaporation of the liquid and the heat required for the heating of the liquid itself, that is,

$$q_{cd} + q_{cv} + q_{ra} = \frac{\pi d^2}{4} v_l \rho_l L_V + c_{pl}\left(M_l - \frac{\pi d^2}{4} v_l \rho_l\right)(T_l - T_\infty) \quad (5\text{-}13)$$

Where, ρ_l is the density of the liquid; L_V is the latent heat of vaporization of the liquid; v_l is the rate of descent of the liquid level; c_{pl} is the specific heat of the liquid; M_l is the total mass of the liquid in the oil pool; T_∞ is the initial temperature of the liquid. The descending speed of the liquid level can be expressed as:

$$v_l = \frac{q_{cd} + q_{cv} + q_{ra} - c_{pl} M_l (T_l - T_\infty)}{\frac{\pi d^2}{4} \rho_l [L_V - c_{pl}(T_l - T_\infty)]} \quad (5\text{-}14)$$

Substitute the above three formulas into formula (5-14) to obtain:

$$v_l = \frac{1}{\rho_l [L_V - c_{pl}(T_l - T_\infty)]} \Big[\frac{4k}{d}(T_F - T_l) + h(T_F - T_l) + \sigma(\varepsilon_F \varphi_F T_F^4 - \varepsilon_l T_l^4) - c_{pl} M_l (T_l - T_\infty) \Big] \quad (5\text{-}15)$$

When d is very small, the first term at the right end of equation (5-15) is relatively large, so there is an inverse relationship between v_l and d. When d is very large, the first term at the right end of equation (5-15) is relatively small, so v_l is approximately independent of d. These justify the results of Fig. 5-4.

In addition, it can be seen that in order to prevent the spread of such fires, it is necessary to control the heat exchange process between the outside and the liquid. Therefore, the use of foam extinguishing agent to generate a layer of foam on the liquid surface can reduce the heat flow and prevent the evaporation of liquid, which is a better method to prevent the spread of fire and extinguish fire.

If there is water in the oil pool, the water generally sinks to the bottom of the oil pool, but the boiling point of water (100℃) is much lower than that of oil. According to the previous introduction, when the flame transfers heat to the fuel, the fuel and the pool wall will also transfer heat to the water, so the water temperature deposited at the bottom of the oil pool will continue to rise. When the water temperature rises to the boiling point of water, the water will boil, and there is a layer of oil on the top of the water surface. The uppermost part of the oil layer is in a state of evaporation and combustion. Therefore, the boiling water vapor will boil with the evaporated and burned oil, which may lead to extremely dangerous boil over phenomenon. The boiling phenomenon causes the fire to expand rapidly. The splashing height and scattering area of fuel droplets carried by water vapor have an important influence on the spread of fire. The results show that the splashing height and scattering area diameter are related to the thickness of oil layer and the diameter of oil pool. Generally, the ratio of scattering area diameter (D) to oil pool diameter (d) is more than 10, that is, $\frac{D}{d}>10$. Because the sprayed fuel must pass through the burning pool fire, the pool fire ignites the sprayed fuel, and with the improvement of atomization conditions and oxygen supply conditions, the sprayed fuel burns more fiercely than the oil in the pool. Cause the fire to expand rapidly, and if there are other combustibles around the oil pool, they will be ignited quickly. If there are personnel and equipment engaged in fire fighting around the oil pool, it will cause great casualties and losses. Therefore, for the oil pool fire, we must avoid the occurrence of boiling phenomenon, study the characteristics before boiling, do a good job of forecasting, and prevent the spread and expansion of the fire.

5.4.2 Liquid surface fire

Oil tanker accidents at sea often lead to surface fires. Therefore, it is of great significance to study the spread law of liquid surface fire for extinguishing this kind of fire. The results show that the properties of flammable liquid and the surrounding conditions have a great influence on the spread speed of liquid surface fire.

In a static environment, the initial temperature of the liquid has a significant effect on the speed of fire spread. Fig. 5-5 shows the relationship between the spread rate of methanol liquid surface fire and the initial temperature of methanol. At the beginning, the spread rate of methanol liquid surface fire increases with the increase of the initial temperature of methanol. When the temperature exceeds a certain value, the rate of spread of a liquid surface fire tends to some constant. This is because the flash point of methanol is 11℃. When the temperature reaches 20℃, methanol steam with a certain concentration is formed above the liquid level of the methanol, the steam is mixed with air to form premixed combustible gas with a certain mixing ratio, and the propagation speed of the premixed combustible gas is certain. It shows that the spread rate of methanol liquid surface fire tends to a constant. This constant is the laminar flame propagation speed of the mixture of the maximum methanol concentration and the air. The flame propagation speed is different, the flame shape is also different, and the flame structure of the methanol liquid surface fire photographed by the schlieren method is even more different. Fig. 5-6 shows different initial temperatures of methanol. Schlieren photograph at a given time interval (from fire). It can be seen from Fig. 5-6 that the faster the flame propagates, the greater the inclination of the flame surface.

The above results show that the flame propagation speed is related to the temperature, which must be related to the propagation process. When the initial temperature of methanol is lower than the flash point (11℃), a diffusion flame is formed. To maintain the spread of the liquid surface fire, the methanol in front of the flame must be heated to ensure a certain vapor velocity. In this way, heat must be transferred to the liquid methanol in front of the flame, so that there is a temperature difference between the liquid methanol in front of the flame and the liquid methanol directly below the flame, which causes a surface tension difference. The surface flow of liquid phase methanol is generated under the action of surface tension difference, so that the liquid phase methanol with high temperature flows to the front of the flame, as shown in Fig. 5-7 (a). The periodic variation of the flame (see Fig. 5-7 (b)) is caused by the variation of the surface flow due to the difference in surface tension.

The spreading characteristics of methanol liquid surface fire are also applicable to other flammable liquids and are universal. Fig. 5-8 shows the spread of a liquid surface fire in the

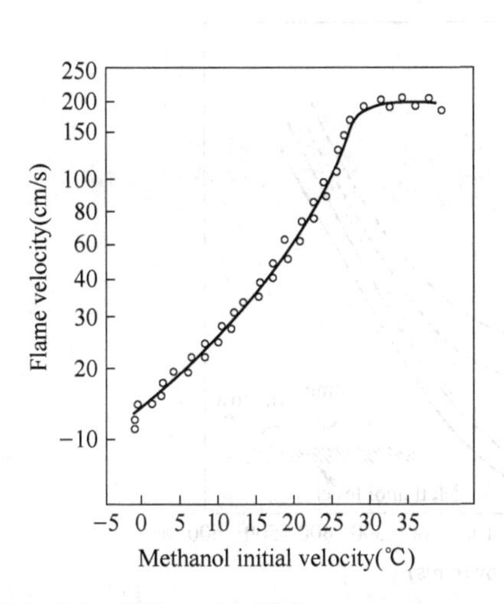

Fig. 5-5 Effect of initial temperature on liquid level fire spreading speed

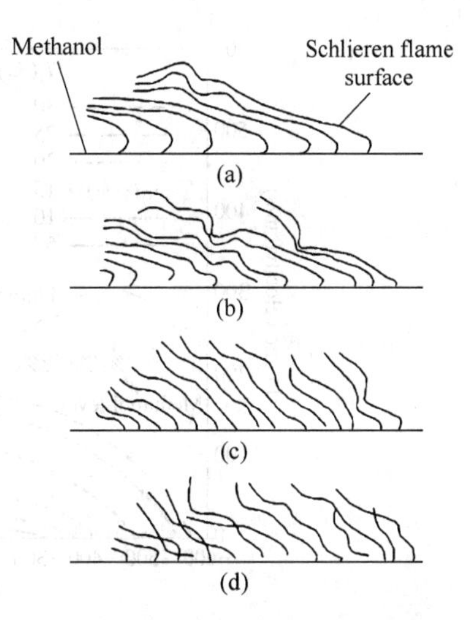

Fig. 5-6 Schlieren photos of methanol liquid surface fire at different initial temperatures

(a) $T = 26.0℃$, $\Delta t = 21ms$; (b) $T = 17.0℃$, $\Delta t = 21ms$; (c) $T = 6.0℃$, $\Delta t = 21ms$; (d) $T = 2.3℃$, $\Delta t = 104ms$

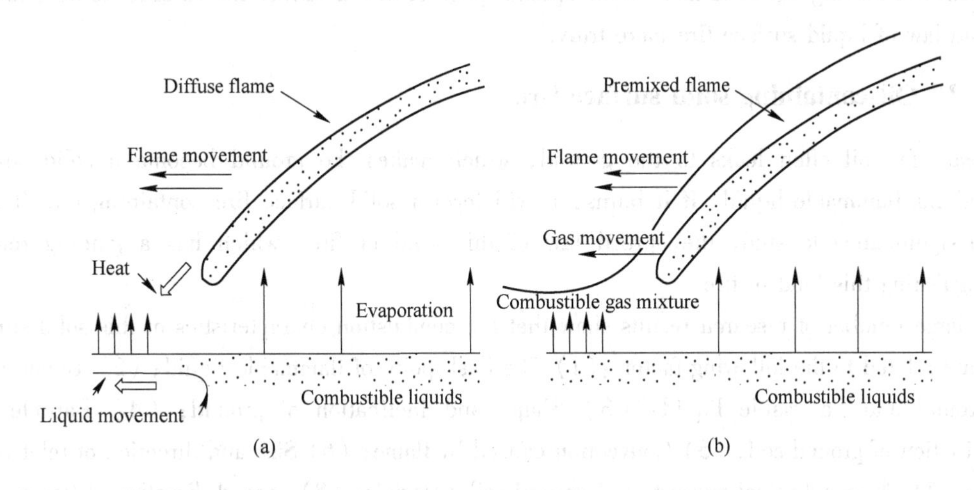

Fig. 5-7 Effect of liquid temperature on heat transfer process

(a) Flammable liquid temperature below flash point; (b) Flammable liquid temperature above flash point

presence of relative velocity. Under the condition of upwind, the initial temperature of methanol has a significant effect. Under downwind conditions, the initial temperature has little effect and is mainly affected by the wind speed. This result is of course related to the evaporation rate of methanol. To study the evaporation problem, we must study the heat transfer problem, because the heat transfer process at the liquid surface plays a very important role in evaporation. In addition, it can be seen that if the reverse wind speed is several times larger than the spread speed of the liquid surface fire, the liquid surface fire can be extinguished. When putting out the liquid level fire, do not blow along the direction of the flame, otherwise the fire will burn more and more vigorously.

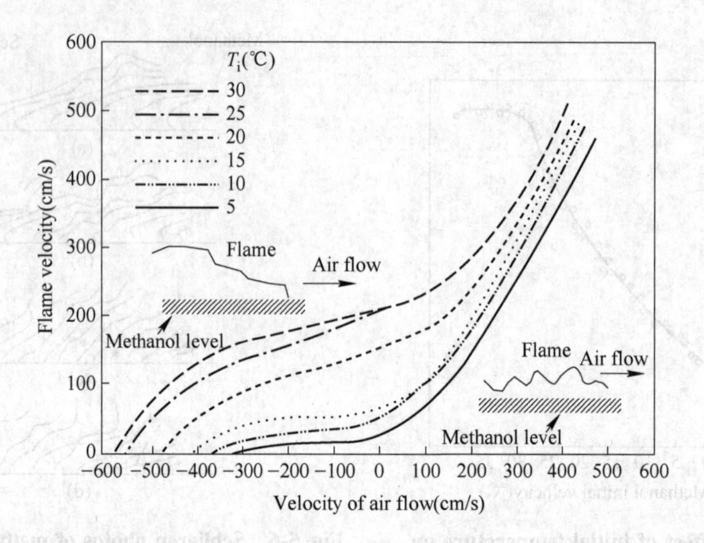

Fig. 5-8 Effect of relative wind speed on liquid surface fire spread rate

In actual fire, the liquid surface is not static, so it is of great practical value to study the influence of moving liquid surface on the spread speed of liquid surface fire in order to describe the spread law of liquid surface fire more truly.

5.4.3 Oil-containing solid surface fire

In real life, oil often leaks to the ground, which makes the ground become a solid surface containing flammable liquid. If it burns, it will form a solid surface fire containing oil. It is of great significance to study the spread law of this kind of fire, which has a guiding role in extinguishing this kind of fire.

A large number of research results show that the combustion characteristics of this solid surface fire are related to the following factors: (1) The flash point of flammable liquid; (2) Temperature of ground and flammable liquid: (3) Shape and inclination of ground; (4) Particle size distribution of ground soil; (5) Convection caused by flame; (6) Size and direction of relative air flow; (7) Thermophysical properties of ground soil materials; (8) Spread direction of flame, etc.

In order to deeply study the influence of the above factors on the combustion of solid surface fire, the experimental device shown in Fig. 5-9 is designed. The fuel container is a 60 × 12 × 1 (cm^3) (length × wide × high) rectangular container, the whole container is placed in a constant temperature bath to maintain a certain temperature (adjustable). The fuel container is placed with the incubation bath into a 60 × 45 (cm^3) wind tunnel, the effect of wind speed on combustion velocity was studied. Add sand with different particle sizes into the fuel container. For example, when the average particle size is 220, the average density of sand is 2.68g/cm^3, and the gap between sands is 0.32cm^3/g (about 46vol%), and then fill the whole fuel container with kerosene with a flash point of 50℃.

For the convenience of future experiments, it is necessary to calibrate the flow field above the fuel container under cold conditions. Fig. 5-10 shows the distribution of the average velocity and turbulence intensity of the air flow at different parts above the fuel container when the mainstream velocity is 300cm/s. The coordinate selection is shown in Fig. 5-10, where U is the mainstream velocity, u is the average velocity in the x direction, u' is the velocity variation in the x direction.

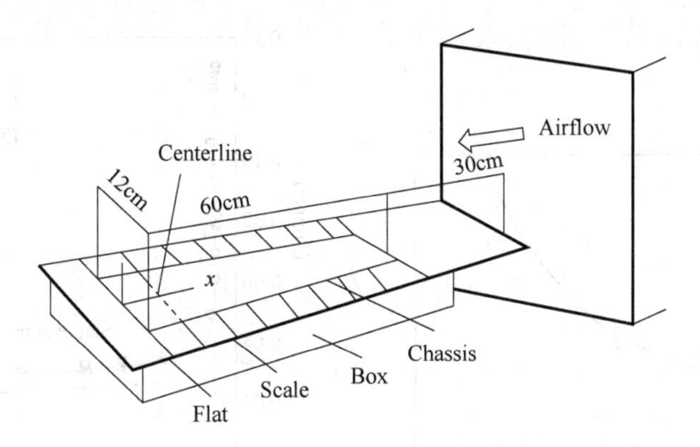

Fig. 5-9 Experimental device of oil-containing solid surface fire

The kerosene was ignited with alcohol cotton yarn from one end, the whole combustion process was recorded with a camera, and the temperature distribution of the sand layer at the center x = 30cm of the combustion vessel was measured with a thermocouple installed in the sand layer in advance.

It can be seen from Fig. 5-11 that when the particle size is very small, the spread speed of sand surface fire is close to a constant, and the spread speed of sand surface fire decreases with the increase of particle size.

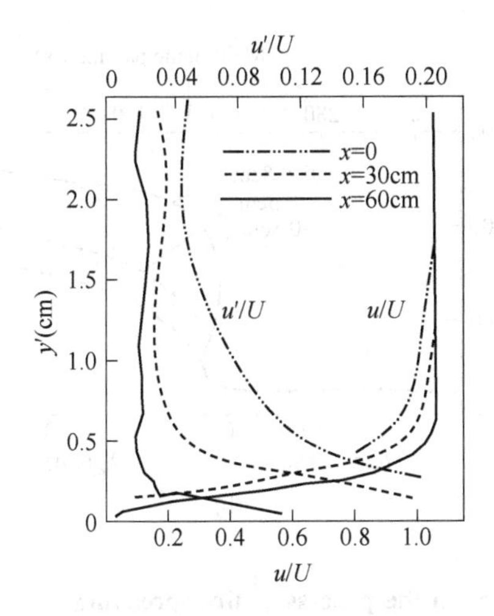

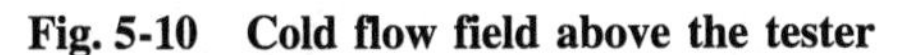

Fig. 5-10 Cold flow field above the tester

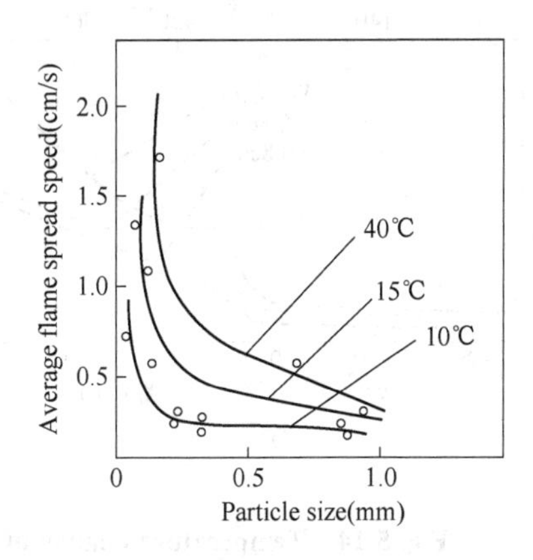

Fig. 5-11 Effect of particle size on fire spread rate of sand surface

When there is no relative wind speed, the initial temperature also has a significant impact on the spread speed of sand surface fire, as shown in Fig. 5-12. The higher the initial temperature is, the faster the sand surface fire spreads. When the relative wind speed increases, the spread speed of sand surface fire decreases, as shown in Fig. 5-13. When the relative wind speed reaches a certain value, there is a sharp decline in the spread speed (fire extinguishing).

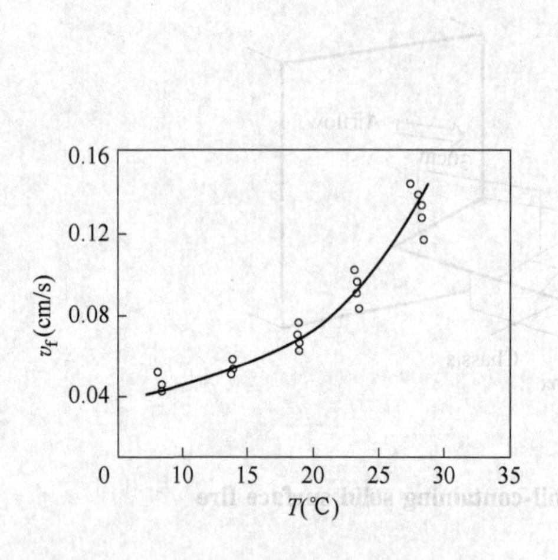

Fig. 5-12 Effect of initial temperature on fire spread rate of sand surface

Fig. 5-13 Influence of relative wind speed on fire intensity

At the same time of the experiment, the temperature change of the sand layer was also measured. Fig. 5-14 shows the measured results. The results show that the temperature of the sand layer in front of the flame is higher, and the higher the relative wind speed is, the higher the temperature is. The temperature of the sand layer behind the flame is basically unchanged.

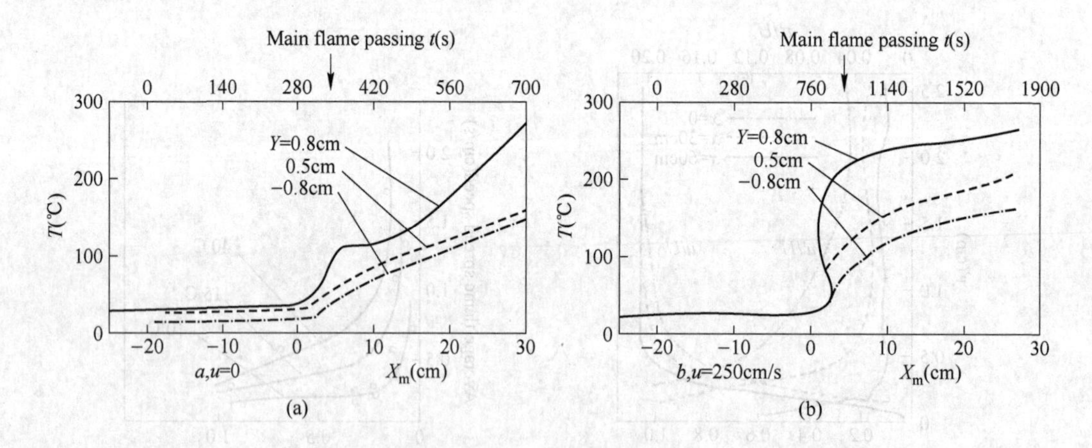

Fig. 5-14 Temperature change of sand layer in the process of fire spreading

5.4.4 Fire spread in liquid mist

Fire spread in liquid mist often occurs in drilling blowout fires and fires after the rupture of liquid fuel containers. In this case, due to the poor spray conditions, the atomization quality is not high, the droplets are large, and the specific gravity of the large droplets is also high. The liquid mist flame is mostly liquid group diffusion flame. In order to understand the characteristics of this kind of flame, some explanation of the spray flame is required first.

The liquid spray flame is generally divided into four types: (1) Pre-evaporation gas combustion. For example, when the ambient temperature is higher and the atomization is finer, the combustion at a greater distance from the nozzle outlet approaches this type. Obviously, it has the characteristics of premixed combustible gas combustion. (2) Droplet group diffusion combustion. For example, when the ambient temperature is low and the liquid mist is coarse, the combustion near the nozzle outlet is close to this type. Therefore, the evaporation of droplets plays an important role in the whole process. (3) The combination of pre-evaporation and droplet group diffusion combustion. When the small droplets have been evaporated before entering the combustion zone, the premixed combustible gas with a certain concentration is formed, while the large droplets have not been evaporated, and the droplet group diffusion combustion is carried out. (4) The combination of pre-evaporation combustion and droplet group diffusion evaporation. The small droplets have evaporated completely before entering the combustion zone, forming a premixed combustible gas with a certain concentration, while the larger droplets entering the combustion zone have not evaporated completely. Because the droplet diameter is too small to ignite, it can only continue to evaporate, thus forming a composite type of pre-evaporation combustion and droplet group evaporation. Obviously, in the fire, due to the limitation of conditions, the diffusion combustion of droplet group is the main form. Other types will also show more or less.

In general, the droplets are large and fall continuously during the combustion process. The droplets may fall on the ground, forming a solid surface containing flammable liquid and causing a fire spreading on the flammable solid surface. In the case of an offshore drilling rig, a combustible liquid surface may form on the water surface, and a fire may spread along the combustible liquid surface. If the flammable liquid collects somewhere. Oil pool fire may also occur, so we must pay attention to the comprehensive effect at the same time.

In order to explain the basic characteristics of droplet diffusion combustion, the model of droplet diffusion combustion can be simplified as Fig. 5-15. i. e. An initial drop diameter is uniform. One-dimensional spray flame without relative motion between droplet and gas flow. If the ambient temperature of the initial gas flow is not too high, but is higher than the droplet temperature, the preheating effect on the mist should be considered. Here, the high temperature gas side also has a preheating effect on the liquid mist, thus forming a preheating evaporation zone for the droplet group. Obviously, the evaporation characteristics of the liquid itself and the ambient temperature have a great impact on this area. If the temperature rises above a certain temperature, the mixture of evaporated steam and air may ignite and form a premixed flame. Then the droplet catches fire again, a diffusion flame is for. It can be seen that there are different multiphase combustion mechanisms with different conditions.

If the premixed flame cannot be formed, only the droplet group diffuses and burns, which is

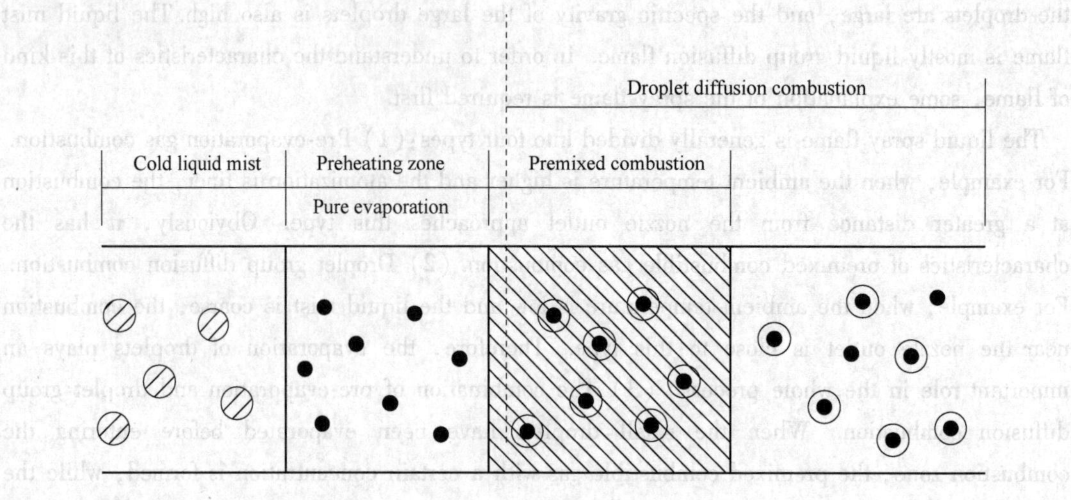

Fig. 5-15 Simplified droplet group diffusion combustion model

mainly discussed here. At this time, although there is no premixed combustion, evaporation has an impact on the gas flow, but it can still be assumed that there is no relative motion between the droplet and the gas flow, and the droplet diameter is uniform.

The overall continuity equation for such a one-dimensional two-phase flame is:

$$\rho_{\varepsilon} u = m = \text{constant} \tag{5-16}$$

Or

$$(\rho_{g} + \rho_{l})u = m = \text{constant}$$

Where ρ_g is the density of the gas phase; ρ_ε is the total density of the gas and liquid phases, u is the average velocity of the gas and liquid phases, and m is the mass flux of the two phases.

The gas-phase continuity equation:

$$\frac{\mathrm{d}(\rho_{g} u)}{\mathrm{d}x} = \bar{\rho}_{l} \frac{\pi d}{4} k_{f} N \tag{5-17}$$

Set

$$z = \frac{\rho_{g}}{\rho_{\varepsilon}}$$

Then

$$\frac{\mathrm{d}z}{\mathrm{d}x} = \frac{\bar{\rho}_{l} \frac{\pi d}{4} k_{f} N}{m} \tag{5-18}$$

Where, $\bar{\rho}_l$ is the average density of the liquid phase, d is the droplet diameter, k_f is the evaporation constant for diffusion combustion, and N is the number of droplets per unit volume.

The energy equation for a two-phase one-dimensional flame is:

$$\frac{\mathrm{d}}{\mathrm{d}x}(\rho_{g} u h_{g} + \rho_{l} u h_{l}) = \frac{\mathrm{d}}{\mathrm{d}x}\left(\lambda \frac{\mathrm{d}T}{\mathrm{d}x} - \sum h_{i} \rho_{g} Y_{i} v_{i}\right) \tag{5-19}$$

Both ends are the same divided by $\rho_\varepsilon u = m$, and integrate to get:

$$zh_g + (1 - z)h_l + \frac{1}{m}\left(-\lambda\frac{dT}{dx} - \sum h_i\rho_g Y_i v_i\right) = \text{constant} \tag{5-20}$$

Where the subscript i denotes the i-th component; h_i is the enthalpy of the i-th component; Y_i is the mass percentage of the i-th component; v_i is the velocity of the i-th component. So there is:

$$h_g = \sum Y_i h_i$$

$$h_i = h_{i,0} + c_p(T - T_{g,0})$$

$$h_l = h_{l,0} + c_{p,l}(T_l + T_{l,0})$$

$$T_{g,0} = T_{l,0} = \text{Standard temperature}$$

This pattern (5-20) can also be written as:

$$\frac{\lambda}{m}\left(\frac{dT}{dx}\right) = z(c_pT - Q') - z_0(c_pT_{g,0} - Q_0') \tag{5-21}$$

Where $$Q' = Q_F + q_v$$

Where Q_F is the heat of combustion; q_v is the latent heat of evaporation; subscript 0 indicates the standard state.

If the above formula is made dimensionless, it can be finally solved:

$$m \propto \frac{1}{d_0}\frac{1}{\sqrt{\rho_l}}\frac{\lambda}{c_p}\sqrt{p\overline{M}_0} \tag{5-22}$$

Where $\overline{M}_0$ is the average molecular weight of the gas mixture in the standard state.

The above results show that the mass spread velocity of droplet group diffusion combustion flame increases with the decrease of droplet size, and other physical parameters and ambient pressure also have great influence on the mass spread velocity. If we want to consider the spatial distribution of droplet diameter, we can imagine that the results will be more reasonable and realistic, but it will also be more complex. However, from the perspective of estimation, it is feasible to adopt the simplest model. More accurate models will not be introduced here. Readers are invited to refer to the relevant literature. Here we focus on the ideas and methods of dealing with this problem.

5.5 Oil tank fire burning

5.5.1 Stable combustion of liquid

Once the flammable liquid ignites and completes the propagation process on the liquid surface, it enters a stable combustion state. Stable combustion of liquids is generally in the form of "pool" combustion in horizontal planes, and some are in the form of "flow" combustion. This section mainly studies the combustion velocity of pool combustion and its influencing factors, and the flame characteristics of liquid stable combustion.

5.5.1.1 Burning rate of liquid

A Liquid burning rate expression method

There are two ways to express the burning velocity of a liquid, namely the gravimetric velocity and the linear velocity.

(1) Combustion linear velocity (v): thickness of liquid layer burned in unit time. It can be expressed as:

$$V = \frac{H}{t}(\text{mm/h})$$

Where H——thickness of liquid burned, mm;

t——time required for liquid combustion, h.

(2) Weight burning rate: the weight (kg) of the liquid burned per unit area (m^2) per unit time (h) can be expressed as:

$$G = \frac{g}{s \cdot t}[\text{kg/(m}^2 \cdot \text{h)}]$$

B Measurement of liquid combustion speed

Fig. 5-16 are schematic view of a liquid combustion rate measuring apparatus. During the measurement, both the container and the burette are filled with flammable liquid, and the liquid gradually decreases due to combustion, but the excess liquid from the gradually rising burette can be used to supplement the burned liquid, so that the liquid level is always kept on the 0-0 line. The burn time and the volume of burette rise were recorded. The combustion speed of the combustible liquid can be calculated.

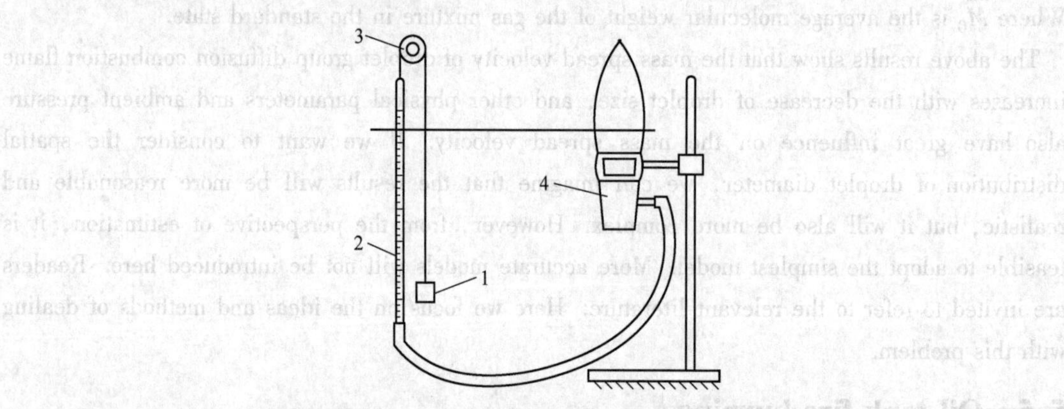

Fig. 5-16 Device for measuring liquid burning

1—Heavy hammer; 2—Buret; 3—Pulley; 4—The diameter is 62mm quartz vessel

C Factors affecting the burning rate of liquid

a Influence of initial temperature of liquid

The mass rate of liquid combustion, G, can be expressed as:

$$G = \frac{Q''}{L_V + \bar{c}_p(t_2 - t_1)} \tag{5-23}$$

Where Q''——heat received by the liquid surface, kJ/($m^2 \cdot$ h);

G——the mass velocity of surface combustion, kg/($m^2 \cdot$ h);

L_V——heat of vaporization of the liquid, kJ/kg;

$\bar{c}_p$——T the average heat capacity of the liquid, kJ/(kg · ℃);

t_2——liquid level temperature during combustion, ℃;

t_1——initial temperature of the liquid, ℃.

It can be seen from the formula (5-23) that the initial temperature t_1 rise and burn faster. This is because when the initial temperature is high, less heat is needed to preheat the liquid to the point where more heat is available for evaporation of the liquid.

b Influence of vessel diameter

The liquid is usually contained in a cylindrical vertical vessel, and its diameter has a great influence on the combustion rate of the liquid (as shown in Fig. 5-17). It can be seen from the figure that the flame has three combustion states: when the diameter of the liquid pool is less than 0.03m, the flame is in a laminar flow state, and the combustion speed decreases with the increase of the diameter; when the diameter is greater than 1m, the flame is in a fully developed turbulent state, and the combustion velocity is constant. Not affected by changes in diameter; diameter between from 0.03m to 1.0m, with the increase of diameter, the combustion state gradually transits from laminar state to turbulent state, and the combustion velocity reaches the minimum value at 0.1m, and then the combustion velocity gradually rises to the constant value of turbulent state with the increase of diameter.

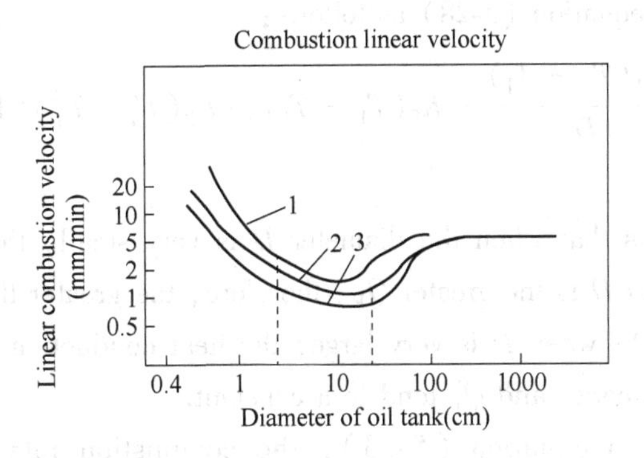

Fig. 5-17 Variation of liquid combustion velocity with tank diameter

1—Gasoline; 2—Kerosene; 3—Light oil

The relationship between the burning velocity of the liquid surface and the diameter can be explained by the change of the relative importance of each of the three mechanisms of heat transfer from the flame to the liquid surface at different stages. If there is no external heat source, equation (5-23) Q_F is the heat transferred from the flame to the liquid surface Q_F. The entire liquid surface receives the heat flux of the flame Q_F can be expressed as the sum of the three heat fluxes of conduction, convection and radiation, namely

$$Q_F = q_{cond} + q_{conv} + q_{rad} \tag{5-24}$$

Heat conduction term q_{cond} represents the amount of heat transferred through the wall of the vessel and can be expressed as:

$$q_{cond} = K_1 \pi D(T_F - T_1) \tag{5-25}$$

In the formula T_F and T_l indicating the temperature of the flame and the liquid surface respectively; K_1 is a heat transfer coefficient considering the heat transfer from the flame to the wall, the heat transfer in the wall and the heat transfer from the wall to the liquid. D is that diameter of the vessel.

The convective heat transfer term can be expressed as:

$$q_{conv} = K_2 \frac{\pi D^2}{4}(T_F - T_l) \tag{5-26}$$

In the formula K_2 is the convective heat transfer coefficient.

The reradiation of the liquid surface is included in the radiative heat transfer term, so it can be expressed as:

$$q_{rad} = K_3 \frac{\pi D^2}{4}(T_F^4 - T_l^4)[1 - \exp(-K_4 D)] \tag{5-27}$$

In the formula K_3 is a constant, including Stefan-Boltzmann constant σ and the angle coefficient of flame radiation to the liquid surface; and $1-\exp(-K_4 D)$ is the emissivity of the flame, where K_4 is a constant that takes into account the average ray path radiated by the flame to the liquid surface and the concentration and emissivity of the radiation particles in the flame.

Add equation (5-25) ~ equation (5-27) and divide by the liquid surface area to obtain the specific expression of in equation (5-28) as follows:

$$Q''_F = \frac{4\sum q}{\pi D^2} = \frac{4K_1(T_F - T_l)}{D} + K_2(T_F - T_l) + K_3(T_F^4 - T_l^4)[1 - \exp(-K_4 D)] \tag{5-28}$$

Equation (5-28) shows that when the diameter D is very small, the heat conduction term is dominant, and the smaller D is the greater Q''_F, therefore, the greater the combustion velocity, as shown in equation (5-23); when D is very large, the heat conduction term approaches 0, while the radiation term is dominant, and Q''_F tend to a constant.

Therefore, according to equation (5-23), the combustion rate is constant; during the transition phase, heat conduction, convection and radiation work together, and because the transition of combustion from laminar flow to turbulent flow enhances the heat transfer from the flame to the liquid surface, the combustion velocity decreases rapidly to a minimum value with the increase of diameter, and then increases with the increase of diameter until it reaches a maximum value.

Burgess et al. 's experimental results on the combustion of hydrocarbon liquids in a large circle pool (D>1m) also show that radiation heat transfer is dominant, and the limiting linear burning velocity of liquids under radiation dominant effect is obtained v_∞, as shown in Table 5-10. And the linear combustion speed v_t is expressed by the following equation:

$$v_t = v_\infty(1 - e^{-KD})$$

Where K is a constant and D is the diameter of the pool.

It can be seen from Table 5-10 that although there are cryogenic liquids (liquefied petroleum gas), their limit burning rates are relatively close. However, the ultimate combustion rate of methanol is very small. This is due to the large latent heat of vaporization and the low emissivity of the flame.

Table 5-10 Limiting velocities for stable combustion of some pool liquids

Name of the liquid	v_∞ (mm/min)
Liquefied petroleum gas	6.6
N-butane	7.9
Hexane	7.3
Xylene	5.8
Methanol	1.7

c Influence of liquid height in thc container

The height of the liquid in the container refers to the height of the liquid level from the edge of the upper opening of the container. As the liquid level in the vessel drops, the linear burning velocity decreases accordingly. This is because as the liquid level decreases, the distance from the liquid surface to the bottom of the flame increases, so the heat transfer rate from the flame to the liquid surface decreases.

d Influence of water content in liquid

When the liquid contains water, the burning rate of the liquid is reduced because a portion of the heat transferred from the flame is consumed by evaporation of the water. Moreover, the more water content, the slower the combustion rate.

e Influence of wind

Wind facilitates the mixing of air and liquid vapor, which can accelerate the combustion rate. Fig. 5-18 shows the burning rate of the three petroleum products as a function of wind speed. It can be seen from this figure that the wind speed has a great influence on the combustion speed of gasoline and diesel, but has little influence on the combustion speed of heavy oil; if the wind

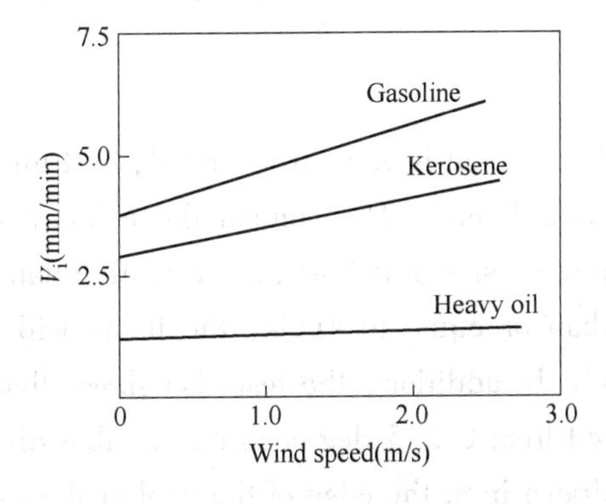

Fig. 5-18 Relationship between combustion speed and wind speed

speed increases beyond a certain level, the burning rate of almost all liquids will tend to some fixed value. This can be explained by the fact that the radiative heat flux from the flame to the liquid surface is influenced by both the radiation intensity of the flame and the inclination of the flame. When the wind speed increases, the radiation intensity of the flame increases with the increase of the combustion speed, but the inclination of the flame also increases. This reduces the radiation angle factor from the flame to the liquid surface. Combining the effects of the two factors on the radiative heat flux, the heat flux obtained at the surface of the liquid tends to be constant, so the burning rate tends to be constant.

When the combustion test is carried out in a small diameter tank, the combustion rate of some liquid fuels may be slowed down with the increase of wind speed. A similar phenomenon also occurred in the simulated fire test of a large diameter ground oil pool. It is believed that the former is mainly due to the small diameter of the tank and the unstable combustion caused by the wind, while the latter is due to the insufficient oxygen supply caused by the flame surrounded by layers of smoke.

5.5.1.2 Flame characteristics of oil tank combustion

A Combustion state of flame

As mentioned earlier, when the pool diameter D is less than 0.03m, the flame is laminar. At this time, the air diffuses to the flame surface, and the combustible liquid vapor also diffuses to the flame surface, so the main mode of combustion is diffusion combustion. When $0.03\text{m}<D<1.0\text{m}$, the combustion changes from laminar flow to turbulent flow; When $D>1.0\text{m}$, the flame develops into a turbulent state. The shape of the flame is changed from a conical shape in a laminar flow state to a turbulent flame with an irregular shape.

Most actual liquid fires are turbulent flames. In this case, the evaporation rate of oil surface is large, and the flame combustion is intense. Due to the buoyancy movement of the flame, a negative pressure area is formed between the bottom of the flame and the liquid surface, so that a large amount of air is sucked to form an upper and lower airflow mass which is violently rolled, the flame is pulsed, and the smoke column generates mushroom-shaped entrainment movement. Cause a large amount of air to be drawn in. Fig. 5-19 is a schematic diagram of a turbulent flame.

B Inclination of flame

The flame of the oil in the liquid pool is generally conical, and the bottom of the cone is equal to the area of the burning liquid pool. The conical flame has a certain angle of inclination under the action of wind, and the size of this angle is directly related to the wind speed. When the wind speed is greater than or equal to 4m/s, the flame will be inclined to the leeward direction by about 60° ~ 70°. In addition, the test also shows that under the condition of no wind, the flame will be tilted from 0 to 5 degrees in a variable direction. This may be due to the imbalance of air being drawn in at the edge of the pool or the asymmetry of the flame being drawn into the air.

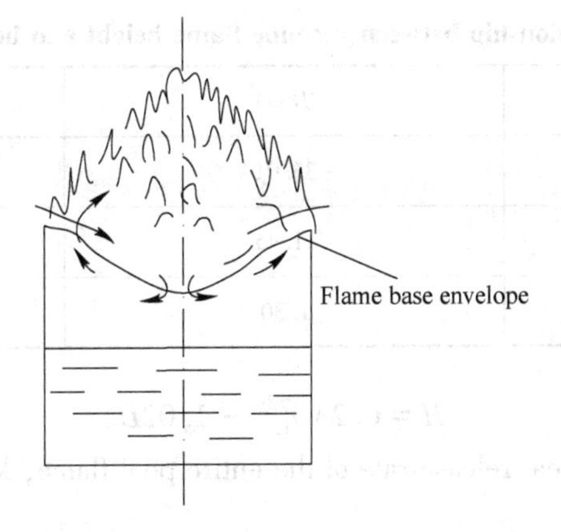

Fig. 5-19 Turbulent buoyant diffusion flame

C Height of flame

The flame height usually refers to the height of the top of the column composed of visible luminous carbon particles, which depends on the diameter of the liquid pool and the type of liquid. If the diameter of the round pool D is taken as the abscissa, and the ratio H/D of the flame height H to the diameter D of the round pool is taken as the ordinate, the test results shown in Fig. 5-20 can be obtained.

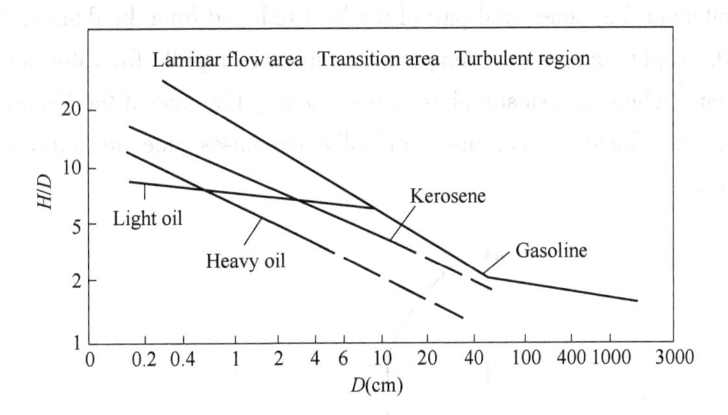

Fig. 5-20 Flame height of petroleum products

It can be seen from the figure that in the laminar flame region, H/D decreases with the increase of D, while in the turbulent flame region, H/D is basically independent of D. In general, there are the following relationships:

$$\text{Laminar flame zone}: H/D \propto D^{-0.1\sim0.3} \tag{5-29}$$

$$\text{Turbulent flame zone}: H/D \approx 1.5 \sim 2.0 \tag{5-30}$$

The relationship between the height of the gasoline flame and the diameter of the liquid pool obtained from the test is listed in Table 5-11, and the data in the table is basically consistent with the formula (5-29) and formula (5-30).

Hesdestad mathematically processed the extensive experimental data and obtained the following formula for the flame height:

Table 5-11 Relationship between gasoline flame height and liquid pool diameter

D(m)	H(m)	H/D
22.30	35.01	1.56
5.40	11.45	2.12
0.38~0.44	1.30	3.25

$$H = 0.23Q_C^{2/5} - 1.02D \tag{5-31}$$

In the formula Q_C is the heat release rate of the entire pool flame, kW; the units of H and D are in m.

Equation (5-31) is in good agreement with the experimental results in the range of $7\text{kW}^2/\text{m} < Q_C^{2/5}/D < 700\text{kW}^2/\text{m}$. For large pool flames (e.g., $D > 100\text{m}$), the above equation is not applicable because the flame breaks into small flames.

D Temperature characteristics of flame

The flame temperature mainly depends on the type of combustible liquid. Generally, the flame temperature of petroleum products is between 900℃ and 1200℃. The temperature distribution of the flame along the longitudinal axis is shown in Fig. 5-21. There is a vapor zone from the oil surface to the bottom of the flame, and part of the heat radiated from the flame to the liquid surface is absorbed by the vapor zone, so the temperature increases rapidly from the liquid surface to the bottom of the flame. There is a steady phase after reaching the base of the flame; when the height increases again, the flame temperature gradually decreases due to outward heat loss and entrainment of air.

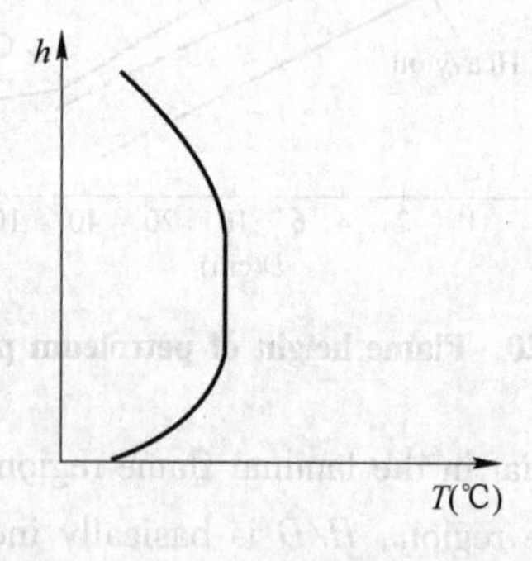

Fig. 5-21 Temperature distribution along longitudinal and transverse direction of flame

E Flame radiation

Another characteristic of a flame is that it transfers heat by radiation to the objects surrounding the liquid pool. The radiant heat flux from the flame to the object depends on the temperature and thickness of the flame, the concentration of the radiant particles in the flame, and the geometric relationship between the flame and the object being radiated. It is necessary to calculate the flame radiation to determine the fire safety distance between oil tanks and design the fire sprinkler system. Two approximate calculation methods are described below.

a Point source method

As shown in Fig. 5-22, the flame height is approximately calculated by the following equation:

$$H = 0.23Q_C^{2/5} - 1.02D \tag{5-32}$$

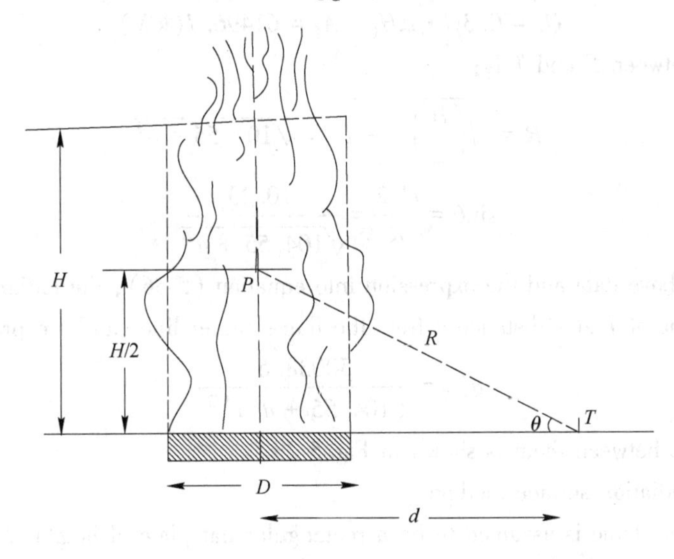

Fig. 5-22 Schematic diagram of oil tank fire radiation

Heat release rate of liquid pool for:

$$Q_C = G \cdot \Delta H_C \cdot A_f \tag{5-33}$$

In the formula A_f is the area of the liquid surface and G is the evaporation rate per unit area of the liquid surface.

Assuming that 30% of the total heat is transferred outward as radiant energy, the rate of radiant heat is:

$$Q_r = 0.3G \cdot \Delta H_C \cdot A_f$$

The so-called point source method is the assumption Q_r is emitted from a point source at a height of $H/2$ from the liquid surface on the central axis of the flame. Therefore, the radiative heat flux at a distance R from the point source is:

$$q''_r = 0.3G \cdot \Delta H_C \cdot A_f/(4\pi R^2) \tag{5-34}$$

In Fig. 5-22, the following relationship exists:

$$R^2 = (H/2)^2 + d^2 \tag{5-35}$$

Where d is the horizontal distance from the central axis of the flame to the radiated body.

Assuming that the angle between the radiated body and the line-of-sight PT is θ, the radiant heat flux projected onto the surface of the radiation receiver is:

$$q''_r = 0.3G \cdot \Delta H_C \cdot A_f \cdot \sin\theta/(4\pi R^2) \tag{5-36}$$

For example, the gasoline tank has a diameter of 10m and a weight burn rate of 0.058 kg/(m^2 · s). The combustion heat of gasoline is $\Delta H_C = 45$kJ/g. In the event of a fire, the flame heat release rate is:

$$Q_C = G \cdot \Delta H_C \cdot A_f \approx 204989(\text{kW})$$

The flame height is:

$$H = 0.23Q_C^{2/5} - 1.02D = 20.45(\text{m})$$

The total radiation rate of the flame is:

$$Q_r = 0.3G \cdot \Delta H_C \cdot A_f = 61496.7(\text{kW})$$

The distance between P and T is:

$$R = \sqrt{\left(\frac{H}{2}\right)^2 + d^2} = \sqrt{104.55 + d^2}$$

$$\sin\theta = \frac{H/2}{R} = \frac{10.23}{\sqrt{104.55 + d^2}}$$

Substitute the above data and the expression into equation (5-36), the radiant heat flux $q''_{r,T}$ on the horizontal plane at T at a distance d from the flame center line can be expressed as follows:

$$q''_{r,T} = \frac{50038.6}{(104.55 + d^2)^{3/2}}$$

The relationship between them is shown in Fig. 5-23.

b Rectangular radiation surface method

In this method, the flame is assumed to be a rectangular flat plate of height H and width D, the heat is radiated from both sides of the deck, and the radiation force on both sides is:

$$E = \frac{1}{2}[0.3G \cdot \Delta H_C \cdot A_f/(H \cdot D)] \quad (5\text{-}37)$$

The radiant heat flux at point T in Fig. 5-23 is:

$$q''_{r,T} = \Phi E \quad (5\text{-}38)$$

Where Φ is the angle coefficient of the horizontal element where T is located facing the rectangular surface of the flame. The flame center line divides the flame plane into two rectangles, and the angle coefficient Φ' of the horizontal infinitesimal plane where T is located for each rectangle is half of Φ, that is

$$\Phi = 2\Phi' \quad (5\text{-}39)$$

The relationship between $q''_{r,T}$ and d obtained by Method b is also shown in Fig. 5-23. It can be seen from the figure that for the same value of d, the radiant heat flux at T calculated by Method b is higher than that obtained by Method a, because it discusses the radiator as an amplification source and ignores the temperature non-uniformity in the flame and the shielding effect of smoke on radiation.

5.5.2 Boil-over and splashing during combustion of crude oil and heavy petroleum products

The vapor of the flammable liquid is mixed with air on the liquid surface while burning, and the heat released by the burning will spread inside the liquid. Due to the different characteristics of liquids, the propagation of heat in liquids has different characteristics. Under certain conditions,

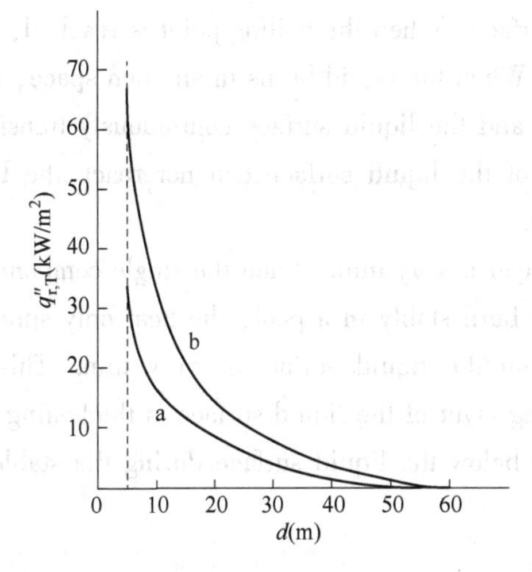

Fig. 5-23 Calculated radiant heat flux of 10m diameter gasoline tank fire

a, b—Point source method

the propagation of heat in liquids the propagation will form a thermal wave and cause boiling and splashing of the liquid, make the fire more violent.

5.5.2.1 Basic concepts

(1) Boiling point: The temperature at which the lightest hydrocarbon in crude oil boils and is the lowest boiling point in crude oil.

(2) Final boiling point: The temperature at which the heaviest hydrocarbon in crude oil boils and is the highest boiling point in crude oil.

(3) Boiling range: The temperature range of the lowest and highest boiling points at which all fractions with different specific gravities and boiling points are converted into vapor. Various single-component liquids have only boiling points but no boiling range.

(4) Light components: a small part of hydrocarbon components with the lightest specific gravity and the lowest boiling point in crude oil.

(5) Heavy components: a small part of hydrocarbon components with the largest specific gravity and the highest boiling point in crude oil.

5.5.2.2 Characteristics of heat transfer in the liquid layer during the combustion of a single liquid component

Single component liquid (such as methanol, acetone, benzene, etc.) and mixed liquid with narrow boiling range (such as kerosene, gasoline, etc.) form stable combustion in a very short time when burning on the free surface, and the combustion rate is basically unchanged. The combustion of such substances has the following characteristics:

(1) The liquid surface temperature is close to but slightly lower than the boiling point of the liquid. When the liquid burns, the heat transferred from the flame to the liquid surface raises the

temperature of the liquid surface. When the boiling point is reached, the temperature of the liquid surface will no longer rise. When the liquid burns in an open space, the evaporation is carried out in a non-equilibrium state, and the liquid surface continuously transfers heat to the inside of the liquid, so the temperature of the liquid surface can not reach the boiling point, but is slightly lower than the boiling point.

(2) The liquid heating layer is very thin. When the single component oil and the mixed oil with a very narrow boiling range burn stably in a pool, the heat only spreads to the shallow oil layer, that is, the heating layer on the liquid surface is very thin. This is not consistent with our imagination that "the heating layer of the liquid surface is thickening with time". Fig. 5-24 shows the temperature distribution below the liquid surface during the stable combustion of butanol and gasoline.

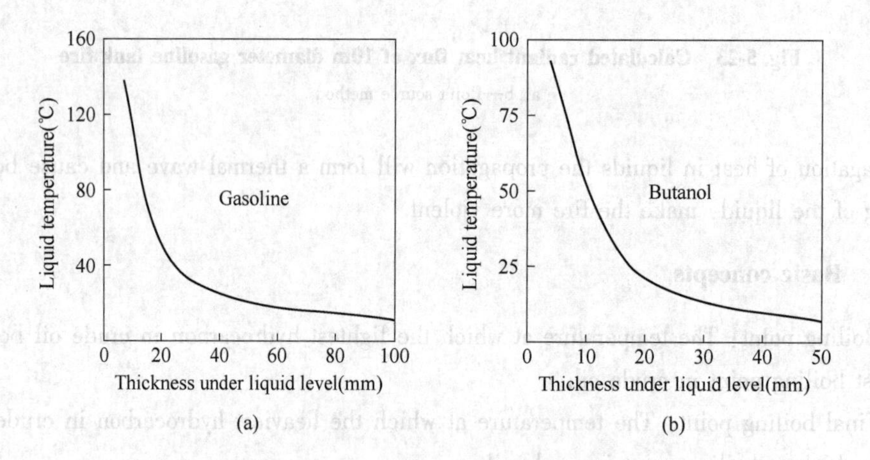

Fig. 5-24 Temperature distribution under liquid surface during stable combustion of butanol and gasoline

(a) Gasoline; (b) Butanol

When the liquid burns steadily, the evaporation rate of the liquid is certain, the shape of the flame and the heat release rate are certain, so the heat transferred from the flame to the liquid surface is also certain. This heat is used on the one hand to vaporize the liquid and on the other hand to heat the liquid layer downward. If the heating thickness becomes thicker and thicker, then according to the Fourier heat conduction law, less and less heat is transferred to the liquid through the liquid surface, and more and more heat is used to evaporate the liquid, thus intensifying the flame combustion. Obviously, this is inconsistent with the premise of stable combustion of liquid. Therefore, when the liquid is burning stably, the temperature distribution under the liquid surface is constant.

5.5.2.3 Characteristics of heat transfer in liquid layer during crude oil combustion

The mixed liquid with a wide boiling range is mainly some heavy oil products, such as crude oil, residual oil, wax oil, asphalt, lubricating oil and so on. Because there is no fixed boiling point, in the combustion process, the heat transferred from the flame to the liquid surface first makes the low boiling point components evaporate and enter the combustion zone for combustion, while the heavy part with a higher boiling point carries the heat received on the surface and sinks to the deep liquid. A hot front is formed that propagates deep into the liquid, gradually penetrating and heating the cold liquid layer. This phenomenon is called the thermal wave property of the liquid, and the thermal front is called the thermal wave.

The initial temperature of the thermal wave is equal to the temperature of the liquid level, which is equal to the boiling point of the lightest component in the crude oil at that time. With the continuous combustion of crude oil, the boiling point of the evaporation components on the liquid surface is getting higher and higher, and the temperature of the thermal wave will gradually rise from 150℃ to 315℃, which is much higher than the boiling point of water. The speed at which the thermal wave moves downward in the liquid layer is called the thermal wave propagation speed. It is faster than the linear burning speed of the liquid (that is, the falling speed of the liquid level), as shown in Table 5-12. After the thermal wave propagation velocity of a certain oil is known, the thickness of the high temperature layer inside the liquid can be estimated according to the combustion time, and then the boiling over and splashing of the heavy oil containing water can be judged. Therefore, the propagation velocity of heat wave is an important parameter to extinguish the heavy oil fire.

Table 5-12 Comparison of thermal wave propagation velocity and linear combustion velocity

Type of oil		Thermal wave propagation speed (mm/min)	Linear burning speed (mm/min)
Light oil product	Water content<0. 3%	7~15	1. 7~7. 5
	Water content>0. 3%	7. 5~20	1. 7~7. 5
Heavy fuel oil and fuel oil	Water content<0. 3%	~8	1. 3~2. 2
	Water content>0. 3%	3~20	1. 3~2. 3
Primary distillate (light component of crude oil)		4. 2~5. 8	2. 5~4. 2

5. 5. 2. 4 Boiling and splashing of heavy oil product

Heavy petroleum products with moisture and high viscosity, such as crude oil, heavy oil, asphalt oil, etc., may produce boil over and splashing when they are burned.

The viscosity of crude oil is relatively high, and it contains a certain amount of water. The water in crude oil generally exists in the form of emulsified water and water cushion. The so-called emulsified water is formed in the process of crude oil exploitation and transportation when the water in the crude oil is strongly stirred into tiny water droplets suspended in the oil. After being placed for a long time, the oil and water are separated, and the water settles at the bottom to form a water cushion due to the large specific gravity.

When the heat wave moves to the deep layer of the liquid, because the temperature of the heat wave is much higher than the boiling point of water, the heat wave will vaporize the emulsified water in the oil product, and a large amount of steam will pass through the oil layer and float up to the liquid surface. In the process of moving up, gas-in-oil bubbles are formed, that is, part of the oil forms foam containing a large number of steam bubbles. In this way, the volume of the liquid must be expanded and overflowed outward. At the same time, part of the oil that has not formed foam is thrown out of the tank by the expansion force of the steam below, which makes the liquid surface boil violently, just like "running pot". This phenomenon is called boiling over.

The boil over process shows that the formation of boil over must have three conditions:

(1) Crude oil has the characteristics of forming heat wave, that is, wide boiling range and large difference in specific gravity;

(2) The crude oil contains emulsified water, which turns into steam when exposed to heat waves;

(3) The viscosity of crude oil is high, which makes it difficult for water vapor to pass through the oil layer from bottom to top. If the viscosity of crude oil is low, water vapor can easily pass through the oil layer, and it is not easy to form boil over.

With the progress of combustion, the temperature of the heat wave gradually rises, and the downward transmission distance of the heat wave also increases. When the heat wave reaches the water cushion, a large amount of water in the water cushion evaporates, and the vapor volume expands rapidly, so that the liquid layer above the water cushion is thrown into the air and sprayed outside the tank. This phenomenon is called splashing.

In general, boil over occurs much earlier than splashing. The time of boil over is related to the type and moisture content of crude oil. According to experiments, oil containing 1% water will boil over after 45 to 60 minutes of combustion. The occurrence time of splashing is related to the thickness of oil layer, the moving speed of heat wave and the linear velocity of oil combustion. It can be approximately calculated by the following formula:

$$\tau = \frac{H - h}{v_0 + v_t} - KH \qquad (5\text{-}40)$$

Where τ——the time when splashing is expected to occur, h;

H——height of oil level in storage tank, m;

h——height of water cushion in the storage tank, m;

v_0——linear combustion velocity of crude oil, m/h;

v_t——thermal wave propagation velocity of crude oil, m/h;

K——advance coefficient, h/m, take 0 when the oil storage temperature is lower than the ignition point, and take 0.1 when the temperature is higher than the ignition point.

Oil surface creep and swelling usually occur before splashing in oil tank fire, and the flame increases, shiny, white; oily foam appear 2 ~ 4 times; smoke color changes from thick to light and there is a violent "hissing" "Hiss" sound, etc.. The metal tank wall will tremble, accompanied by strong noise (caused by the intense boiling of the liquid surface and the deformation of the metal tank wall), the smoke will be reduced, and the flame will be brighter. The size of the flame is larger, and the flame is shaped like a rocket.

In case of splashing in case of oil tank fire, the fuel can be thrown out for 70 ~ 120m. It not only makes the fire develop violently, but also seriously endangers the life safety of firefighters. Therefore, evacuation should be organized in time to reduce casualties.

5.6 Burning and explosion of oil storage tank in gas station

The oil storage tank is the most important and dangerous part of the gas station. It will cause the combustion and explosion of the oil storage tank when it encounters open fire, static electricity or the failure of the safety accessories of the oil tank, improper protective measures for unloading oil from the tank truck and so on. Once the explosion occurs, the consequences will be very serious. Therefore, it is very important to study the combustion and explosion characteristics of the oil in the storage tank for eliminating the combustion and explosion disasters of the gas station.

Gasoline and diesel are the main commodities of the gas station. Gasoline is the third category of flammable liquid, flash point<-18℃, and the number of dangerous goods is 31001. When it is out of control, fire, explosion and poisoning accidents will occur. The diesel is not listed in the list of hazardous chemicals. The flash point of -35# and -50 # diesel is not lower than 45℃, and the flash point of -20#, -10#, 0#, 5# and 10# diesel is not lower than 55℃, which belongs to combustible chemicals. It can be seen that the fire risk of gas stations is mainly gasoline.

5.6.1 Flash point and ignition point of gasoline

Gasoline, as one of the most common flammable liquids in human production and life, can catch fire or explode when heated, exposed to fire or in contact with oxidants. Mixing with air can form explosive mixture, which will cause combustion and explosion when exposed to open fire and high heat energy. Its vapor is heavier than air and can diffuse to a considerable distance at a lower place, causing backfire when it meets a fire source. In case of high heat, the internal pressure of the container increases, and there is a risk of cracking and explosion. It can accumulate static electricity and ignite its vapor.

Gasoline is the main oil product stored in gas stations. It is generally a colorless or yellowish transparent liquid with a density of 730kg/m^3, which is lighter than water and insoluble in ice. According to the classification of oil hazard level, the flash point below 45℃ is flammable, while the flash point of gasoline is much lower than room temperature, about -50 ~ 20℃. Gasoline is easy to evaporate at room temperature, the vapor is heavier than air, and it is an explosive mixed gas with air. In case of a very small ignition source, it can explode. At normal temperature, gasoline can volatilize a large amount of steam, and it is easy to burn or explode with open fire or electric spark. See Table 5-13 for the flash point and auto-ignition point of gasoline and diesel.

Table 5-13 Flash point and auto-ignition point of gasoline and diesel oil (℃)

Oils	Flash point	Spontaneous ignition point
Gasoline	<-28	510~530
Light diesel oil	45~120	350~380
Heavy diesel	>120	300~330

5.6.2 Analysis of the causes of gas station explosion

5.6.2.1 Ground explosion

(1) Leakage type explosion. Leakage type explosion refers to the explosion caused by the leakage of combustible gas, steam and dust into the atmosphere due to the rupture of containers, machinery or other equipment for handling, storing or transporting combustible substances for some reason, which meets the fire source when the explosive concentration limit is reached. An important factor causing leakage-type explosion is that the mixture of oil and gas reaches the explosion limit. Open unloading is a direct way to cause the mixing of oil and gas. The other is oil leakage, which is mainly caused by equipment leakage and human factors. The main facilities and equipment of gas stations are oil tanks, refueling machines, pumps, pipelines and valves. Leakage often occurs in various situations, such as weld crack, pipeline aging, flange gasket wear, poor sealing and so on, resulting in the formation of explosive mixed gas by oil and gas mixing. Leakage caused by human factors refers to off-duty, misoperation, illegal operation, overfilling, blind command and poor maintenance of equipment operation. For example, on March 12, 2003, a staff member of a gas station left his post without authorization when unloading oil, the pipe joint fell off and disconnected, the oil splashed out everywhere from the oil pipe, and the volatile gas met the heat source on the engine of the tank truck and exploded.

(2) Disequilibrium explosion. The destructive equilibrium explosion refers to the explosion caused by the increase of steam pressure and expansion of the overfilled or closed oil tank after being heated. When the flammable liquid storage tank is baked, the temperature of the liquid in the container rises and the phase pressure increases. If the container is equipped with a safety valve, the safety valve will open when the pressure in the container reaches the set pressure of the safety valve. If the relief area of the safety valve is large enough and the set pressure is small enough, all the liquid will leak out of the container and an accident will occur. In practice, the set pressure of the safety valve is higher, and because the container is exposed to fire, the tensile strength of the container material decreases sharply under the influence of high temperature, so that the container can not bear the set pressure of the safety valve. In this way, even if the safety valve is opened to exhaust rapidly, the container will crack and then cause a steam explosion. A test to measure heat transfer from an external fire source to an automotive tank showed that an uninsulated $130m^3$ automotive tank experienced a vapor explosion after approximately 25 minutes

of exposure to a fire, despite continued discharge through a safety valve set at 1.86MPa. When the sun shines or the temperature is very high in hot summer, it will also increase the pressure in the sealed oil tank, expand the volume and cause explosion. If the spray cooling system fails, a destructive explosion is more likely to occur.

(3) Other types of fire and explosion. Illegal refueling, such as refueling plastic barrels and refueling motorcycles directly, can cause fires or explosions. Sometimes, several types of explosions occur one after another, causing two or three explosions. The harm caused by it is often more and more serious.

5.6.2.2 Underground oil tank explosion

(1) Non-directly buried underground oil tank. Non-directly buried underground oil tank mainly refers to the oil tank located in the basement or with underground operation room. Because the basement is a closed or confined space, if the oil tank, oil pump, pipeline and valve leak or leak due to corrosion, wear, rupture, aging and poor sealing, the oil vapor will mix with the air to reach the explosion limit, and fire and explosion will occur when it meets the fire source.

(2) Explosion of directly buried oil tank. Directly buried underground oil tank explosion means that the oil tank is directly buried in the soil, the oil vapor in the tank mixes with the air, reaches the explosion limit, and explodes when it meets the fire source. It can be divided into the following two situations:

1) There is oil in the tank. When the oil tank is filled with oil and the mixed gas in the non-liquid part of the tank reaches the explosion limit, it will explode when it meets the fire source. The fire source here refers to fire, lightning sparks, impact sparks, etc. Lightning sparks are caused by the failure of lightning protection facilities, resulting in the accumulation of static electricity and lightning strikes. Impact spark here refers to the spark generated by collision with the edge of the tank cover or orifice when opening the tank cover or orifice or measuring the oil with a steel ruler, and the friction between the steel ruler and the edge of the oil measuring orifice when it is extended and pulled out.

2) There is no oil in the tank. There is no oil in the oil tank, and the tank is filled with the same kind of oil without being cleaned; oil tanks should be cleaned during maintenance and annual inspection. If the oil tank is not cleaned or is not thoroughly cleaned, the mixed gas reaching the explosion limit will be produced. At this time, when the welding fire is encountered, sparks will be produced when the tank cover or orifice is opened, or after the oil tank is washed, the sedimentation of water from a certain volume of oil will produce electrostatic sparks, etc., which will cause an explosion. On December 13, 2001, an oil depot in Hunan prepared to clean a 5000m^3 earth-covered oil tank. When opening the manhole of the oil tank, the oil and gas leaked from the tank and filled the tank channel. The blower with poor explosion-proof performance was used to ventilate in the pipeline, which led to an explosion. The earth-covered tank roof and steel tank roof were opened and collapsed, and the residual oil in the tank was burned. Four people were burned in the accident. A 5000m^3 steel tank was burned down, and 42.5 tons of No. 90 gasoline was burned down.

5.6.3 Fire source analysis

According to the statistical analysis of 100 cases of fire and explosion accidents in gas stations, the ignition sources mainly include 8 types, such as "electrical appliances, open fire, welding, static electricity, lightning, engine, smoking and others", as shown in Table 5-14 below. Among them, there were 45 cases of electrical appliances, lightning and static electricity, accounting for 45%; There were 37 cases of open fire, welding and smoking, accounting for 37%; Engine 8 cases, accounting for 8%; Others 10 cases, accounting for 10%. That is to say, In order to prevent fire and explosion accidents in gas stations, the ignition source must be strictly controlled to be out of control. For some uncontrollable ignition sources, such as lightning, static electricity and other ignition sources, scientific methods should be adopted to guide and prevent them in strict accordance with the technical requirements of lightning protection and static electricity prevention. For the ignition source caused by human factors, we should strengthen the ideological education of personnel and strive to create a safe and cultural environment for gas stations. Start from me, start from small things, and put an end to man-made ignition sources.

Table 5-14 Statistics of ignition sources of 100 cases of fire and explosion accidents in gas stations

Project	Number of cases	Proportion(%)	Project	Number of cases	Proportion(%)
Power source	17	17	Lightning strike	4	
Switch	8		Induction	1	
Wires	4		Engine	8	8
Other	5		Electric appliance	2	
Open fire	14	14	Mars	3	
Lighter	2		Hot face	1	
Stove	9		Other	2	
Other	3		Smoking	14	14
Welding	9	9	Match	8	
Welding	8		Lighter	6	
Cut	1		Other	10	10
Static electricity	23	23	Unknown	7	
Thunderbolt	5	5	Collision	3	

Exercises

1. What are the main hazardous characteristics of flammable liquids, give examples of each and explain what parameters are used to evaluate each of these characteristics?
2. What is called flashover, why do flammable liquids flash?

3. How does the rate of combustion of an oil change with the diameter of the vessel ? Combined with the mechanism of flame propagation to the liquid surface, the reason for this change is explained.
4. What is called a heat wave, what is boiling over? What is spatter, briefly describe the formation of boil over and splash respectively, and explain how they are similar and different?
5. According to the formation conditions of heat wave, boiling overflow and spatter, whether the heavy oil that can form heat wave can definitely form boiling overflow or spatter during fire combustion is explained?
6. How to judge the explosion risk of vapor of combustible liquid at room temperature by explosion temperature limit?
7. Is the concept of explosion temperature limit of flammable liquid the same as that of flammable gas?
8. Why is it unsafe to store flammable liquids in refrigerators?

Chapter 6 Combustion and Explosion of Combustible Solid

扫码获得
数字资源

6.1 Overview of solid combustion

6.1.1 Form of solid combustion

According to the combustion mode and combustion characteristics of various combustible solids, the forms of solid combustion can be roughly divided into five types.

(1) Evaporation and combustion. Sulfur, phosphorus, potassium, sodium, candles, rosin, asphalt and other combustible solids, when heated by the fire source, first melt and evaporate, and then the vapor and oxygen combustion reaction. This form of combustion is generally referred to as evaporative combustion. Camphor, naphthalene and other easily sublimable substances do not undergo melting process during combustion, but their combustion phenomenon can also be regarded as an evaporative combustion.

(2) Surface combustion. The combustion reaction of combustible solids (such as charcoal, coke, iron, copper, etc.) is caused by the direct action of oxygen and substances on their surfaces, which is called surface combustion. This is flameless combustion, sometimes referred to as out-of-phase combustion.

(3) Decomposition and combustion. Combustible solids, such as wood, coal, synthetic plastics, calcium materials, etc., undergo thermal decomposition when heated by a fire source, and then the decomposed combustible volatiles react with oxygen. This form of combustion is generally called decomposition combustion.

(4) Fumigation burns (Smoldering). Under the conditions that the air does not circulate, the heating temperature is low, the combustible volatiles decomposed are less or escape faster, and the water content is more, the combustion phenomenon of only smoking without flame often occurs, which is fumigation combustion, also known as smoldering.

(5) Power combustion (Explosion). It refers to the explosive combustion of combustible solids or the combustible volatiles separated from their components when they meet fire sources, mainly including combustible dust explosion, explosive explosion, flashover and other situations. Among them, flashover refers to the release of combustible gas from combustible solids due to thermal decomposition or incomplete combustion, when it is mixed with air in an appropriate proportion and then meets the fire source. Explosive premixed combustion occurs. For example, celluloid, which can precipitate carbon monoxide, and polyurethane, which can precipitate hydrogen cyanide, often produce flashover when a large amount of accumulation burns.

6.1.2 Parameters for evaluating solid flammability

The solid combustion process is relatively complex, and the parameters for evaluating the combustion characteristics mainly include:

6.1.2.1 Melting point flash point and fire point

The initial temperature at which a solid becomes a liquid is called the solid's melting point; the flash point is the lowest temperature at which some low-melting flammable solids will flash heating a combustible solid to a certain temperatures? The minimum temperature at which a solid continues to combust in the presence of an open flame is called the solid ignition point.

Melting point, flash point and ignition point are the indexes of solid combustion characteristics, and also the important parameters to evaluate the fire risk of solid. Generally, the lower the melting point of the flammable solid, the lower the flash point and ignition point, and the greater the fire risk.

6.1.2.2 Thermal decomposition temperature

Solid thermal decomposition temperature refers to the initial temperature at which combustible solid decomposes when heated, which is one of the main parameters for evaluating the fire risk of solid decomposed by thermal energy. The basic rule is that the lower the thermal decomposition temperature of combustible solids, the lower the ignition point, and the greater the fire risk. Thermal decomposition temperature and ignition point of several combustible solids are shown in Table 6-1.

Table 6-1 Thermal decomposition temperature and ignition point of several combustible solids

Solid name	Thermal decomposition emperature (℃)	Ignition point (℃)	Solid name	Thermal decomposition temperature (℃)	Ignition point (℃)
Nitrocellulose	40	180	Cotton	120	210
Celluloid	90~100	150~180	Wood	150	250~295
Ma	107	150~200	Silk	235	250~300

6.1.2.3 Spontaneous ignition point

The self-ignition point of a combustible solid is the lowest temperature at which it can burn automatically. The lower the spontaneous ignition point of the solid, the easier it is to burn, so the greater the fire risk. Autoignition points of common macromolecular substances are shown in Table 6-2.

Table 6-2 Spontaneous ignition points of common macromolecular substances

Name of substance	Self-ignition point (℃)	Name of substance	Self-ignition point (℃)	Name of substance	Self-ignition point (℃)
Cotton	255	Polyethylene	349	Polyamide	424
Newspaper	230	Polyvinyl chloride	454	Cellulose acetate	475
White Pine	260	Plexiglass	450~462	Cellulose nitrate	141

6.1.2.4 Specific surface area

Specific surface area is the surface area of a solid per unit volume. For the same combustible solid, the greater the specific surface area, the greater the fire risk. As far as combustible dust is concerned, the specific surface area has a very important influence on the lower explosion limit, the minimum detonation energy, the maximum explosion pressure and other parameters. Generally speaking, with the increase of the specific surface area of dust, its lower explosive limit decreases. The minimum detonation energy decreases and the maximum explosion pressure increases.

6.1.2.5 Oxygen index

Oxygen index refers to the volume percentage of the minimum oxygen content in the mixed gas when maintaining the combustion of substances under specified conditions. Oxygen index (loi) is an indicator of the relative combustibility of a substance, and it is also an important indicator for evaluating the fire risk of combustible solids (especially polymers). The smaller the oxygen index is, the greater the fire risk is. Combustible materials are generally considered to have an oxygen index of less than 22. Those between 22 and 27 belong to flame retardant materials; greater than 27 is a highly flame retardant material. After flame retardant treatment, the oxygen index of the material will be improved in different degrees. The oxygen index of some common polymers is shown in Table 6-3.

Table 6-3 Oxygen index of common high polymer

Name of substance	Oxygen Index	Name of substance	Oxygen Index	Name of substance	Oxygen Index
Polystyrene	18	Polybenzimidazole	41	Neoprene	26
Polyvinyl alcohol	22	Polyamide	41	Silicone rubber	26~39
Polyvinyl chloride	45	Polyalditol	31	Acetal copolymer	15
Polyphenylene oxide	28	Phenolic resin	35	Polycarbonate	27
Polysulfone	32	Epoxy resin	20	Polytetrafluoroethylene	>95

In addition to the above parameters, for combustible dust and explosives, there are other important parameters to assess the risk of fire and explosion, such as the lower limit of explosive concentration of dust, the sensitivity of explosives, etc.

6.2 Solid ignition and combustion theory

In the actual fire, the most common combustible solid is the solid that can release combustible gas when it is heated. This section mainly discusses the ignition and combustion of this kind of solid.

6.2.1 Solid ignition conditions and ignition time

Whether a solid that can release combustible gas when heated can be ignited depends on whether the combustible gas released can maintain a certain concentration, which can also be judged by the equilibrium equation, that is,

$$(\varphi \cdot \Delta H_C - L_V) \cdot G_{cr} + Q_E - Q_l = S \tag{6-1}$$

Where, φ is the heat of combustion of a solid at the point of ignition (ΔH_C) the number of copies delivered to its surface; L_V is the amount of heat required by the solid to release the combustible gas; G_{cr} is the critical mass flow rate of the combustible gas released by the solid at the ignition point; Q_E and Q_l the heating rate and the heat loss rate of the fire source per unit solid surface, respectively; S is the net heat gain rate per unit solid surface.

Q_E it can be determined by calculation. ΔH_C and L_V available in the literature. For an infinite solid of a certain thickness, Q_l it can be estimated by the following formula:

$$Q_l = \varepsilon \sigma T_i^4 + k \cdot \frac{T_S - T_0}{\sqrt{\alpha t}} \tag{6-2}$$

Where, ε is the emissivity of a solid; σ is the Stefan-Boltzmann constant; T_i, T_S, T_0 respectively the ignition point of the solid, the surface temperature at the ignition point and the ambient temperature; k and α are the thermal conductivity and diffusivity of the solid, respectively; t is the time that the solid is heated by the heat source.

G_{cr} and φ there are the following relationships:

$$G_{cr} = \frac{h}{c} \cdot \left(1 + \frac{3000}{\varphi \cdot \Delta H_C}\right) \tag{6-3}$$

Where, h is the convective heat transfer coefficient between the flame and the solid surface; c is the heat capacity of air.

If measured by experiment, G_{cr} it can be estimated according to the formula (6-3). Of some polymers. G_{cr} and φ values are shown in Table 6-4.

Table 6-4 Some high polymer G_{cr} and φ value

Name of substance	G_{cr} (g/(m^2 · s))	φ	Name of substance	G_{cr} (g/(m^2 · s))	φ
Polyoxymethylene	3.9	0.45	Phenolic foam (GM-57)	4.4	0.17
Polymethyl methacrylate	3.2	0.27	Polyethylene-42% Cl	6.5	0.12
Polyethylene	1.0	0.27	Polyurethane foam	5.6	0.11

Continuted Table 6-4

Name of substance	G_{cr} (g/(m^2 · s))	φ	Name of substance	G_{cr} (g/(m^2 · s))	φ
Polypropylene	2.2	0.26	Polyisocyanate foam	5.4	0.11
Polystyrene	3.0	0.21	Polyethylene-25% Cl	6.0	0.19

In equation (6-1), if $S< 0$, the solid cannot be ignited or can only be flashed; if $S>0$, the heat received by the solid surface can not only maintain continuous combustion, but also the excess part. This part of heat can further improve the release rate of combustible gas and create better conditions for the continuous combustion of solids. $S=0$ is the critical condition under which a solid can be ignited.

【Example 6-1】 An organic glass plate with a thickness of 50mm was irradiated with a flame at a temperature of 1300℃. If the surface temperature reached the ignition point (about 6 seconds), the flame was removed immediately to determine whether the glass plate could be ignited.

【Explanation】 According to relevant data $\alpha=1.1\times10^{-7}\,\text{m}^2/\text{s}$, $k=0.19\,\text{W}/(\text{m}\cdot\text{K})$, $\Delta H_C=26.2\,\text{kJ/g}$, $L_V=1.62\,\text{kJ/g}$, $G_{cr}=3.2(\text{g}/(\text{m}^2\cdot\text{s}))$, $\varphi=0.27$, with the given condition $T_S=T_i=270+273=543$ (K), $t=6$s.

Given $T_0=20+273=293$ (K), $\varepsilon=0.8$, from formula (6-2):

$$Q_t=0.8\times5.67\times10^{-8}\times543^4+0.19\times\frac{543-293}{\sqrt{1.1\times10^{-7}\times6}}$$
$$=6.24\times10^4(\text{W/m}^2)$$
$$=62.4(\text{kW/m}^2)$$

Since the flame is removed as soon as the surface temperature reaches the ignition point, $Q_E\to0$. From formula (6-1):

$$S=(0.27\times26.2-1.62)\times3.2-62.4=-44.93\ (\text{kW/m}^2)$$

Because $S<0$, so the plexiglass plate cannot be ignited.

If a similar calculation is performed for a polyurethane foam whose surface always reaches the ignition temperature at 0.2s, the result is $S=18.4(\text{kW/m}^2)>0$. Therefore, the polyurethane foam is easily ignited and burns strongly.

Under the continuous action of fire source, the ignition time of combustible solid is related to the type, shape and size of combustible material, the intensity of fire source, heating mode and other factors. In this paper, the ignition time of thin objects such as curtains and curtains with small B_i number is estimated by using the "lumped hot melt analysis method".

Assume that the thickness, density, heat capacity, and convective heat transfer coefficient between a thin body and its surroundings are τ, ρ, c, and h, respectively, the ignition point of the thin body, and the ambient temperature. (Or the initial temperature of the object) are T_i and T_0, respectively. When both sides of a thin body are heated by a hot gas flow with temperature T_∞, the energy balance equation can be written as follows during the time interval dt:

$$2A \cdot h \times (T_\infty - T) \cdot dt = (\tau A) \cdot \rho \cdot c \cdot dT$$

Where, A is the heated area of the thin object; T is the temperature of the thin object at time t; dt is the change in temperature of the thin object after dt.

The above equation can be changed to:

$$dt = \frac{\tau\rho c}{2h} \cdot \frac{dT}{T_\infty - T} \tag{6-4}$$

Integrating the above equation from T_0 to T_i, the ignition time t_i is:

$$t_i = \frac{\tau\rho c}{2h} \cdot \ln\left(\frac{T_\infty - T_0}{T_\infty - T_i}\right) \tag{6-5}$$

In the same way, if one side of the object is heated and the other side is insulated, the ignition time is:

$$t_i = \frac{\tau\rho c}{h} \cdot \ln\left(\frac{T_\infty - T_0}{T_\infty - T_i}\right) \tag{6-6}$$

If one side of the object is heated and the other side is not adiabatic, then:

$$t_i = \frac{\tau\rho c}{2h} \cdot \ln\left(\frac{T_\infty - T_0}{T_\infty + T_0 - 2T_i}\right) \tag{6-7}$$

When the heat flux on one side of the object is Q''_r when the other side is adiabatic, assuming that the absorption rate of the body is a, during the time interval dT, the energy balance equation can be written as:

$$A \cdot (aQ''_r) \cdot dt - h \cdot A \cdot (T - T_0) = \tau A\rho c dT$$

or

$$dt = \frac{\tau\rho c}{aQ''_r - h \cdot (T - T_0)} dT \tag{6-8}$$

Integrating this equation from T_0 to T_i, the ignition time is:

$$t_i = \frac{\tau\rho c}{h} \cdot \ln\left[\frac{aQ''_r}{aQ''_r - h \cdot (T - T_0)}\right] \tag{6-9}$$

If one side is heated by radiation and the other side is not insulated, then:

$$t_i = \frac{\tau\rho c}{2h} \cdot \ln\left[\frac{aQ''_r}{aQ''_r - 2h \cdot (T - T_0)}\right] \tag{6-10}$$

It is rare that both sides of an object are heated by radiation at the same time.

【Example 6-2】 For a curtain with a thickness of 0.8 mm, the density, heat capacity and convective heat transfer coefficient between it and the surrounding air are 0.3g/cm³, 1.2kJ/(kg · K) and 15W/(m² · K), respectively, the initial temperature is 20℃, and the ignition point is 260℃. When the curtain is hung vertically in hot air at 300℃, the ignition time of the curtain is calculated.

【Solution】 Given $\tau = 8 \times 10^{-4}$m, $\rho = 300$kg/m³, $C = 1.2 \times 10^3$J/(kg · K), $H = 15$ (W/m²), $T_\infty = 573$ K, $T_0 = 293$K, $T_i = 533$K.

Using equation (6-5), the required ignition time is:

$$t_i = \frac{8 \times 10^4 \times 300 \times 1.2 \times 10^3}{2 \times 15} \cdot \ln\left(\frac{573 - 293}{573 - 533}\right) \approx 19\text{s}$$

If one side of the curtain is heated by radiation with a heat flux of 20 kW/m^2, the two sides lose heat, and the absorptivity of the curtain is 0. 8, the ignition time can be obtained from equation (6-10):

$$t_i = \frac{8 \times 10^4 \times 300 \times 1.2 \times 10^3}{2 \times 15} \cdot \ln\left[\frac{0.8 \times 20 \times 10^3}{0.8 \times 20 \times 10^3 - 2 \times 15 \times (533 - 293)}\right] \approx 6\text{s}$$

6. 2. 2 Solid flame propagation theory

Once a combustible solid is ignited, the flame will spread on its surface or shallow layer. In the fire scene, the speed of flame propagation and the size of combustible area determine the speed of fire development. Therefore, the flame propagation characteristics of solids are an essential element of fire development.

In the theory of solid flame propagation, the concept of "combustion initiation surface" is used to unify all types of flame propagation or fire spread (including premixed flame propagation, smoldering propagation, dispersed fuel bed flame propagation, forest fire spread, etc.). Combustion initiation surface refers to the interface between the burning flame and the unburned material as the solid flame propagates. The rate of heat transfer across this interface determines the rate of flame propagation or fire spread. According to the energy conservation equation, the "basic equation of flame propagation" is:

$$\rho \cdot v \cdot \Delta h = Q \tag{6-11}$$

from this:

$$\rho = \frac{Q}{v \cdot \Delta h} \tag{6-12}$$

Where, v is the flame propagation speed; Q is the rate of heat transfer across the interface; ρ is the density of the solid; Δh is the enthalpy change of a unit mass of solid when it rises from the initial temperature T_0 to the ignition temperature T_i.

6. 2. 3 Factors affecting ignition and combustion of solids

The process from ignition to stable combustion (including flame propagation) of combustible solids is affected by many factors.

6. 2. 3. 1 External fire source or external heat source

Generally, the ignition source must be within the flow of combustible volatiles to ignite the solid, and the heating rate per unit surface area of the solid, (Q_E), the easier the solid is to be ignited. Table 6-5 lists some critical heating rates for solid ignition (Q_{Ecr}).

The external heating source will accelerate the stable combustion rate of the solid and the flame propagation rate on its surface, mainly because the external heating preheats the unburned part of the material in front of the flame front, and also accelerates the combustion rate behind the flame front, thus providing an additional forward propagation and intensifying the whole combustion process.

Table 6-5 Critical heating rates for ignition of some solid materials

Name of substance	Q_{Ecr}(kW/m^2)	Name of substance	Q_{Ecr}(kW/m^2)
Wood	12	Flexible polyurethane foam	16
Chip board	28	Polyoxymethylene	17
Hardboard	27	Polymethylene	12
Plexiglass	21	Polyethylene-42% Cl	22

The solid burning rate can be expressed by the surface linear burning rate V_S or the weight burning rate G_S, and the latter is more often used in practice. The gravimetric burn rate, G_S, can be calculated from:

$$G_S = \frac{Q_F - Q_l}{L_V} \tag{6-13}$$

In the formula, Q_F is the heat flux provided by the combustion flame to the solid surface, which is composed of radiative heat flux and convective heat flux, and the share of the two varies with the size of the combustion area. In addition to those solids (such as polyformaldehyde) whose combustion flame is not bright, in the combustion of large area (diameter greater than 1m), the heat transfer from the flame to the surrounding surface is mainly radiation.

Under certain external conditions, Q_F and Q_l the size depends on the nature of the solid itself, some combustible solid Q_F and Q_l see Table 6-6 for values.

Table 6-6 Some solid Q_F and Q_l value

Solid name	Q_F(kW/m^2)	Q_l(kW/m^2)
FR phenolic foam	25.1	98.7
Polyoxymethylene	38.5	13.8
Polyethylene	32.6	26.3
Polycarbonate	51.9	74.1
Wood (American fir)	23.8	23.8
Polystyrene	61.5	50.2
Phenolic plastic	21.8	16.3
Polymethyl methacrylate	38.5	21.3
FR polyisocyanurate foam	50.2	58.5
Polyurethane foam	68.1	57.7
FR polystyrene foam	34.3	23.4
Flexible polyurethane	51.2	24.3

It can be seen that for some solid $Q_F \leqslant Q_l$, They cannot achieve stable combustion only with the heat provided by the combustion flame. For it to burn stably, heat must be supplied to its surface from the outside. It is assumed that the heat externally supplied to the solid surface is still used. Q_E represents, and $Q_E+Q_F>Q_l$, then the weight burning rate of the solid can be calculated from:

$$G_S = \frac{Q_E + Q_F - Q_l}{L_V} \tag{6-14}$$

If $Q_E = Q_l$ that is, the externally supplied heat is fully used to balance the heat loss, the resulting combustion rate is the ideal combustion rate, that is,

$$G_{Si} = \frac{Q_F}{L_V} \tag{6-15}$$

In real solid fire, due to $Q_E \leqslant Q_l$, so G_{Si} in fact, it is the maximum combustion velocity that a solid can achieve in stable combustion. The average burning velocities of some combustible solids under simulated experimental fire conditions are listed in Table 6-7.

Table 6-7 Average burning velocities of some solids under simulated fire conditions

Solid name	$\overline{G_S}$(g/(m^2 · s))
Wood (14% water)	13. 8
Natural Rubber	8. 3
Cloth electric bakelite	8. 9
Phenolic plastic	2. 8
Cotton (water content 6% ~ 8%)	2. 4
Polystyrene resin	8. 3
Newspaper	6. 7
Plexiglass	11. 5

6. 2. 3. 2 Properties of solid materials

The lower the melting point, thermal decomposition temperature and "heat of gasification" (L_V), and the higher the combustion heat, the faster the combustible solid releases combustible gas, the easier it is to be ignited, and the faster the stable combustion speed after ignition. Table 6-8 lists the average burning velocities for some solid materials.

Table 6-8 Average burning rate of several common solid substances (G/(m^2 · s))

Name of substance	Average burning speed	Name of substance	Average burning speed
Wood (14% moisture)	13. 9	Cotton (moisture 6% ~ 8%)	2. 5
Natural Rubber	6. 7	Paper	6. 7
Cloth electric bakelite	8. 9	Plexiglass	11. 5
Phenolic plastic	2. 8	Man-made fiber (moisture 6%)	6

In addition, the thermal inertia (K、ρ、C) of solid materials has an important effect on their ignition and combustion performance, and for thick solid materials, this effect may play a major role. Materials with low thermal inertia are easily ignited and burn rapidly.

Another important property that affects the stable combustion of combustible solids is the rate of heat release from combustion, q_c, which can be calculated from the following equation:

$$q_c = G_S \cdot \Delta H_C \cdot A_F \cdot \mu \tag{6-16}$$

Where, A_F is the surface area of the burning solid; μ is the heat release coefficient, and the μ values of some combustible solids are shown in Table 6-9.

Table 6-9 Exothermic coefficient of some combustible solids

Solid name	Q_E(kW/m²)	μ	μ_c	μ_r
Cellulose	52.4	0.716	0.351	0.365
Polyoxymethylene	0	0.755	0.607	0.148
Polymethyl methacrylate	0	0.867	0.622	0.245
	39.7	0.710	0.340	0.370
Polypropylene	0	0.752	0.548	0.204
	39.7	0.593	0.233	0.360
Polystyrene	0	0.607	0.385	0.222
	39.7	0.464	0.130	0.334
Polyvinyl chloride	52.4	0.357	0.148	0.209

The net heat flux received by the surface of a combustible solid is assumed to be Q_{net}, then there is

$$Q_{net} = Q_E + Q_F - Q_l$$

Combining equations (6-14) and (6-16),

$$q_c = Q_{net} \cdot A_F \cdot \mu \cdot \frac{\Delta H_C}{L_V} \tag{6-17}$$

This equation shows that the solid combustion heat release rate is proportional to the ratio $\Delta H_C/L_V$ the relationship is very close. And $\Delta H_C/L_V$ or L_V comparison, $\Delta H_C/L_V$ and can better reflect the stable combustion characteristics of the solid. Of a hydrocarbon polymer than of a corresponding oxygen-containing derivatives $\Delta H_C/L_V$. Therefore, the stable combustion performance of the former is better than the latter. Table 6-10 lists some of the solid material $\Delta H_C/L_V$ values, in an order that generally corresponds to the actual stable combustion characteristics of these materials.

Table 6-10 For some solid materials $\Delta H_C/L_V$ value

Name of material	$\Delta H_C/L_V$	Name of material	$\Delta H_C/L_V$	Name of material	$\Delta H_C/L_V$
Red oak	2.96	Nylon	13.10	Polypropylene	21.37
Polyoxymethylene	6.37	Epoxy (FR) glass fiber	13.38	Polystyrene	20.04
Polyvinyl chloride	6.66	Polymethyl methacrylate	15.46	Polyethylene	24.84

6.2.3.3 Shape, size and surface location of solid materials

The same material with large specific surface area tends to be easily ignited and has good stable combustion performance, because the larger the specific surface area, the more opportunities for the material to contact with oxygen in the air, and the easier and more common the oxidation. Because of the strong ability of heat conduction from the surface to the interior of the thin object, the pre-heating effect of the unfired part is better when heated. Therefore, thin objects are easier to catch fire than thick objects.

The same material, under the same external conditions, is easier to be ignited by the fire source when it is upside down than when it is downward, and the stable burning speed of the vertical surface is faster than that of the horizontal surface. Vertically upward (+90°) is the fastest, and the vertical downward (−90°) is the slowest. This is mainly due to the different positions of the solid surface. The degree of pre-heating of the unburned solid fraction by the flame and the thermal products is different. Fig. 6-1 illustrates flame propagation in different orientations and the interaction of the flame and heat products with solids.

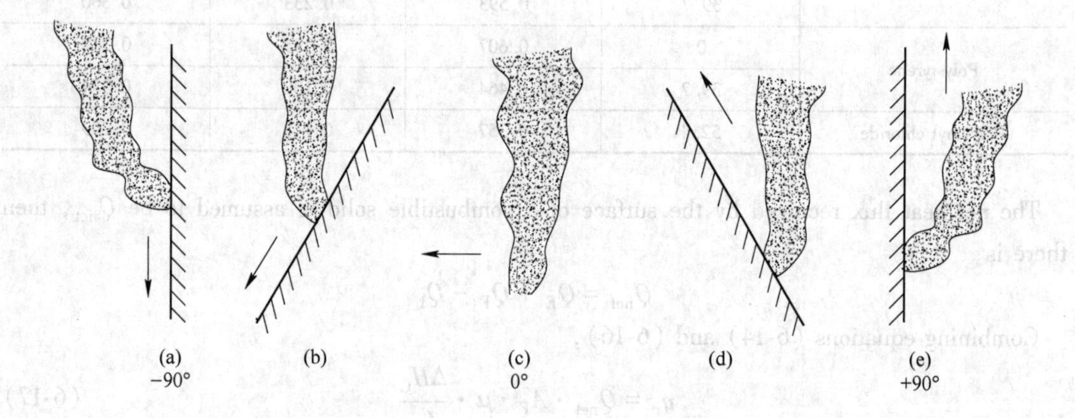

Fig. 6-1 Flame propagation in different directions
(a) ~ (e) Comparison of stable combustion velocity on different surfaces

6.2.3.4 External environmental factors

Figs. 6-2 and 6-3 depict the effects of wind velocity and oxygen concentration and pressure and oxygen concentration, respectively, on the horizontal flame propagation velocity of the hardboard. It can be seen from the figure that the oxygen concentration in the external environment increases, and the ignition and combustion ability of the substance is significantly improved. This is because the flame temperature increases as the oxygen concentration increases, and a higher temperature flame transfers more heat to the combustible surface.

Because the external air flow can promote the mixing of combustible volatiles and air at the flame front, and the flame inclination caused by wind increases the rate of forward heat transfer, it helps the combustion of substances, but too high wind speed can blow out the flame.

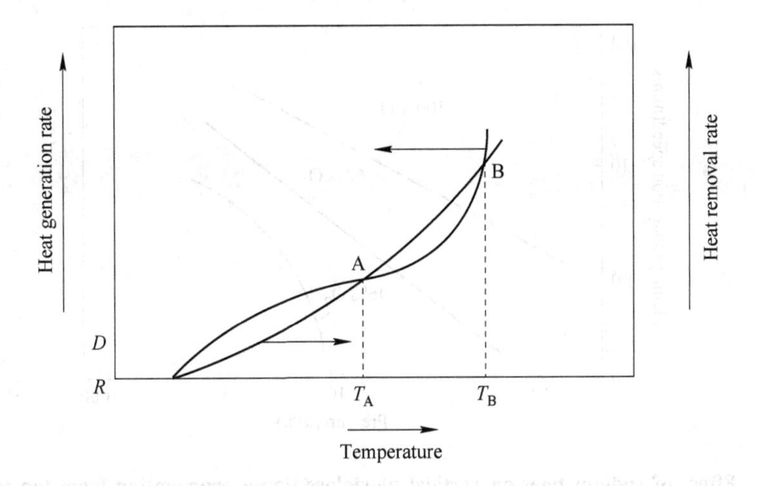

Fig. 6-2 Influence of wind speed and oxygen concentration on flame propagation speed of hardboard

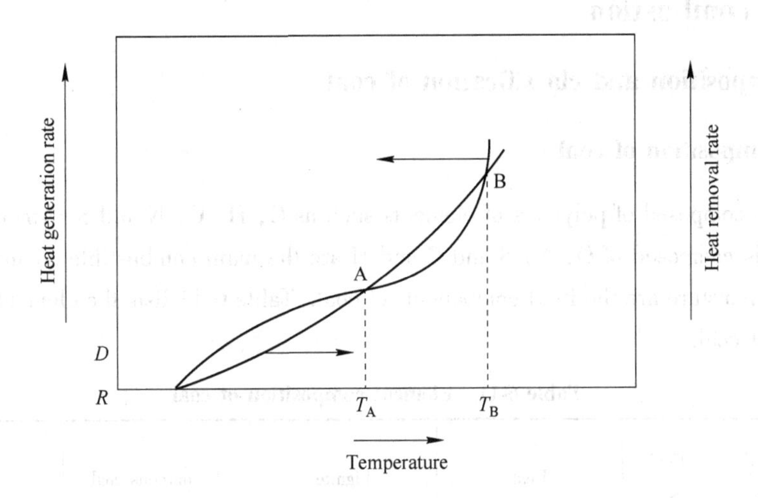

Fig. 6-3 Effect of pressure and oxygen concentration on flame propagation velocity of hard fiberboard

Increasing the ambient pressure will result in a faster combustion rate because the higher pressure helps the flame to stably adhere to the surface of the material.

In addition, as the ambient temperature (i. e., the initial temperature of the substance) increases, the rate of combustion increases. This is because when the initial temperature of the material is higher, less heat is needed for the temperature of the unburned part of the material in front of the flame front to rise to the ignition point, and the external radiation heat will cause an increase in the burning rate, as shown in Fig. 6-4. This is primarily because radiant heat preheats the unburned portion of the material ahead of the flame front, at the same time, the additional radiation heat accelerates the burning rate of the material behind the flame front, which provides an additional forward heat transfer and intensifies the whole combustion process.

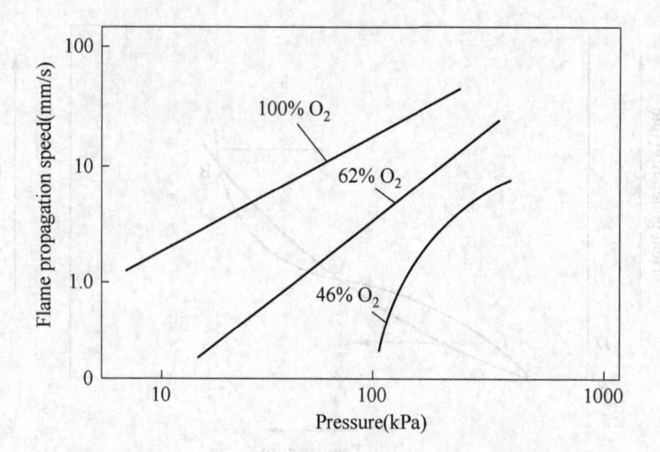

Fig. 6-4 Effect of radiant heat on vertical plexiglass flame propagation from top to bottom
(Figures are applied radiant heat flux, kW/m^2)

6.3 Coal combustion

6.3.1 Composition and classification of coal

6.3.1.1 Composition of coal

Coal is mainly composed of polymers of elements such as C, H, O, N and S. Among them, C, H and compounds composed of O, N, S and C and H are the main combustible components of coal, while ash and moisture are the inert components of coal. Table 6-11 lists the elemental composition of each type of coal.

Table 6-11 Element composition of coal

Type content (%) / Coal / Element	Peat	Lignite	Bituminous coal	Anthracite
C	60~70	70~80	80~90	90~98
H	5~6	5~6	4~5	1~3
O	25~35	15~25	5~15	1~3
N	1~3	1.3~1.5	1.2~1.7	0.2~1.3
S	0.3~0.6	0.2~3.5	0.4~3	0.4

6.3.1.2 Classification of coal

According to the degree of carbonization, coal can be divided into four categories: peat, lignite, bituminous coal and anthracite. Table 6-12 and Table 6-13 show the elemental composition and moisture content of these four types of coal.

Table 6-12 Element content of different coals (%)

Coal type	C	H	O	N	S
Peat	60~70	5~6	25~35	1~3	0.3~0.5
Lignite	70~80	5~6	15~25	1.3~1.5	0.2~0.35
Bituminous coal	80~90	4~5	5~15	1.2~1.7	0.3~0.4
Anthracite	90~98	1~3	1~3	0.2~1.3	0.4

Table 6-13 Moisture content of coal (%)

Coal type	Peat	Lignite	Bituminous coal	Anthracite
Moisture content of raw coal	60~90	30~60	4~15	2~4
Moisture content after air drying	40~50	10~40	1~8	1~2

(1) Peat. Peat is the youngest and least carbonized coal. It is the unique product of ancient marsh environment. Under the condition of abundant water and lack of air, it is a soft organic accumulation layer after deaths. The content of nitrogen and ash elements in this peat is low, the water holding capacity is as high as 40%, and the ventilation is good. Peat is a valuable natural resource because of its unique light weight. In recent years, it has been widely used in horticulture and production of green organic compound fertilizers in China and the world because of its water retention, air permeability, rich organic matter, irreplaceable role of other materials and moderate price.

(2) Lignite. Lignite, also known aswood coal, is a brownish-black, dull low-grade coal between peat and bituminous coal. It is formed by further carbonization of peat and has a bulk density of 750~800kg/m^3. It has high carbon content, strong chemical reactivity, easy oxidation and spontaneous combustion. It's easy to get weathered and fragmented in the air, and is not easy to store and transport.

(3) Bituminous coal. When bituminous coal burns, the flame is longer and comes with smoke. It is a kind of coal with high degree of coalification. Compared with lignite, it has less volatile matter, larger density of about 1.2~1.5, lower water absorption and higher carbon content, of about 75%~90%, therefore its calorific value is higher. Most of it has cohesive properties and is often used as a raw material for daily coking, oil refining, vaporization, low-temperature dry distillation and chemical industries, and of course it can also be used directly as fuel. It can also be used as fuel cells, catalysts or carriers, soil conditioners, filters, building materials, and adsorbents to treat wastewater, etc.

(4) Anthracite. Anthracite is the most highly coalized coal, and it is also the oldest coal. Anthracite has high fixed carbon content, low volatile matter, high density and high hardness and hight ignition point, and it produces no smoke when burning. Generally, its carbon content is more than 90% and the volatile matter is less than 10%. Sometimes the very large volatile matter content is called semi-anthracite, and the very small is called high anthracite.

Anthracite's calorific value is very high, which can reach the 25000 of 32500kJ/kg. The release temperature of the volatile matter is higher. Its coke is not sticky, and it is difficult to catch fire and burn out. Anthracite Lump coal's main application is for fertilizer (nitrogen fertilizer, synthetic ammonia), ceramic industry and manufacturing and forging industries. Anthracite powder is mainly used for blast furnace injection in the metallurgical industry (coal for blast furnace injection mainly includes anthracite, lean coal and gas coal).

6.3.2 Combustion process of coal

Usually, when the coal particles are heated, there is always a part of the gaseous substances decomposed and separated out. Generally, this part of the gaseous substances is called volatile matter, and the remaining part is called coke. If the released volatile matter meets a certain amount of air and has a sufficient temperature, it will ignite and burn. Because coke is more difficult to ignite than volatile, coke generally starts to burn after part or all of the volatiles are burned.

The combustion process of coal can be divided into the following steps. First, the coal is heated and dried, and the volatiles begin to separate out. At this time, if the temperature in the furnace is high enough with the existence of oxygen, the volatiles will burn and form a bright flame. Because oxygen is consumed in the combustion of volatile matter, it can not reach the surface of coke, so the coke is still dark, while the temperature in the center of the coke is only 600 ~ 700℃. In this way, the combustion of the volatiles acts as a hindrance to the combustion of the coke. But on the other hand, the volatile burns near the coal particles, and the coke is heated, so that after the volatile burns out, the coke can burn violently, so the volatile can promote the later combustion of the coke. The lower the carbonization degree of coal is, the lower the temperature at which the volatiles begin to separate out, and the easier it is to catch fire. However, coal with less volatile matter, such as lean coal, is difficult to catch fire although it does not contain much moisture.

Coke remains after the release of volatile matter from the coal, of which fixed carbon is the combustible component. Fixed carbon accounts for a large proportion of the combustible components of coal. Although the ignition of carbon particles is relatively late, its combustion time is very long, so the combustion of carbon is the most decisive part of the coal combustion process.

The combustion reaction on the surface of carbon particles includes the following five processes:

(1) Diffusion of oxygen to the carbon surface;

(2) Adsorption of oxygen on the carbon surface;

(3) Chemical reactions of the adsorbed oxygen with the carbon to form a product which is adsorbed on the surface of the carbon;

(4) Desorption of the product from the carbon surface;

(5) Diffusion of desorbed products into the surrounding environment.

The above processes' occurrence is continuous. According to relevant studies, the chemical reaction between oxygen adsorbed on the carbon surface and carbon atoms is not a simple process.

$$C + O_2 \longrightarrow CO_2 \quad \text{or} \quad 2C + O_2 \longrightarrow 2CO$$

The above two equations can only simply express the material balance relationship at the beginning and after the completion of the chemical reaction, but can not explain how the whole chemical reaction is carried out.

In fact, the carbon combustion reaction is quite complex, which is divided into two stages: the primary reaction and the secondary reaction.

During the primary reaction, the carbon atom reacts with the oxygen adsorbed on its surface to form a carbon-oxygen complex C_3O_4, which then undergoes a dissociation reaction under the impact of other oxygen molecules or a thermal decomposition reaction under high temperature conditions to form CO_2 and CO, i. e.

$$3C + 2O_2 \longrightarrow C_3O_4$$
$$C_3O_4 + C + O_2 \longrightarrow 2CO_2 + 2CO$$
$$C_3O_4 \longrightarrow 2CO + CO_2$$

The CO_2 and CO produced by the primary reaction will continue to react with carbon and oxygen in the following secondary reaction.

Sccondary reactions include:

$$CO_2 + C \longrightarrow 2CO$$
$$2CO + O_2 \longrightarrow 2CO_2$$

There are two chemical reactions, in which CO_2 is an endothermic reduction reaction on the carbon surface, and CO and oxygen undergo an exothermic oxidation reaction in the space around the carbon.

6.3.3 Influencing factors of coal combustion process

6.3.3.1 Effect of volatile matter on combustion process of coal particles

Volatile matter has dual effects on the combustion of coal particles, both positive and negative. Because the ignition temperature of the mixture of the volatile matter and the air is very low, the mixture ignites and burns before the coke, and an envelope flame is formed around the coal particles, so that the temperature of the coke is increased, and favorable conditions are provided for the ignition and combustion of the coke. The increase of coke temperature also promotes the release of volatiles. In addition, after the volatile matter is separated out, a large number of cavities will be formed inside the coke, thus increasing the total surface area of the coke reaction and improving the combustion rate.

The adverse effects of volatile matter on the combustion of coal particles are mainly manifested in the following aspects: the combustion of volatile matter consumes a large amount of oxygen, resulting in a significant reduction in the oxygen diffused to the surface of coke, thus reducing the combustion rate of coal particles, especially in the early stage of combustion, the inhibition effect of volatile matter combustion on the whole combustion of coal particles is particularly obvious.

6.3.3.2 Effect of ash content on combustion process of coal particles

Ash is generally divided into intrinsic ash and extrinsic ash. External ash is the mineral impurities mixed in the process of coal mining and transportation. Its content varies greatly, but it can be removed by coal washing and other measures. Intrinsic ash refers to the minerals already present in the coal during its formation, which are generally evenly distributed in the combustible matter of the coal. Post-treatment measures such as coal washing can not remove the inherent ash of coal, so when coal is burned, these inherent ash will affect the combustion process to a certain extent.

Inherent ash is more evenly distributed in the combustible material. If the combustion temperature is lower than the softening temperature of ash, a gradually thickening ash crust will be formed on the outer surface of coke particles with the progress of combustion; if the combustion temperature is higher than the melting temperature of ash, the ash layer of large coal particles will melt and fall, so no ash shell will be formed on the surface of coke particles. But when large coal particles are accumulated and burned in layers, the slag of the ash will block the ventilation pores between the coal seams. Both conditions interfere with the combustion of the coal.

6.3.3.3 Effect of coke on combustion process of coal particles

Because coke accounts for the largest share of coal, the calorific value of coke accounts for the main part of the calorific value of coal, and its ignition is the latest and the burnout time is the longest (about 90% of the total burnout time), so the combustion of coke plays a decisive role in the combustion process of coal particles. For most coals, the proportion of calorific value contained in coke is generally more than 50%.

6.4 Solid smoldering

Smoldering is the slow burning of some solid substances without visible light, usually producing smoke and accompanied by temperature rise. In terms of the combustion performance test of substances, smoldering is defined as the continuous, smoky and flameless combustion of substances under the specified test conditions. The main difference between smoldering combustion and flaming combustion ispresence or absence of flame, and the main difference between smoldering combustion and flameless combustion is whether combustible gas can be decomposed by heat. Under certain conditions, smoldering combustion can be changed into flaming combustion.

6.4.1 Occurrence conditions of smoldering

Smoldering is a unique combustion form of solid materials, but whether it can occur depends entirely on the physical and chemical properties of solid materials themselves and the external environment.

Smoldering mainly occurs when the solid material is in the condition of no air circulation, such as the smoldering inside the solid stack, and the smoldering of the solid in the room with good sealing. However, there are also cases of smoldering on the surface of the solid dust layer exposed to the external heat flow. In either case, the occurrence of smoldering requires a heat source with appropriate heating intensity. Becausewhen the heating intensity is too small, the solids can not ignite; when the heating intensity is too big, the solids will burn with flames. In porous materials, common heat sources that cause smoldering include:

(1) Natural heat source. The smoldering in the solid stack is mostly the result of spontaneous combustion, and the basic characteristic of the spontaneous combustion of the stacked solid is that it starts with a smoldering reaction inside the stack and then spreads slowly outward until it turns into a flaming combustion on the surface of the stack.

(2) The smoldering itself becomes the heat source. The ongoing smoldering of a solid may become the source of ignition and lead to the smoldering of another solid, such as the smouldering of cigarettes, which often causes the smouldering of carpets, bedding, sawdust, vegetation and so on, and then a vicious fire occurs.

(3) Smoldering after the flame has been extinguished. For example, after the external flame of solid stack flaming combustion is extinguished by water, the water flow does not completely enter the inside of the stack, where it is still in a hot state, so smoldering may occur; indoor solids are in the process of flaming combustion. When the air is consumed to a certain extent, the flame will be extinguished, and then the solid combustion will exist in the form of smoldering.

In addition, asymmetric heating and hot spots inside the solid may cause smoldering.

6.4.2 Propagation theory of smoldering

The horizontal smoldering phenomenon of columnar cellulose materials can well explain the propagation of smoldering. Studies have shown that if one end of the material is properly heated, smoldering may occur, and then it propagates along the unburned area to the other end. The smoldering structure is divided into three regions, as shown in Fig. 6-5.

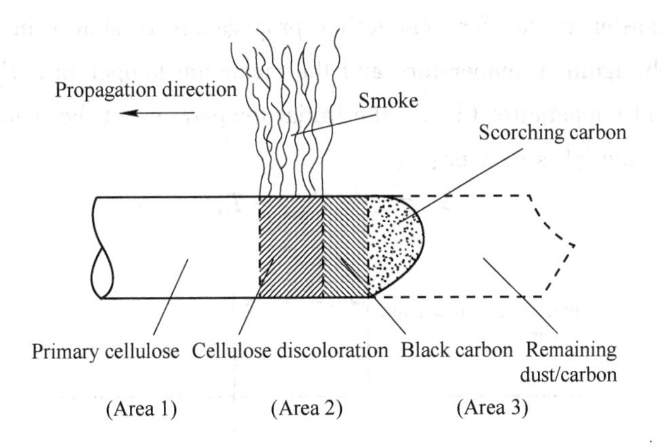

Fig. 6-5 Schematic diagram of smoldering of cellulose rod along horizontal direction

Zone Ⅰ: pyrolysis region. In this zone the temperature rises sharply and fumes are volatilized from the starting material. For the same solid material, the smoke produced in smoldering is quite different from that produced in flaming combustion. Because smoldering usually does not cause obvious oxidation, the smoke contains flammable gases, high boiling point liquids condensed into suspended particles and tar, so it is flammable. There have been smoke explosions caused by the smoldering of latex pads.

Zone Ⅱ: carbonization zone. In this zone, the surface of the char is oxidized and exothermic, and the temperature rises to a maximum. In still air, cellulosic materials smolder at typical temperatures in this region of 600~750℃. Part of the heat generated in this area enters the raw material through conduction, so that its temperature rises and pyrolysis occurs, and the pyrolysis product (smoke) volatilizes, leaving the carbon. For most organic materials, the temperature required to complete this decomposition and carbonization process is greater than 250~300℃.

Zone Ⅲ: residual ash/char zone. In this zone, the glowing combustion no longer takes place and the temperature slowly decreases.

Because smoldering propagation is continuous, there is no clear boundary but a gradual transition stage between the above regions. Whether smoldering can propagate and how fast it propagates depends on the stability of region Ⅱ and its forward heat transfer.

In order to explain the propagation speed of smoldering theoretically, the interface between region Ⅰ and region Ⅱ is defined as the combustion initiation surface. Since the rate of heat transfer across this interface determines the speed of smoldering propagation, in still air, from equation(6-12), we have the following equation:

$$v_{ag} = \frac{q}{\rho \cdot \Delta h} \tag{6-18}$$

Where, v_{ag} is the propagation speed of smoldering, q is the net heat transfer across the combustion initiation surface; ρ is the density of the solid material (packing) and Δh is the change of enthalpy per unit mass of material when it rises from ambient temperature to ignition temperature.

A simple heat transfer model for smoldering propagation is shown in Fig. 6-6. When the difference between the ignition temperature and the maximum temperature T_{max} of Zone Ⅱ is not too large, the ambient temperature (i. e., the initial temperature of the material) is T_0, and the heat capacity of the material is c, then:

$$\Delta h = c \cdot (T_{max} - T_0) \tag{6-19}$$

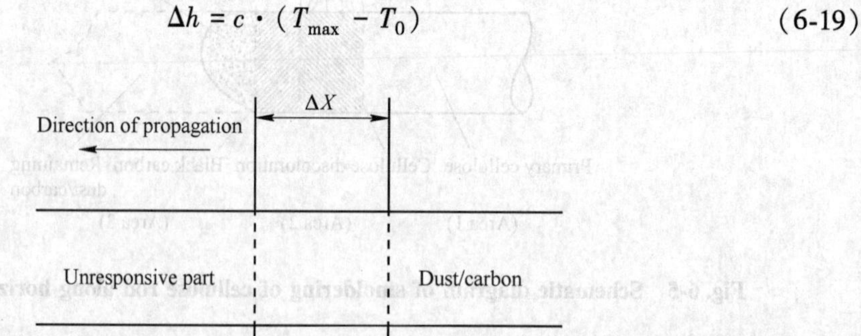

Fig. 6-6 Simple heat transfer model for smoldering propagation

Assuming that the heat transfer is by conduction and is quasi-steady state, we have:

$$q \approx \frac{k \cdot (T_{\max} - T_0)}{x} \tag{6-20}$$

Where, k is the thermal conductivity of the material and x is that heat transfer distance.

Substitute equations (6-19) and (6-20) into equation (6-18) to obtain:

$$V_{ag} \approx \frac{k}{\rho c x} = \frac{\alpha}{x} \tag{6-21}$$

Where, α is the thermal diffusivity.

It is found that the heat transfer distance is about 0.01m. For insulating fiberboard, α is about 8.6×10^{-3}mm/s.

Although the propagation velocity of smoldering determined by equation (6-21) is rough, its order of magnitude is reliable. For example, the order of magnitude of the actual smoldering propagation velocity of the insulating fiberboard is 10^{-2}mm/s, which is basically consistent with the above calculation results.

6.4.3 Influencing factors of smoldering

Smoldering is a very complex combustion phenomenon, which is affected by many factors. These factors mainly include.

6.4.3.1 Properties and dimensions of solid materials

Experiments show that the material with soft texture, fine and less impurities has good smoldering performance. This is because the thermal insulation performance and thermal insulation performance of this kind of material are better, and the heat is not easy to dissipate. Cotton is a typical representative of this kind of material.

The influence of the size (mainly the diameter) of a single material on smoldering is very complex, and it is difficult to draw a unified conclusion. The effect of dust layer size on smoldering can be explained from two aspects of thickness and particle size. For the fine dust layer, the propagation velocity of smoldering increases with the decrease of the thickness in a certain range, but the propagation velocity of smoldering decreases when the thickness decreases to a certain extent. Moreover, there is a lower limit for maintaining the smoldering thickness of the dust layer, as shown in Table 6-14. This effect can be explained as follows: the air is difficult to enter the smoldering zone when the thickness is large; when the thickness is too small, there is too much heat loss.

Table 6-14 Lower limit of smoldering thickness of cork powder with different particle sizes

Particle size (mm)	0.5	1.0	2.0	3.6
Lower limit of thickness (mm)	~12	~36	~47	~36

It can be seen from Table 6-14 that the lower limit of thickness increases with the increase of particle size, but when the particle size increases to a certain extent, the lower limit of thickness decreases due to burning. For a certain thickness of dust layer, the propagation velocity of smoldering increases slowly with the decrease of particle size. Although it is more difficult for the air to enter the smoldering zone due to the reduced particle size, the adiabatic condition is improved, the heat loss is reduced, and the behavior of the smoldering dust layer shows that the latter effect is slightly dominant, so the propagation speed is slightly increased.

By the way, when the density of the dust layer decreases, the propagation speed of smoldering also increases. This conclusion can also be explained by following the above.

6.4.3.2 Apply air flow (wind) speed

The test shows that the lower limit of smoldering thickness of the dust layer under the action of the external air flow is significantly reduced, as shown in Fig. 6-7. With the increase of the air flow velocity, the smoldering propagation velocity increases obviously, especially when the air flow direction is consistent with the smoldering propagation direction. This is in addition to the fact that the air flow promotes the transport of oxygen to the smoldering zone, but also because of the increased heat transfer from region Ⅱ to region I. This effect is more significant for coarse particle size dust. If the air flow velocity is too high, the smoldering will change to flaming combustion.

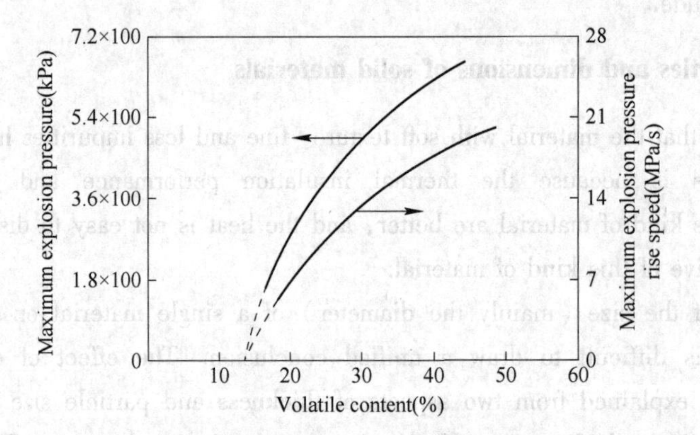

Fig. 6-7 Variation of the lower thickness limit with air flow

(Beech sawdust, average particle size 0.48mm)

With the increase of oxygen concentration in the environment, the propagation speed of smoldering also increases significantly, which is also due to the enhanced diffusion rate of oxygen into the smoldering zone. Since the maximum temperature in the combustion zone is directly related to the oxygen concentration, i.e., the higher the oxygen concentration, the higher the temperature in the combustion zone. Therefore, the above effect of the applied air flow or ambient oxygen concentration on the smoldering also indicates the effect of the maximum temperature in the combustion zone on the smoldering. The test results also show that the propagation speed of smoldering increases with the increase of the maximum temperature in region Ⅱ, as shown in Fig. 6-8. In the formula (6-21), many practical factors affecting the smoldering are ignored. Therefore, the relationship between the propagation speed of smoldering and the maximum temperature in region Ⅱ is not reflected.

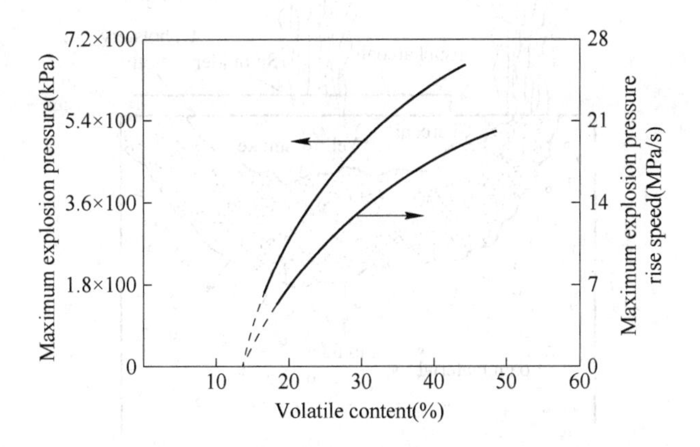

Fig. 6-8 Relationship between smoldering propagation speed and maximum temperature in region Ⅱ

(Horizontal smoldering of cellulose rod)

6. 4. 3. 3 Direction of smoldering propagation

It is found that the smoldering speed of upward propagation is the fastest, the smoldering speed of horizontal propagation takes the second place, and the speed of downward propagation is the slowest for the same solid material under the same environmental conditions, which indicates that the upward propagating smoldering state is more dangerous. The general explanation is as follows: for upwardly propagating smoldering, the products of combustion or pyrolysis flow by buoyancy to the unburned portion of the material, and the former preheats the latter, in which case, oxygen is less hindered from entering region Ⅱ. On the contrary, there is no such "preheating" effect in the downward propagating smoldering, and this situation is unfavorable to the diffusion of oxygen to the region Ⅱ; the horizontal propagating smoldering is in the middle.

6. 4. 3. 4 Smoldering of binary material system

Some polymeric foams, such as flexible polyurethane foams with high elasticity, are difficult to smolder when they are alone, but they can smolder if they are combined with many materials such as fabrics to form a binary material system. This shows that some materials which are easy to smolder play a decisive role in the smoldering of other materials which are difficult to smolder, as is shown in Fig. 6-9. If the foam smoldering occurs in still air, the maximum temperature reached in Zone Ⅱ does not exceed 400℃, which is significantly lower than the maximum temperature in Zone Ⅱ of cellulose smoldering (600℃), which may be the main reason why some foams are difficult to smolder when they are alone. Even in the case shown in Fig. 6-9, the rate of foam smoldering propagation is relatively slow. It has also been suggested that the mechanism of smoldering propagation of these foams involves radiative heat transfer through a sparse mesh structure.

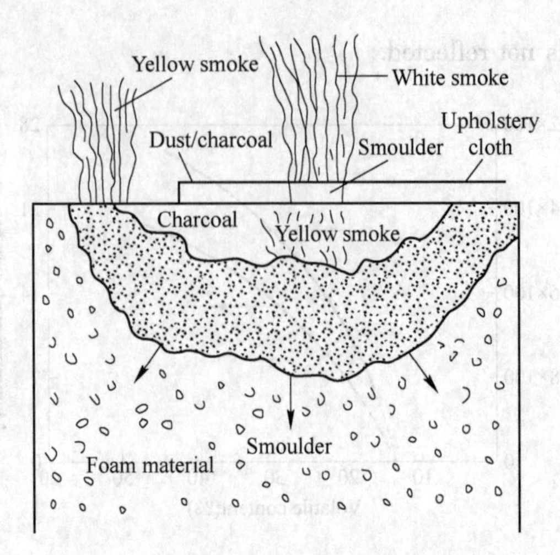

Fig. 6-9 Schematic diagram of the interaction of smoldering of fabric-foam system

In addition to the above factors, the smoldering characteristics of solid materials are also affected by impurities in them. In addition, humidity is unfavorable to smoldering, because humidity increases the heat capacity of the unburned part of the material, increases the heat demand for thermal decomposition, and limits the propagation of smoldering

6.4.4 Transition from smoldering to flaming combustion

The transition from smoldering to flaming combustion is one of the important contents of smoldering research. The above factors that are favorable for smoldering are also favorable for the transition from smoldering to flaming combustion, for example, the external air flow is favorable for this transition; the upward-propagating smoldering is more likely to transition to flaming combustion than the downward-propagating one. Soft slight smoldering like cotton can easily turn into flaming combustion.

Generally speaking, when the temperature of region Ⅱ increases, the temperature of region Ⅰ rises due to heat conduction, the pyrolysis rate accelerates, the volatile matter increases, and the concentration of combustible gas in the space near region Ⅰ increases. When the temperature continues to rise, it can also ignite spontaneously. This completes the transition from smoldering to flaming combustion. Because this transition is an unstable process, it is difficult to determine the transition temperature accurately.

Generally speaking, the transition from smoldering combustion to flaming combustion mainly includes the following situations:

(1) The smoldering changes to flaming combustion as it propagates from the inside to the outside of the material stack. Inside the material stack, only smoldering can occur due to lack of oxygen. However, as long as the smoldering does not interrupt its propagation, it will eventually develop to the outside of the stack, and because it is no longer anoxic, it is likely to turn into flaming combustion.

(2) The heating temperature is increased, then the smoldering changes to flaming combustion. When the smoldering solid material is affected by external heat, the release rate of volatiles in region I increases with the increase of heating temperature. When this rate exceeds a certain critical value, smoldering will develop into flaming combustion. This transformation can also occur within the material stack.

(3) The smoldering of materials in a confined space is converted into flaming combustion(even a flashover). In a confined space, due to insufficient oxygen supply, the solid materials in the space are smoldering, and a large number of incomplete combustion products are generated to fill the whole space. At this time, if some parts of the space are suddenly opened, flammable mixed gases are formed in the space due to the entry of fresh air, and then flaming combustion occurs, which may also lead to a flashover. This sudden transition from smoldering to flashover is very dangerous.

6.5 Dust explosion

Dust refers to solid substances in a dispersed state. The phenomenon of dust explosion makes dust much more dangerous and harmful than the original large solid. Therefore, people should pay attention to the study of this problem.

6.5.1 Conditions of dust explosion

Generally, dust explosion shall meet three conditions:

6.5.1.1 The dust itself is combustible

Combustible dust includes organic dust and inorganic dust. However, under normal conditions, not all combustible dusts can explode. See Table 6-15 for common explosive dust.

Table 6-15 Types of common explosive dust

Type	For example
Carbon products	Coal, charcoal, coke, activated carbon, etc.
Fertilizer	Fish meal, blood meal, etc.
Food category	Starch, sugar, flour, cocoa, milk powder, cereal flour, coffee powder, etc.
Xyloid	Wood flour, cork flour, lignin flour, paper flour, etc.
Synthetic products	Dye intermediates, various plastics, rubber, synthetic detergents, etc.
Agro-processing category	Pepper, pyrethrum, tobacco, etc.
Metals	Aluminum, magnesium, zinc, iron, manganese, tin, silicon, ferrosilicon, titanium, barium, zirconium, and the like

6.5.1.2 Dust is suspended in the air at a certain concentration

Dust that is deposited (in the aerogel state) is not explosive. Only dust in suspension (aerosol state) can explode. Whether the dust can be suspended in the air and the suspension time depend on the dynamic stability of the dust, which is mainly related to the particle size, density, ambient temperature and humidity of the dust.

Suspended dust can explode only when its concentration is within a certain range. This is because the dust concentration is too small, the combustion heat release is too small, it is difficultto form a continuous combustion and can not explode; if the concentration is too high, the oxygen concentration in the mixture is too low to cause an explosion.

6.5.1.3 The presence of an ignition source sufficient to cause a dust explosion

Dust combustion first requires heating, or melting and evaporation, or thermal cracking to release combustible gas, so dust explosion requires more energy. Its minimum ignition energy is about 10~100MJ, which is $10^2 \sim 10^3$ times larger than that of the combustible gas.

The reason why the dust with the above conditions can explode is that the combustible dust suspended in the air forms a highly dispersed system, and its surface area and surface energy (reflected by adsorption and activity) are greatly increased; at the same time, the interface between dust particles and oxygen in the air is enlarged, and the oxygen supply is more than abundant. Thus, upon ignition by an ignition source of sufficient energy, the reaction rate is greatly accelerated and explosive.

6.5.2 Process and characteristics of dust explosion

Dust explosion generally goes through the following processes: firstly, the surface temperature of dust particles receiving the energy of the fire source rises rapidly, so that the dust particles are decomposed or distilled rapidly, and the generated combustible gas is released into the gas phase around the particles; secondly, the mixture of combustible gas and air is ignited by the fire source and then burns with flame. The onset of such combustion is usually localize, and the combustion heat is transferred by radiation and convection to make the flame spread and diffuse. Third, the flame is in the process of propagation the generated heat promotes the decomposition or dry distillation of more and more dust particles, releases more and more combustible gas, accelerates the combustion cycle step by step, and finally leads to dust explosion.

It should be pointed out that the above process is for dust explosion that can release combustible gas. This kind of dust releases combustible gas, some through thermal decomposition (such as wood powder, paper powder, etc.), some through melting evaporation or sublimation (such as camphor powder, naphthalene powder, etc.). Essentially, the explosion of this kind of dust is the explosion of combustible gas, but the combustible gas is "stored" in the dust. The dust is not released until it is heated.

Charcoal, coke and some metal dusts do not release combustible gas in the process of explosion. After receiving the heat energy of the fire source, they directly react violently witho xygen in the air and catch fire. The reaction heat produced makes the flame spread. In the process of flame propagation, the hot dust or its oxides heat the surrounding dust and air, making the high-temperature air expand rapidly. Resulting in a dust explosion.

Compared with gas explosion, dust explosion has two characteristics:

(1) More ignition energyand detonation time is required for a dust explosion than for a gas explosion, and the process is complex. This is because dust particles are much larger than gas molecules, and dust explosion involves a series of physical and chemical processes such as decomposition and evaporation. The detonation time of dust explosion is long, which makes it possible to detect the precursor of explosion with rapid devices, and curb the development of explosion (explosion suppression technology).

(2) The maximum explosion pressure of the explosion is slightly less than that of the gas explosion. However, the explosion pressure of the former rises and falls slowly, so the product of pressure and time (that is, the energy released by the explosion) is larger, and the dust particles disperse while burning, so the destructiveness of the explosion and the degree of burning damage to the surrounding combustibles are also more serious.

In terms of hazards, dust explosion has the following characteristics:

(1) The air wave generated by the initial explosion of the dust will raise the deposited dust. Formation of explosive concentration in the new space, resulting in a secondary explosion. In addition, at the initial explosion site of dust, the air and combustion products are heated and expanded, and the density becomes thin. After a very short time, a negative pressure area is formed, and the fresh air flows back to the explosion site, which promotes the secondary impact of air (referred to as "return wind"). A secondary explosion may also occur if dust and fire are still present at the explosion site, which tends to be more stressful and destructive than primary explosions. In the continuous production system, this kind of secondary explosion may occur continuously, forming a chain explosion, and some may reach the level of detonation, resulting in very large damage.

(2) Some dust explosion accidents not only show the characteristics of explosion continuity Moreover, with the continuation of the explosion, the reaction speed and explosion pressure continue to accelerate and rise, and show a leap-forward development, thus showing the characteristics that the farther away from the detonation point, the more serious the damage. Especially when there are obstacles or bends in the way of explosion propagation, the explosion pressure will rise sharply. Table 6-16 lists the explosion pressure test data for coal dust. When there are obstacles, the propagation of dust explosion is blocked, the explosion shock wave is reflected back, and the pressure increases exponentially.

Table 6-16 Coal dust explosion pressure test data

Distance from explosion point (m)	Explosion pressure (Pa)	
	No obstructions	There are obstacles
91.5	2.91×10^4	1.58×10^5
120.9	4.51×10^4	5.72×10^5
137.2	1.10×10^5	1.05×10^6

(3) Dust the explosion (especially the dust of organic matter) is easy to cause incomplete combustion. A large amount of incomplete combustion products such as CO will be produced. This will not only cause personnel poisoning, but also may cause gas explosion in closed places.

6.5.3 Important characteristic parameters of dust explosion

6.5.3.1 Explosion pressure and rate of pressure rise

When dust explodes, the volume of the product will mostly exceed the volume of the initial mixture, especially the high temperature produced by the explosion makes the volume of product and air mixture expand rapidly, resulting in a sharp increase in explosion pressure. Generally speaking, the explosion pressure of a certain dust refers to the maximum pressure that the dust can reach when it explodes at a specified concentration, which is expressed by P_m; the maximum explosion pressure of a certain dust in a large concentration range is called the maximum explosion pressure of the dust, which is expressed by P_{max}.

The ratio of explosion pressure to time is called the rate of pressure rise. expressed by $\left(\frac{dP}{dt}\right)_m$ the maximum value of the pressure rise velocity reached by a certain dust in a large concentration range, which is called the maximum explosion pressure rise velocity of the dust, expressed by $\left(\frac{dP}{dt}\right)_{max}$. The pressure rise velocity of dust explosion is an important parameter to measure the intensity of dust explosion.

The existence of explosion pressure and pressure rise speed is the main reason for the destruction of the equipment. The greater the dust explosion pressure, the faster the pressure rise, and the more serious the damage to the equipment.

6.5.3.2 Explosion limit

Dust explosion limit is the minimum concentration (lower limit) or maximum concentration (upper limit) of dust that can explode when the mixture of dust and air meets fire source, which is generally expressed by the mass of dust contained in unit volume space. When the chemical composition and heat of combustion of the dust are known, and certain simplifying assumptions are made, the explosion limit can be calculated. But it is still usually determined with special instruments.

Experiments show that the lower explosive limit of many industrial dusts is $20 \sim 60 g/m^3$, and the upper explosive limit is $2000 \sim 6000 g/m^3$. Because of dust sedimentation and other reasons, it is difficult to reach the upper limit of explosion in the actual situation, so the upper limit of explosion of dust is generally of no practical value, while the lower limit of explosion is of great significance, the lower the lower explosion limit, the greater the risk of explosion.

6.5.3.3 Minimum detonating energy (E_{min})

The minimum detonation energy of dust explosion can also be obtained from the spark discharge energy. Combustible dust can explode when it touches a fire source with energy exceeding its minimum ignition energy.

In addition to the above parameters, the spontaneous ignition point of dust in suspension is also an important characteristic parameter of dust explosion, which reflects that when the dust is in suspension, the risk of fire and explosion caused by spontaneous combustion due to heating is too small.

6.5.4 Influencing factors of dust explosion

The influence of various factors on dust explosion is mainly reflected in the influence of these factors on the characteristic parameters of dust explosion. When analyzing and solving the explosion problem of actual dust, the following main factors should be considered.

6.5.4.1 Physical and chemical properties of dust

The more combustible volatile dust contains, the greater the risk of explosion, and the higher the explosion pressure and pressure rise speed, as shown in Fig. 6-10. This is because this kind of dust releases more combustible gas when it is heated, and a large amount of combustible gas mixes with air to form an explosive mixture, which makes the system react more easily and violently. For example, 1kg of coking coal containing 20%~26% of volatile matter, it can release 290~350L combustible gas at high temperature, so its dust is easy to explode and form a higher explosion pressure (0.4~0.6MPa).

The combustion heat is related to the amount of combustible gas released by the dust, the dust with high combustion heat is prone to explosion, as shown in Fig. 6-11, and the explosion power is also large. In addition, the dust with fast oxidation speed, such as magnesium, ferrous oxide and dye, is easy to explode, and the maximum explosion pressure is high; the dust that is easy to electrify is also easy to explode .

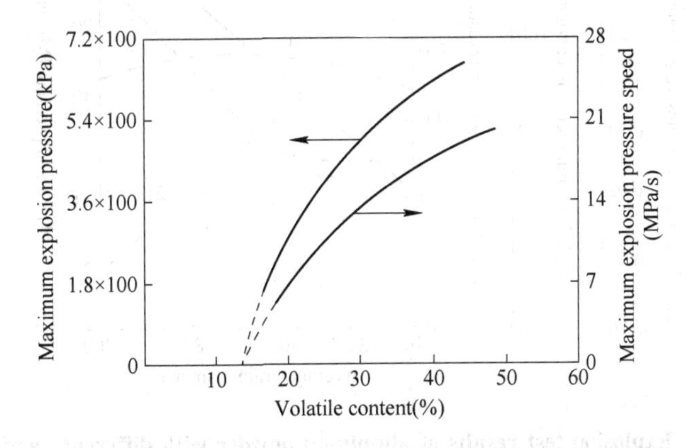

Fig. 6-10 Relationship between dust explosion pressure, pressure rise rate and volatile content

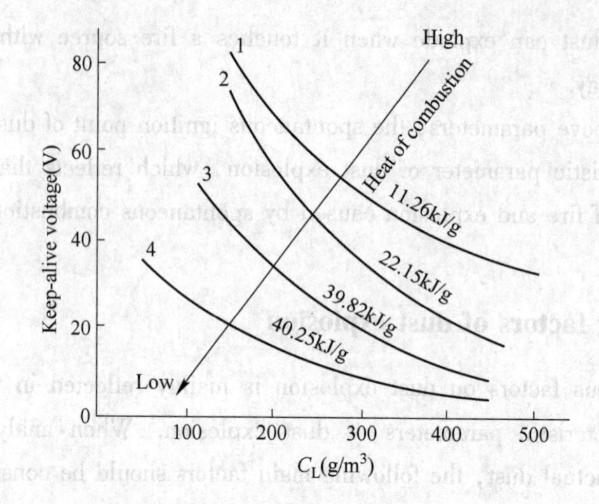

Fig. 6-11 Influence of dust combustion heat on explosion performance

1—2,4,6-Trinitrophenol; 2—Trinitronaphthalene; 3—Anthracene; 4—Naphthalene

6.5.4.2 Particle size and concentration of dust

Particle size is an important factor of dust explosion. The smaller the particle size of the dust, the greater the specific surface area, the greater the degree of dispersion in the air and the longer the suspension time; the stronger the activity of oxygen adsorption, and the faster the oxidation reaction rate, the easier it is to explode, that is, the smaller the minimum ignition energy and lower limit of explosive concentration is, and the greater maximum explosion pressure and maximum pressure rise rate correspondingly are, This is illustrated by the results of the aluminum dust explosion test shown in Fig. 6-12.

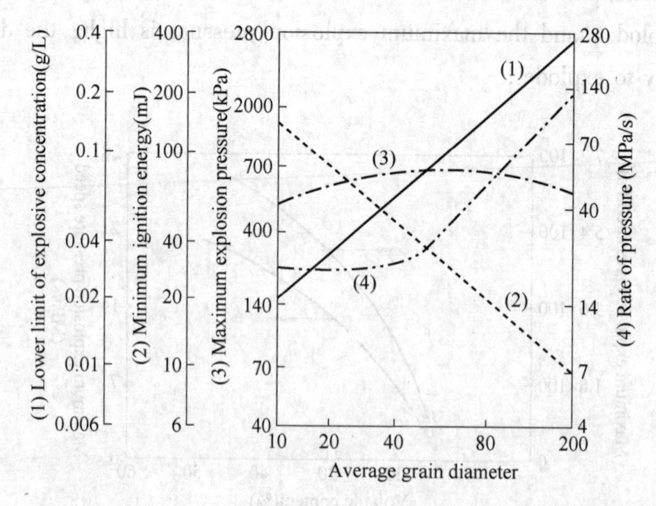

Fig. 6-12 Explosion test results of aluminum powder with different particle size

If the particle size of the dust is too large, it will lose its explosive properties. For example, dusts such as polyethylene, flour and methyl cellulose with particle sizes of more than 400μm can not explode, while most coal dusts can only explode with particle sizes of less than 1/10~1/15mm. When the coarse dust larger than the explosion critical particle size is mixed with a certain amount of fine explosible dust, it may be called an explosive mixture.

Combustible dust can explode only when its concentration is within the limit of explosive concentration, and its most easily detonated concentration is generally 2 to 3 times higher than its stoichiometric concentration for complete combustion.

The maximum values of dust explosion pressure and pressure rise velocity also occur when the dust concentration is 2 to 3 times higher than the stoichiometric concentration, but the concentrations at which the two reach the maximum values are not necessarily the same (as shown in Fig. 6-13), and the dust concentration at which the maximum explosion pressure or pressure rise velocity occurs is also different from the concentration when the dust is most likely to be ignited. However, the relationship between these two concentrations is roughly as follows:

$$C_{\mathrm{Emin}} = \left| C_{\mathrm{Pmax}} - \left(C_{(\mathrm{d}P/\mathrm{d}t)\max} - C_{\mathrm{Pmax}} \right) \right| \tag{6-22}$$

In the formula, C_{Emin}, C_{Pmax}, $C_{(\mathrm{d}P/\mathrm{d}t)\max}$ are the dust concentrations at which the minimum ignition energy, the maximum explosion pressure, and the maximum pressure rise velocity occur, respectively.

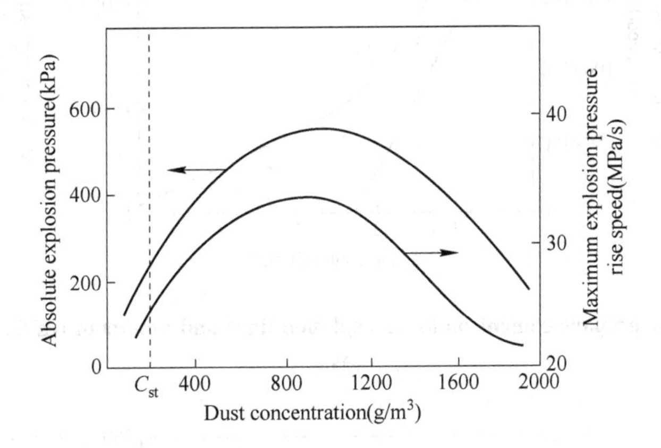

Fig. 6-13 Effect of concentration on explosion pressure and pressure rise rate (Benzoic acid dust)

Under the condition of a certain particle size, the higher the dust concentration is, the lower the ignition temperature is, but this effect gradually weakens with the increase of particle size, as shown in Fig. 6-14.

6.5.4.3 Content of combustible gases and inert components

When a certain amount of combustible gas is mixed into the mixture of combustible dust and air, the explosion hazard of dust increases significantly, which is reflected in the decrease of minimum ignition energy and lower explosion limit, and the increase of maximum explosion pressure and maximum pressure rise rate (as shown in Figs. 6-15 and 6-16). This is because the mixing of the combustible gas makes the mixture easy to ignite and increases the combustion speed.

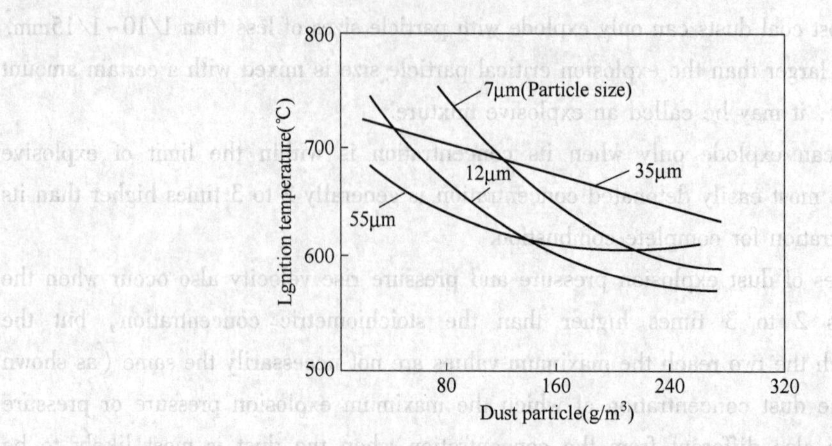

Fig. 6-14 Relationship between ignition temperature and concentration of magnesium powder

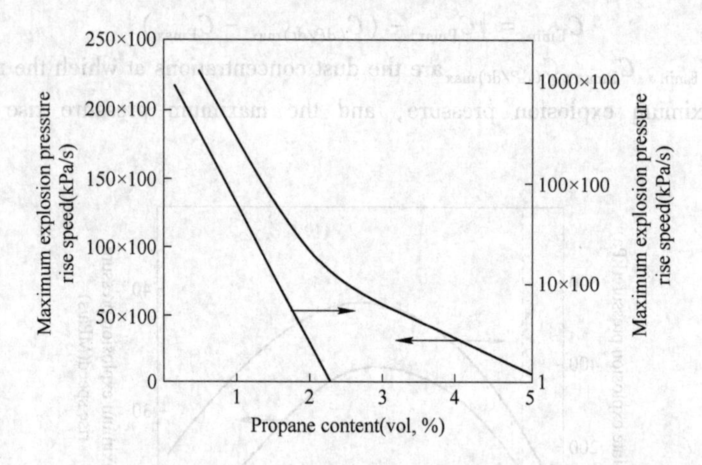

Fig. 6-15 Influence of propane content on lower explosion limit and minimum ignition energy of PVC dust

(125μm)

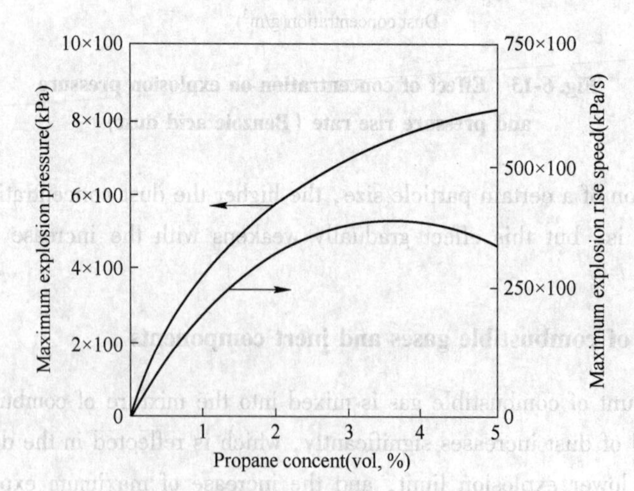

Fig. 6-16 Propane content to PVC dust

(125μm)

The lower explosive limits of the dusts in the combustible dust, air and gas mixture and the concentration of the combustible gas (mainly methane and propane) have the following approximate relationship:

$$C_{mdl} = C_{dl}\left(\frac{L_{GL}}{L_G} - 1\right)^2 \tag{6-23}$$

Where, C_{mdl} and C_{dl} are the lower explosive limits of the dust in the mixture and air; L_{GL} and L_G are the combustible gas contents in the mixture and the lower explosive limits of the combustible gas in air respectively.

When the mixture of combustible dust and air is mixed with a certain amount of inert gas, it will not only reduce the concentration range of dust explosion, but also reduce the pressure and pressure rise speed of dust explosion, as shown in Fig. 6-17. This is mainly because the mixing of inert gases reduces the oxygen content of the dust environment, which reduces or even completely lose the explosive performance of the dust. Table 6-17 lists the critical oxygen contents for some combustible dust atmospheres when inerted with nitrogen.

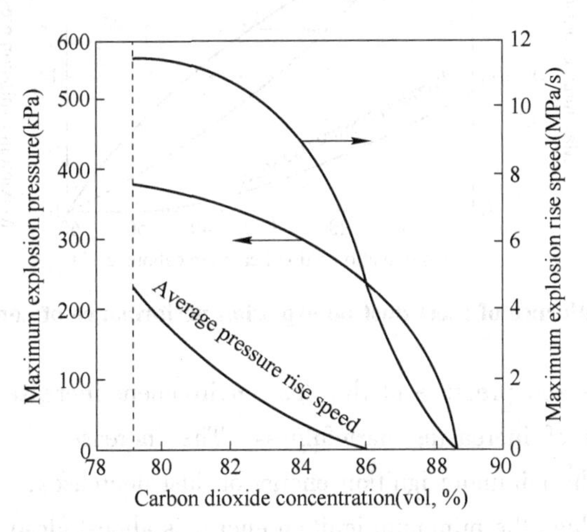

Fig. 6-17 Oxygen concentration, explosion pressure and pressure rise speed and the speed of detonation

Table 6-17 Critical oxygen content (Vol%) of some dusts when inerting with N_2

Dust name	Critical oxygen content	Dust name	Critical oxygen content	Dust name	Critical oxygen content
Coal dust	14.0	Organic pigment	12.0	Rosin powder	10.0
Cadmium laurate	14.0	Calcium stearate	11.8	Methyl cellulose	10.0
Barium stearate	13.0	Wood flour	11.0	Light metal dust	4~6

Inert dust mixed with combustible dust will also weaken or even lose its explosive performance, as shown in Fig. 6-18. This is because inert dusts have a cooling effect, and some inert dusts also have a negative catalytic effect.

6.5.4.4 Ambient conditions for dust explosion

The moisture in the combustible dust environment will weaken the explosion performance of the dust, because the moisture plays the role of adsorption of non-combustible components. Moisture can bind small particles of dust, reduce the dispersion of dust and shorten its floating time; moisture evaporation absorbs a lot of heat, preventing the combustion chemical reaction of the dust; the water vapor occupies the spaces, diluting the oxygen concentration in the environment, which reduces the combustion rate of the dust. The weakening effect of water increases with the increase of its content.

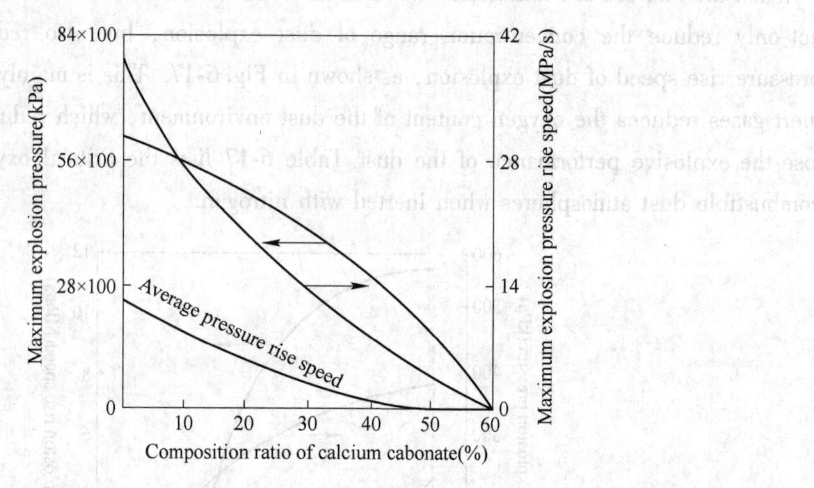

Fig. 6-18 Influence of inert dust on explosion performance of combustible dust

When the temperature and pressure of the dust environment increase, the dust explosion will change in the direction of increasing harmfulness. The increase of temperature promotes the release of volatiles, so the minimum ignition energy of dust decreases, and when the temperature increases to a certain value, the minimum ignition energy is almost close to zero. This temperature value is the ignition temperature of the suspended dust. The ignition temperatures of some dust clouds are shown in Table 6-18.

Table 6-18 Ignition temperature of some dust clouds

Dust name	Cellulose acetate	High pressure polyethylene	Polystyrene	Grain	Peanut shell	Powdered sugar	Skimmed milk	Wheat flour
Ignition temperature (℃)	460	450	500	400	460	370	490	440

Dust explosion has a low pressure limit. Generally, dust cannot explode when the ambient pressure is lower than several kilopascals.

6.5.4.5 Intensity of fire source or ignition method

The higher the temperature of the fire source, the longer the contact time with the combustible dust, air mixture, or the greater its energy, the more likely the dust will explode. The data listed in Table 6-19 show that the lower explosive limit of dust is lower when the fire source is stronger.

Ignition mode has a great influence on the explosion characteristics of dust, which is illustrated in Table 6-19 and furtherreflected in Table 6-20. For example, a weak capacitive discharge (energy of a few tens of mJ) can give a large explosion characteristic value as can a chemical detonator having a higher ignition energy (energy of 10000J).

Table 6-19 Relationship between lower explosion limit of dust and ignition source

Source of ignition / Lower explosive limit (%) / Dust	1200℃ glowing body	33V5A arc light	6.5V3A Induction Coil Spark
Starch	7.0	10.3	13.7
Dust from the wheat barn	10.3	10.3	No fire
Sugar	10.3	17.2	34.4

Table 6-20 Influence of ignition mode on dust explosion characteristics

Dust name	Ignition mode	Ignition Energy (J)	P_{max} ($\times 10^5$Pa)	$(dP/dt)_{max}$ ($\times 10^5$Pa/s)
Lycopodium	Chemical detonator	10000	3.2	186
	Capacitor discharge	0.08	8.3	199
	Fix the spark gap	10	8.4	153
Cellulose	Chemical detonator	10000	9.7	150
	Capacitor discharge	0.04	9.2	147
	Fix the spark gap	10	8.2	63

6.5.4.6 The volume of the container

Like the explosion of combustible gas, the larger the volume of the container, the longer the time of dust explosion, the longer the time from the beginning of explosion to the maximum pressure rise (as shown in Fig. 6-19), and the smaller the maximum pressure rise speed of dust explosion. Numerous dust explosion tests (as shown in Fig. 6-20) have demonstrated that the "Cubic Law" is fully applicable to dust explosions if the container volume is not less than 0.04m^3, i.e.:

$$\left(\frac{dP}{dt}\right)_{max} \cdot V^{\frac{1}{3}} = K_{st} = \text{Constant} \tag{6-24}$$

In addition to the above factors, other factors will be encountered in the actual conditions, such as the turbulence of dust, air mixture, the water content of dust particles, cohesion and thermal conductivity. When analyzing and solving the problem of actual dust explosion, the influence of these factors should be considered comprehensively according to the field conditions.

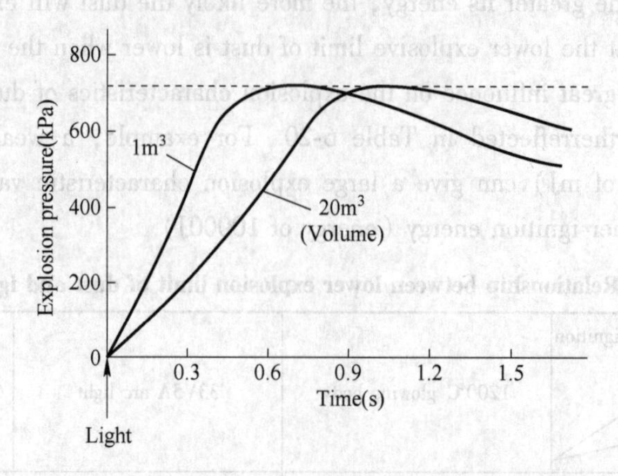

Fig. 6-19 Influence of vessel volume on coal dust explosion time

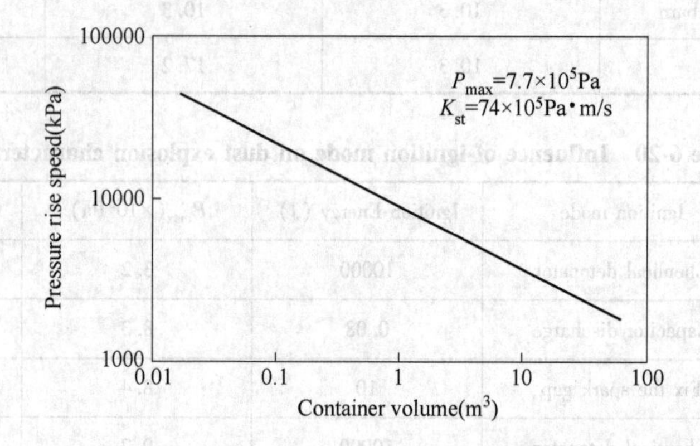

Fig. 6-20 Validation of "Cubic Law" for dust explosion

6. 5. 5 Prevention and control of dust explosion

The basic methods for the prevention and control of flammable gas explosions are also applicable to prevent flammable dust explosions. These methods include strict control or elimination of ignition sources; prevention of formation of explosive gases. Main measures include elimination of dust source and protection with inert gas; suppression of explosion and explosion-proof pressure relief. Compared with the explosion of combustible gas, the detonation time of dust is long, and the explosion pressure and pressure rise speed are small, so the latter two measures to control dust explosion are very important.

6. 5. 5. 1 Suppression of dust explosion

The dust explosion suppression device can quickly spray fire extinguishing agent at the initial stage of dust explosion to extinguish the flame and stop the development of explosion. It consists of an explosion detection mechanism and an extinguishing agent spraying mechanism. The former must react quickly and act accurately in order to quickly detect the precursors of the explosion and send out signals. The latter receives the signals sent by the former., and is immediately activated after passing through the amplified signal, to spray fire extinguishing agent.

The structure and effects of the dust explosion suppression device are shown in Fig. 6-21.

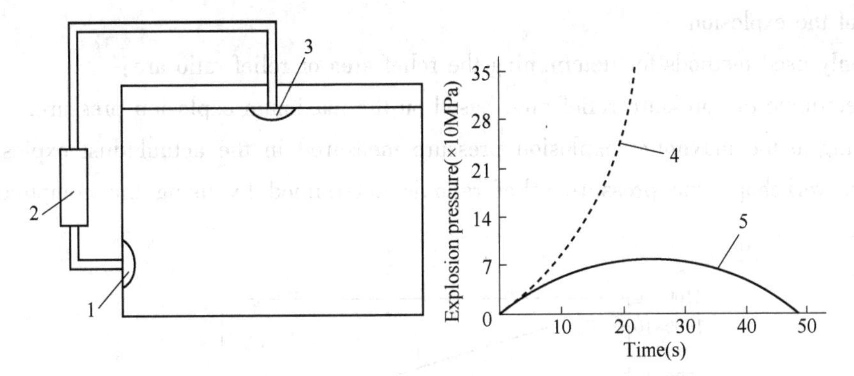

Fig. 6-21 Explosion suppression device

1—Pressure sensor; 2—Amplifier; 3—Suppressor;

4—Normal explosion pressure curve; 5—Suppressed post-explosion pressure curve

6.5.5.2 Set explosion-proof pressure relief device

Weak surfaces (pressure relief surface) shall be set at proper position of equipment or plant to discharge pressure, flame, dust and products at the initial stage of explosion, so as to reduce explosion pressure and explosion loss, as shown in Fig. 6-22.

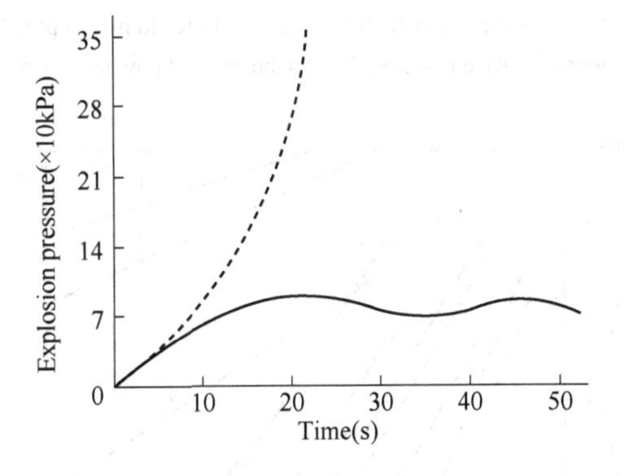

Fig. 6-22 Schematic diagram of explosion-proof pressure relief

When explosion-proof pressure relief techniques are used, the maximum pressure and maximum pressure rise speed of dust explosion must be taken into account. In addition, the volume and structure of equipment or plant, as well as the material, strength, shape and structure of pressure relief surface should be considered. Facilities used as pressure relief surface include blasting plate, side door, hinge window, etc. Materials used as the pressure relief surface include metal foil, waterproof paper, waterproof cloths or plastic boards, rubber, asbestos boards, gypsum board, etc. The most important problem is to determine the size of the explosion-proof pressure relief area.

At present, there are many methods to determine the pressure relief area, but no matter which method is used, the pressure relief area should be increased appropriately, because the explosion of the pressure relief surface material itself may interfere with the dust explosion and increase the intensity of the explosion.

Commonly used methods for determining the relief area or relief ratio are:

(1) Determine the pressure relief ratio based on the maximum explosion pressure.

According to the maximum explosion pressure measured in the actual dust explosion in the production workshop, the pressure relief ratio is determined by using the computation graph Fig. 6-23.

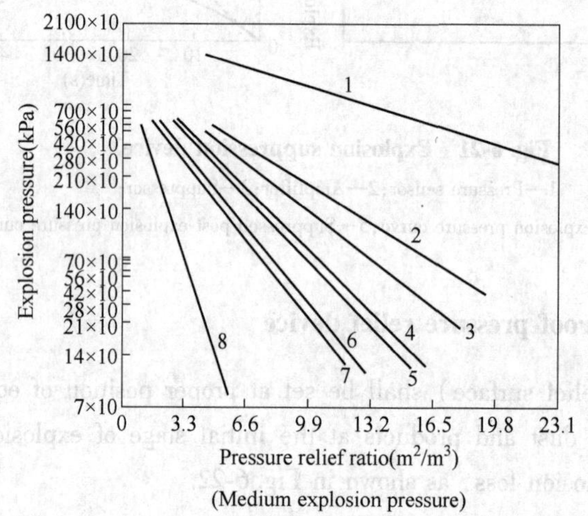

1—Aluminum powder; 2—Coarse magnesium powder; 3—Fine aluminum powder; 4—Phenolic resin; 5—Corn flour; 6—Rice noodles; 7—Pittsburgh coal powder; 8—Soap powder

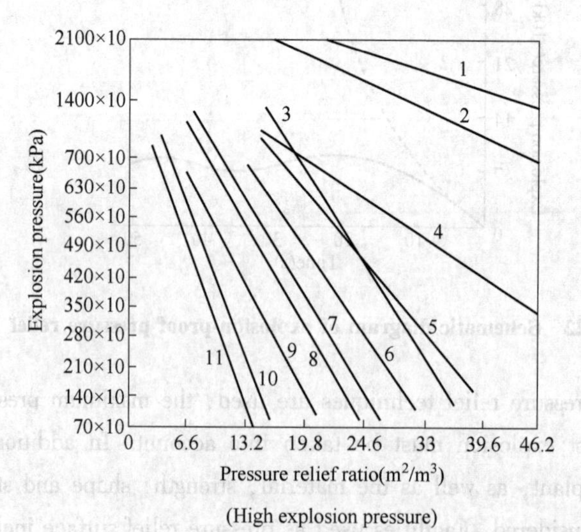

1—Aluminum powder; 2—Magnesium powder; 3—Cork powder; 4—Corn flour; 5—Cellulose acetate; 6—Soybean flour; 7—Granulated sugar; 8—Pittsburgh coal powder; 9—Wood flour; 10—Sodium lignosulfate; 11—Cocoa powder

Fig. 6-23 Pressure relief ratio chart of dust explosion

(2) Determine the pressure relief ratio based on the volume of equipment or building. The specific results are listed in Table 6-21.

Table 6-21 Pressure relief ratio of equipment and building

Type of equipment and building	Relief ratio (m^2/m^3)
Machines and stoves of light construction up to 28.32m^3	0.33~0.11
Machines and stoves up to 28.32m^3 capable of withstanding high pressure	0.11
Rooms, buildings, tanks, vessels, etc. from 28.32 to 707.92m^3 (in this case, the relative position of the explosion point and the pressure relief hole and the volume of possible explosion must be considered)	0.11~0.07
For rooms above 707.92m^3, the hazardous installations are only a small part of the building	
(1) Reinforced concrete wall	0.04
(2) Lightweight concrete, brick or wood structures	0.05~0.04
(3) Simple plate pressing structure	0.07~0.05
Large rooms above 707.92m^3 with hazardous installations occupying the majority of the room	0.33~0.07

(3) Determine the pressure relief area mainly based on the pressure rise speed. Explosion strength (i.e., K_{st}), explosion pressure after pressure relief (i.e., the maximum explosion pressure P_{red} that can be reached in the equipment after pressure relief) and stop action pressure (i.e., the static working pressure P_{stat} when the bursting plate is broken) shall be considered when the pressure relief design is carried out by using this method.

The specific method is to determine the selected computation graph as shown in Fig. 6-24 according to the static action pressure, then, based on the size of the equipment, lead the line vertically upward on the computation graph to intersect with the desired post-relief explosion pressure. Draw a horizontal line from the intersection point to the left intersects with the dust explosion order line determined by the explosion intensity; at last, draw a vertical leader down from intersection, then we obtain the pressure relief area of the equipment.

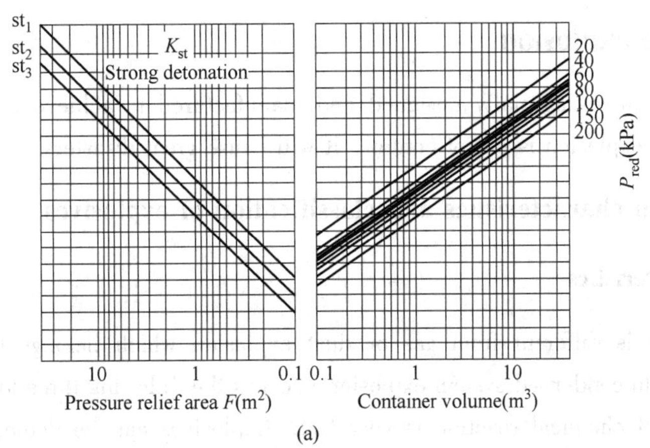

(a)

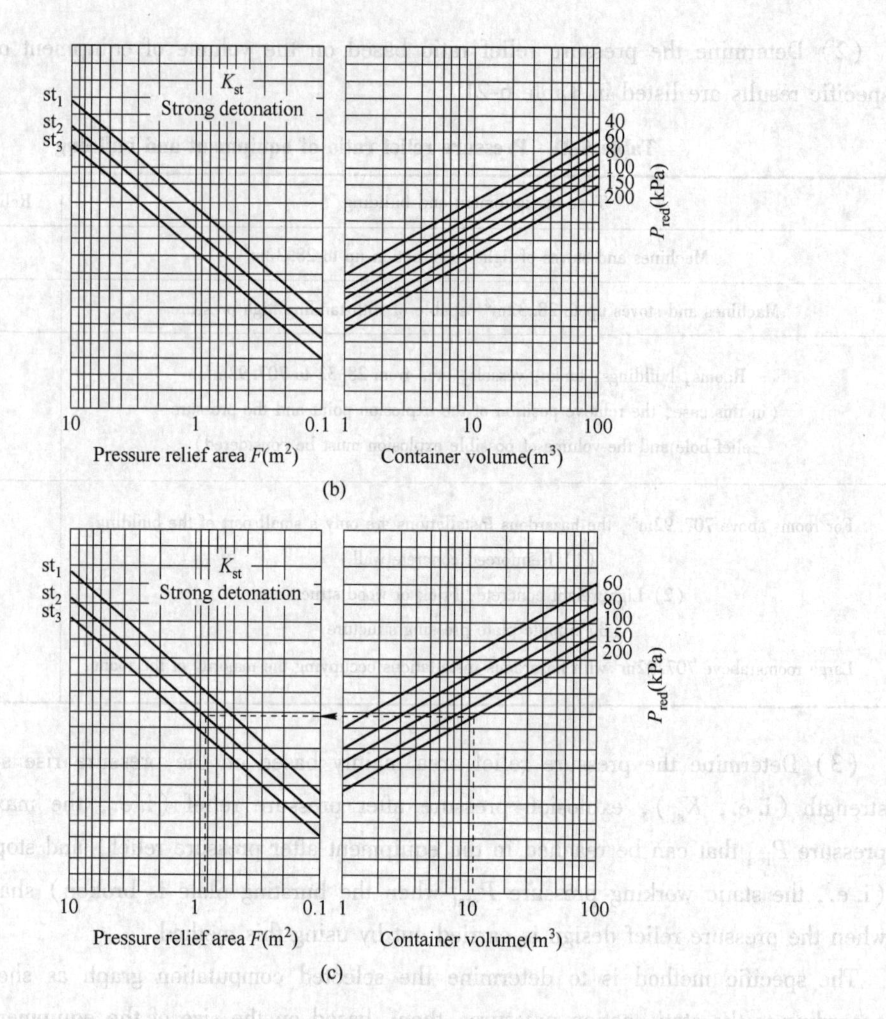

Fig. 6-24 Pressure relief area calculation diagram of dust explosion

(a) $P_{stat} = 10^4$ Pa; (b) $P_{stat} = 2\times10^4$ Pa; (c) $P_{stat} = 5\times10^4$ Pa

(In the figure, st_1, st_2, and st_3 are dust explosion classifications based on the explosion intensity K_{st}. K_{st} ranges for st_1, st_2, and st_3 levels are 0 to 200, 201 to 300, and greater than 300, respectively)

6.6 Explosive explosion

Explosives are substances specially designed and manufactured to achieve controlled explosion. However, once the explosion is out of control, it will cause great disasters.

6.6.1 Explosion characteristics and classification of explosives

6.6.1.1 Characteristics

Explosive explosion is different from gas or dust explosion which belongs to highly dispersed system. It belongs to condensed system explosion and has the following three main characteristics:

(1) The speed of chemical reaction is very fast. Explosives can be detonated in 0.0001s or

even less. For example, the explosion of a 1m long detonating cord can be completed in only 0.00003s.

(2) It can release a large amount of heat. Explosives explode with a heat of reaction of thousands to tens of thousands of joules. The temperature can reach thousands of degrees celsius and produce high pressure. For example, when 1kg of nitroglycerin explodes, it can release 6100 ~ 6620kJ of heat, while the temperature can reach 4250℃ and the pressure can reach 900MPa.

(3) It can produce a large amount of gas products. At the moment of explosion, the explosive rapidly changes from solid state to gas state, which increases the volume by hundreds of times. For example, 1kg of RDX can produce 890L of gas after explosion, and 1kg of TNT can produce 695L of gas after explosion.

The above characteristics show that explosive explosion has great power and great destructive effect.

6.6.1.2 Classification

According to the actual use, explosives can be divided into three categories:

(1) Primary explosives. Such as mercury fulminate, lead azide, lead styphnate, dinitrodiazophenol, etc. Its main characteristic is that it has high sensitivity and can explode under the action of very weak external energy. It is mainly used to make detonators, caps and other detonating materials.

(2) High explosives. Such as TNT, RDX, TAY, gelatinous explosives, etc. Their properties are relatively stable, but they have high power and brisance, and their explosion has a great destructive effect on the surrounding medium. They are the main charges of engineering blasting and military ammunition, and can be divided into single explosives and mixed explosives.

(3) Propellants. Such as black powder, smokeless powder and the like. Their characteristic reaction is deflagration, which can produce high temperature and high pressure in closed and semi-closed environments, and is mainly used as propellant for guns and shells.

6.6.2 Explosive properties of explosives

6.6.2.1 The sensitivity of the explosive

The degrees of difficulty of explosive explosion under the action of external energy are called the sensitivity of explosive. Explosives that are easy to explode is highly sensitive; non-explosive can be seen as less sensitive or insensitive. Sensitivity mainly includes:

(1) Thermal sensitivity. It refers to the degree of difficulty of explosion caused by explosives under the action of heat, which is usually expressed by explosion point and flame sensitivity, which correspond to uniform heating and flame ignition.

(2) Mechanical sensitivity. The mechanical sensitivity of explosives is defined as the degree of difficulty for explosives to explode under mechanical action, which is usually expressed by impact sensitivity, friction sensitivity and needling sensitivity.

(3) Explosion sensitivity. The explosive sensitivity is the degree of difficulty for an explosive to explode under the action of another explosive's explosion.

Table 6-22 lists the sensitivity of some explosives.

Table 6-22 Sensitivity of some explosives

Sensitivity / Measured value / Dynamite	Burst point (℃)		Flame sensitivity (cm) (maximum height for 100% firing)	Friction sensitivity (explosion percentage)	Impact sensitivity (drop height or explosion percentage)	
	5s	5min			H_{100}(cm)	H_0(cm)
Lead stephenate	265	—	54	70	36	11.5
Mercury fulminate	210	178~180	20	100	9.5	3.5
Diazodinitrophenol	176	170~173	17	25	—	17.5
Lead azide	345	305~315	< 8	76	33	10
TNT	475	295~300	—	0	4~8 (explosive percentage)	
Tetryl	257	190	—	24	44~52	
RDX	277	225~230	—	48~52	72~80	
Tai'an	225	210~220	—	92~96	100	
Nitroglycerin	222	200~205	—	—	100	
Ladder/Black (50/50)	—	—	—	4~8	50	

Note: 1. In the above table, the test conditions for friction sensitivity of primary explosive: swing angle 80°, charge amount 0.1g, pressure 7.12×10^4 kPa; the test conditions for friction sensitivity of high explosive: swing angle 90°, pressure 5.93×10^5 kPa.

2. H_{100} is the minimum drop height of 100% primary explosive; H_0 is the maximum drop height of 100% primary explosive without explosion.

6.6.2.2 The stability of explosives

The stability of explosives refers to the ability of explosives to keep their properties unchanged under the influence of temperature, humidity, sunshine and other conditions during long-term use and storage. The more powerful this explosive is, the more stable it is. It mainly includes:

(1) Chemical stability. It refers to the ability of explosives to keep their chemical properties unchanged in the process of use and storage, although they are affected by external conditions. It mainly depends on the chemical structure of explosives, and impurities, temperature and humidity also have a greater impact on them.

(2) Physical stability. It refers to the ability of explosives to maintain mechanical strength without moisture absorption and volatilization.

(3) Thermal stability. The ability of explosives to maintain their physical and chemical properties unchanged under the action of heat becomes the thermal stability of explosives. For

example, under the same conditions, the thermal stability of tetryl is worse than that of TNT and better than that of nitroglycerin.

6.6.2.3 Thermochemical parameters of explosive

The thermochemical parameters of explosives are important indexes to measure and estimate the explosive power and damage.

(1) Explosive volume (or specific volume). It refers to the volume occupied by gas products (including water) under standard conditions after the explosion of unit mass of explosives. The greater the value of the explosive, the stronger the ability to do external work during explosion.

(2) Explosive heat. The heat released by a unit mass of explosive when it explodes is called the explosion heat of the explosive, which is usually expressed by the heat released when the explosive changes at a constant volume. The work capacity of an explosive during explosion depends mainly on its explosion heat.

(3) Explosion temperature. Explosion temperature refers to the heat released by the explosive at the moment of explosion to heat the product to the highest temperature; the greater the value, the stronger the ability to do work when the explosive explodes. The explosion temperature (t) can be calculated by the following formula:

$$t = \frac{-a + \sqrt{a^2 + 4bQ_V}}{2b} \tag{6-25}$$

Where, a and b are the coefficients determined by the average isochoric heat capacity of the explosion product gas; Q_V is the heat of explosion at constant volume of the explosive.

(4) Explosion pressure. The detonation pressure of an explosive is the pressure at which the heat capacity of its gaseous products no longer changes after the explosive explodes in a certain volume. Generally, the higher the detonation pressure, the stronger the explosive power.

Table 6-23 lists the explosion volume, explosion heat and explosion temperature values of several common explosives.

Table 6-23 Thermochemical parameters of several common explosives

Name of explosive	Explosion capacity (m^3/kg)	Explosion heat (kJ/kg)	Explosion temperature (℃)
Black powder	0.28	2512	2615
TNT	0.695	4229	3050
RDX	0.89	6280	3700
Tai'an	0.78	5862	—
Nitroglycerin	0.715	—	4600
Mercury fulminate	0.30	1717	4350
Ammonium nitrate	0.98	1440	4040

6.6.2.4 The power and brute force of an explosive

The power of an explosive is its ability to do work when it explodes; the more powerful the explosive, the greater the extent and volume of the damage in an explosion. The power of an explosive is usually expressed in terms of TNT equivalent (the ratio of the power of an explosive to the power of TNT), which depends mainly on the heat of explosion.

The brisance of an explosive is the ability of an explosive to shatter an object or medium in direct contact with it when it detonates; it is primarily related to the velocity of detonation. The brisance of an explosive is usually expressed by the height to which the lead column is compressed when the explosive explodes.

Table 6-24 lists the power and brisance of several common explosives.

Table 6-24 Power and brisance of several common explosives

Dynamite	TNT	Tetryl	RDX	Tai'an	Nitroglycerin	Picric acid	Ammonium nitrate
Power (9%)	100	118.48	161.96	143.13	78.93	—	84.63
Brightness (mm)	16	19	24	24	22.5~23.5	19.2	—

6.6.2.5 Oxygen balance of explosives

The vast majority of explosives are organic compounds composed of C, H, O, N and other elements, so the essence of explosive explosion reaction is the redox reaction between these elements, in which C and H are combustible elements, O is a combustion-supporting element, and N is an inert carrier. The so-called oxygen balance of explosives refers to the balance relationship between the oxygen in explosives and the oxygen required for the complete combustion of carbon and hydrogen in explosives.

Explosives of the general composition $C_aH_bO_cN_d$ may exhibit the following three oxygen equilibria:

(1) Positive oxygen balance. $c-(2a+b/2)>0$, the oxygen content in the explosive is the remainder of all carbon and hydrogen oxidation.

(2) Zero oxygen balance. $c-(2a+b/2)=0$, the content of the explosive is just enough to oxidize the carbon and hydrogen completely.

(3) Negative oxygen balance. $c-(2a+b/2)<0$, the oxygen content in the explosive is not enough to completely oxidize carbon and hydrogen.

The oxygen balance of an explosive is expressed numerically in terms of the oxygen balance rate. The oxygen balance rate B of the explosive $C_aH_bO_cN_d$ can be calculated from the following equation:

$$B = \frac{\left[c - \left(2a + \frac{b}{2}\right)\right] \times 16}{M} \times 100\% \tag{6-26}$$

Where M is the molar mass of the explosive.

Table 6-25 lists the oxygen balance rates for some common explosives.

Table 6-25 Oxygen balance rates of some common explosives

Name of explosive	Oxygen balance rate (%)
Potassium nitrate	+39.6
Ammonium nitrate	+20.0
Nitroglycerin	+3.5
Nitroglycol	0
Tai'an	−10.1
Tetryl	−47.4
TNT	−74.0

It is of great theoretical and practical significance to study the oxygen balance of explosives.

(1) Explosives with zero oxygen or near zero oxygen balance release the most heat and have the greatest power when they explode.

(2) Explosives with high oxygen balance ratio burn completely when they explode, more CO_2 and less CO are in the product, but with the presence of nitrogen oxides; explosives with a small oxygen balance do not burn completely when they explode with less CO_2 and more CO in the product; explosives with zero oxygen or near zero oxygen equilibrium produce the least toxic gas.

6.6.3 Explosion of explosives and its destruction mechanism

6.6.3.1 Explosion mechanism

Explosives have two outstanding characteristics, one is that explosive molecules contain unstable groups, which make the molecular structure unstable; the other is that most explosives themselves contain oxygen it doesn't need oxygen from the outside to explode. These two characteristics are the internal causes of explosive explosion. But explosives require an external energy source to explode. Explosion mechanisms include:

(1) Thermal explosion mechanism. Under the action of heat energy, explosives will undergo thermal decomposition. At the beginning, the decomposition speed is slow, and an initial reaction center and accumulate active intermediate products are mainly formed; and later, if the decomposition heat release rate is great than the heat dissipation rate to the surrounding environment, heat accumulation can be generated, so that the decomposition speed is accelerated, the intermediate products are increased and collide with each other, and an oxidation-reduction reaction occurs, then a large amount of heat is released and the temperature rises sharply; and so on forth, proceeding according to the chain reaction mechanism. When the temperature of the explosive rises to its explosion point, the thermal decomposition is transformed into an explosion.

(2) Initiation mechanism of mechanical energy. When the explosive is subjected to impact or friction, the tiny area is first heated to the detonation temperature to form a hot core, so that the explosive locally explodes first, and then the explosion rapidly expands to the whole.

The main reasons for the formation of glowing nuclei are as follows: 1) there are tiny bubbles in explosives, which are strongly compressed under mechanical action, and the temperature in them rises sharply; 2) under mechanical action, the particles or thin layers of explosives rub against each other and generate heat, causing some small areas to heat up sharply.

(3) Detonation mechanism of explosive energy. After the explosion of the main explosive, the high temperature, high pressure gas and shock wave produced cause the secondary explosive to be heated by uniform shock (such as homogeneous explosive) or locally heated by the hot core (such as heterogeneous explosive), which causes the rapid chemical reaction of the secondary explosive and explosion.

6.6.3.2 Destructive effect

When the explosive explodes in the air, there are three main destructive effects on the surrounding medium:

(1) Direct action of explosion products. It mainly refers to the direct expansion impact of products with high temperature, high pressure and high energy density.

(2) Effect of air shock wave. Because the explosion products only act in the close range of the explosion center, the air shock wave is the main destructive force when the explosive explodes in the air. Air shock wave is a kind of supersonic pressure wave with huge energy, which is produced by the violent impact of the rapid expansion of high-pressure product gas on the air when explosives explode. The air is severely compressed and the local pressure is suddenly changed.

The closer to the explosion center, the stronger the destructive effect of the air shock wave is, but its action area is smaller; The farther to the explosion center, the weaker the destructive effect of the air shock wave is, but the action area is larger. The destructive effect of shock wave is mainly measured by three characteristic quantities: the pressure on the wave front, the wave duration and the specific impulse (the product of pressure and time). But because about 75% of the energy of the explosive is transferred to the shock wave when it explodes, therefore, the destructive effect of the shock wave is usually measured by the pressure on the shock front. In fact, the destruction of a large area and the whole of the surrounding medium is always caused by the pressure on the shock front (also known as shock overpressure).

(3) Scattering and killing effect of shell fragments. After the explosion of the explosive, the pressure distribution of the damage area and the air shock wave is shown in Fig. 6-25.

6.6.4 Sympathetic detonation of explosives

The phenomenon that the explosion of one charge (main explosive) can cause the explosion of another charge (secondary explosive) at a certain distance from it becomes the sympathetic explosion of explosives, as shown in Fig. 6-26.

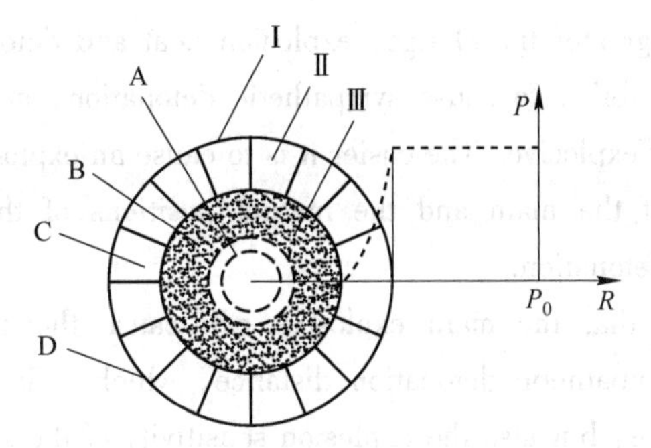

Fig. 6-25 Damage area and shock wave pressure distribution after explosive explosion

A—Zone of action of explosion products; B—Zone of interaction between shock wave and product;

C——Shock wave and fragment dispersion Area; D—Quiescent undisturbed atmospheric;

I —Shock wave front; II —Positive pressure zone; III—Negative pressure area

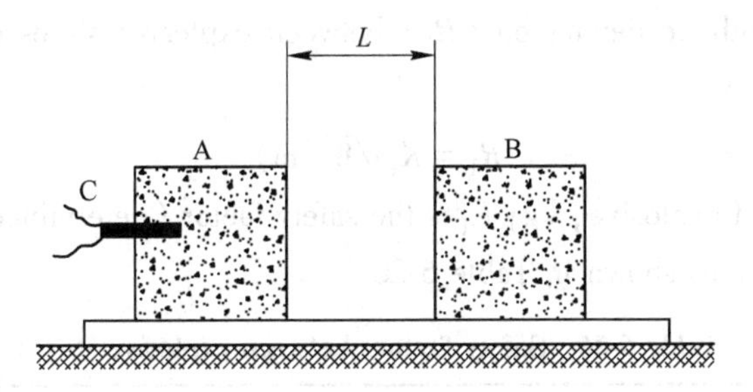

Fig. 6-26 Sympathetic explosion diagram of explosive

A—Main explosive; B—From the explosive; C—Detonator

There are four main reasons for sympathetic detonation:

(1) Shock action of explosive products of main explosive. When the density of the medium (such as air) between the two charges is not very high and the distance between the two charges is close, the explosive products of the main explosive can directly impact the secondary explosive, and the secondary explosive can be triggered to explode under the action of impact and heating.

(2) Impact action of the object thrown by the explosion of the main explosive. The secondary explosive explosion can be caused by the impact of the shell fragments, flying objects and metal jet formed during the explosion of the main explosive.

(3) The shock wave produced by the explosion of the main explosive. When the main explosive explodes, it forms a shock wave in the surrounding medium. When it propagates in the inert medium, its intensity decays continuously. If it still has enough intensity when it propagates to the secondary explosive, it can cause the secondary explosive to explode.

(4) Flame action. If there are combustibles between the main explosive and the secondary explosive, the high temperature and flame generated by the explosion of the main explosive will cause the combustion of these combustibles and the explosion of the secondary explosive.

In the actual sympathetic detonation, two or more factors often act at the same time, but the shock wave is often the main factor.

The test shows that the greater the charge, explosion heat and detonation velocity of the main explosive, the greater the ability to cause sympathetic detonation, and the higher the explosion sensitivity of the secondary explosive. The easier it is to cause an explosion; the nature of the inert medium and the nature of the main and the mutual positions of the explosives have a great influence on sympathetic detonation.

The maximum distance that the main explosive can cause the explosion of the secondary explosive is called the sympathetic detonation distance, which reflects not only the detonation ability of the main explosive, but also the explosion sensitivity of the secondary explosive.

It is of great practical significance to study the sympathetic detonation phenomenon and the sympathetic detonation distance, which can provide the basis for the safe production and storage of explosives. The minimum distance that cannot cause the explosion of the secondary explosive after the explosion of the primary explosive is called the safety distance of sympathetic detonation. The safety distance of sympathetic detonation (R_I) between explosive stores can be estimated by the following formula:

$$R_I = K_I \sqrt{W}\,(\mathrm{m}) \tag{6-27}$$

Where, W is the mass of explosive, kg; K_I is the safety factor (determined by explosive properties and storage conditions), as shown in Table 6-26.

Table 6-26 Safety factors between explosive stores

K_I value / Main explosive	From the explosive	Ammonium nitrate explosive		TNT		Advanced explosives	
		Naked	Bury	Naked	Bury	Naked	Bury
Ammonium nitrate explosive	Naked	0. 25	0. 15	0. 40	0. 30	0. 70	0. 55
	Bury	0. 25	0. 10	0. 30	0. 20	0. 55	0. 40
TNT	Naked	0. 80	0. 60	1. 20	0. 90	2. 10	1. 60
	Bury	0. 60	0. 40	0. 90	0. 50	1. 60	1. 20
Advanced explosives	Naked	2. 00	1. 20	3. 20	2. 40	5. 50	4. 40
	Bury	1. 20	0. 80	2. 40	1. 60	4. 40	3. 20

6. 6. 5 Safety of explosives and safe explosives

Explosive is a kind of important chemical dangerous goods. In order to ensure the safety of its production, use, storage and transportation, it is necessary to strengthen the management, supervision and inspection of explosives and master the dangerous characteristics of various explosives. To improve the evaluation level of explosion hazard and the safety test level of explosives. Accurately master the sensitivity of various explosives; improve the quality inspection level of explosives, etc.

The following characteristics shall be paid attention to during the storage and storage of explosives:

(1) Sensitivity of explosives. When storing, transporting and using explosives, the sensitivity of explosives should be fully understood. Within the range of sensitivity, the explosive will not explode.

(2) Instability of explosives. It means that explosives are not only explosive and sensitive to impact, friction and temperature, but also have the characteristics of acid decomposition, light decomposition and unstable salts in contact with some metals. Lead Azid, for example, can explode when exposed to concentrated sulfuric or nitric acid; TNT is exposed to sunlight? It will improve the sensitivity and easily cause explosion; Picric acid can react with metals to form picrate, and its friction sensitivity and impact sensitivity are higher than those of picric acid. In order to ensure safety, the instability of explosives must be fully understood and used within the specified shelf life.

(3) Sympathetic detonation of explosives. Explosives shall be kept at a sufficient distance to avoid sympathetic detonation.

In addition to the safety distance of sympathetic detonation discussed above, the safety distance of explosive explosion should also include the safety distance from the shock wave in the air to people and buildings.

When the explosive explodes, the minimum allowable distance acting on people, namely the safety distance (R_{II}), can be estimated by the following formula:

$$R_{\mathrm{II}} = 5 \cdot \sqrt{\frac{Q}{Q_{\mathrm{TNT}}} \cdot W}\,(\mathrm{m}) \tag{6-28}$$

The minimum allowable distance, i. e., the safety distance (R_{III}), which causes or prevents damage to the building to an allowable extent, can be estimated by the following equation:

$$R_{\mathrm{III}} = K \cdot \sqrt{\frac{Q}{Q_{\mathrm{TNT}}} \cdot W}\,(\mathrm{m}) \tag{6-29}$$

Where, Q and Q_{TNT} are the explosion heat of the explosive to be estimated and TNT respectively; W is the mass of the explosive to be estimated, kg; K is the safety factor depending on the safety protection level and whether there is an earth embankment outside the explosive warehouse or workshop, see Table 6-27.

Table 6-27 Safety protection level and safety factor *K*

Safety protection level	The extent of possible damage	Safety factor K	
		No earth dike	There is an earthen embankment
1	Completely non-destructive	50~150	10~40
2	Accidental breakage of glass windows	10~30	5~9
3	Glass windows are completely damaged, doors and window frames are partially damaged, wall plastering and interior walls are damaged	5~8	2~4

Continuted Table 6-27

Safety protection level	The extent of possible damage	Safety factor K	
		No earth dike	There is an earthen embankment
4	Damage to interior walls, window frames, doors, wooden houses and sheds, etc	2~4	1.1~1.9
5	Damage to weak masonry and timber structures, toppling of railway vehicles, damage to power lines	1.2~2.0	0.5~1.0
6	Solid brick walls were destroyed, urban and industrial buildings were completely destroyed, and railways were damaged	1.4	

In mines where gas or combustible dust is present, the explosives used shall be safe explosives. If the content of gas or mineral dust in the air reaches a certain concentration, it will form an explosive medium. Under the action of air shock wave pressure, hot solid particles, high temperature gas products and secondary flame formed by the explosion of ordinary explosives, fire and explosion accidents will occur. The explosion hazard concentration and ignition point of common gas and mine dust are listed in Table 6-28.

Table 6-28 Dangerous concentration and ignition point of gas and mine dust

Name of gas and mine dust	Dangerous concentration	Ignition point (℃)
Biogas (CH_4)	5%~15%	645~730
Hydrogen (H_2)	4%~74.2%	550~610
Carbon monoxide (CO)	12.5%~74.2%	650
Sulfur dust (S)	5~1000g/m^3	275~460
Coal dust (C)	10~2500g/m^3	750~1100

However, there is a delay period when the dangerous medium is ignited, but it is related to the temperature, the higher the temperature, the shorter the delay period. For example, when the temperature is 650℃, the ignition delay period of gas is 10s. At 1000℃, the lag period is 1s. When the ignition delay period of the dangerous medium is longer than the action time of the above ignition factors on the medium, the medium will not explode. When the explosive is used for blasting operation in the mine, the temperature of the dangerous medium can rise to about 2000℃, and the ignition delay period (τ) of the medium can be calculated according to the following formula:

$$\tau = K \cdot \exp\left(\frac{E}{RT}\right) \cdot P^{-n} \tag{6-30}$$

Where, K is the reaction kinetic constant, which is generally taken as 10^{-12}; E is the activation energy of the reaction medium, which is taken as $E/R = 3000$; T is the heating temperature of the medium; P is the pressure of the medium; n is the reaction order, which is generally taken as 1.8.

According to the above discussion, the safe explosive used in the mine where there is a danger of gas or combustible mine dust explosion shall meet the following requirements:

(1) Explosion energy is limited. That is to say, the explosion temperature and explosion heat of safe explosives are lower than those of ordinary explosives, which can ensure that the energy and shock wave intensity of explosive explosion products do not exceed a certain range, thus reducing the ignition rate of dangerous media.

(2) Explosion reaction shall be complete. The more complete the explosive reaction is, the less the amount of unreacted hot solid particles and explosive gas in the product is.

(3) The oxygen balance rate should be close to zero. Because the nitrogen oxide and nascent oxygen generated by the explosion of positive oxygen balance explosives are easy to cause the ignition of dangerous media, and the explosion reaction of negative oxygen balance explosives is incomplete. It will increase the solid particles that are not fully reacted, and it is also easy to generate explosive gas and cause secondary flame.

(4) Add flame inhibitor properly. Flame suppressant is a substance with large heat capacity, such as sodium chloride, potassium chloride, etc. It does not participate in the reaction when the explosive explodes, but it can absorb part of the explosion heat and reduce the explosion temperature. In this way, the flame can be reduced and the flame duration can be shortened, and more importantly, the flame suppressant can inhibit the oxidation and combustion reaction of the dangerous medium, thereby preventing the dangerous medium from igniting.

(5) It does not contain metal powder such as aluminum and magnesium. Because such metal powders tend to burn in air, thereby increasing the likelihood of ignition and explosion of hazardous media.

6.7 Combustion and explosion of aluminum powder

Aluminum powder, also known as silver powder, is an important industrial raw material and product, which is widely used in pigments, paints, fireworks, metallurgy, aircraft and shipbuilding industries. Aluminum powder production belongs to the high-risk smelting industry, which will produce a large amount of dust in the production process. Because of the unique characteristics of aluminum powder, such as flammable when wet, low ignition energy, small explosion limit range and so on. It determines that there is a great risk of fire and explosion in the production process of aluminum powder.

6.7.1 Hazardous characteristics of aluminum powder

The flammable and explosive medium in the production process of aluminum powder is aluminum powder. The danger of aluminum powder is mainly reflected in that it belongs to Class B flammable powder and is easy to absorb moisture. Its explosion limit in air is 37 ~ 50mg/m^3, the minimum ignition temperature is 645℃, the minimum ignition energy is 15mJ, the maximum explosion pressure is 0.415MPa, and the minimum explosion oxygen content in nitrogen is 9%.

Aluminum powder mixed with air in the air can form an explosive mixture, when it reaches a certain concentration, it will explode when it meets sparks or a certain amount of electrostatic energy. Contact with acid (such as hydrochloric acid, sulfuric acid, etc.) or strong alkali can produce flammable dangerous gas (hydrogen), which is easy to cause combustion and explosion; mixing with oxidant can form explosive mixture; contact with fluorine, chlorine, etc. causes violent chemical reactions. Therefore, aluminum powder should not be mixed with acids, alkalis, oxidants and other items for storage.

6.7.2 Characteristics of aluminum powder fire

The danger of aluminum powder fire is that it can catch fire and burn when it meets a small open fire in the air; if aluminum powder stained with grease in the air is piled up and stored for a long time, the accumulated heat will not be dissipated, and it is also easy to cause spontaneous combustion or explosion. Moreover, the smaller the particle size of aluminum powder is, the greater the explosion risk is. When its concentration in the air reaches $37\sim50mg/m^3$, it will explode in case of open fire. Therefore, in the management of chemical dangerous goods, aluminum powder is classified as Class Ⅱ flammable goods.

Aluminum powder fire is characterized by high flame temperature, fast combustion speed, strong explosion power and strong radiation heat. When burning, it usually has a green-blue flame and emits a silver-white dazzling light. The explosion pressure can reach $6.3kg/cm^2$, which has great destructive power and harmfulness to the surrounding buildings and personal safety.

6.7.3 Explosion mechanism of aluminum dust

Aluminum dust explosion is an instantaneous chain reaction, the explosion process is more complex, restricted and affected by many factors. The mechanism of aluminum dust explosion is shown in Fig. 6-27.

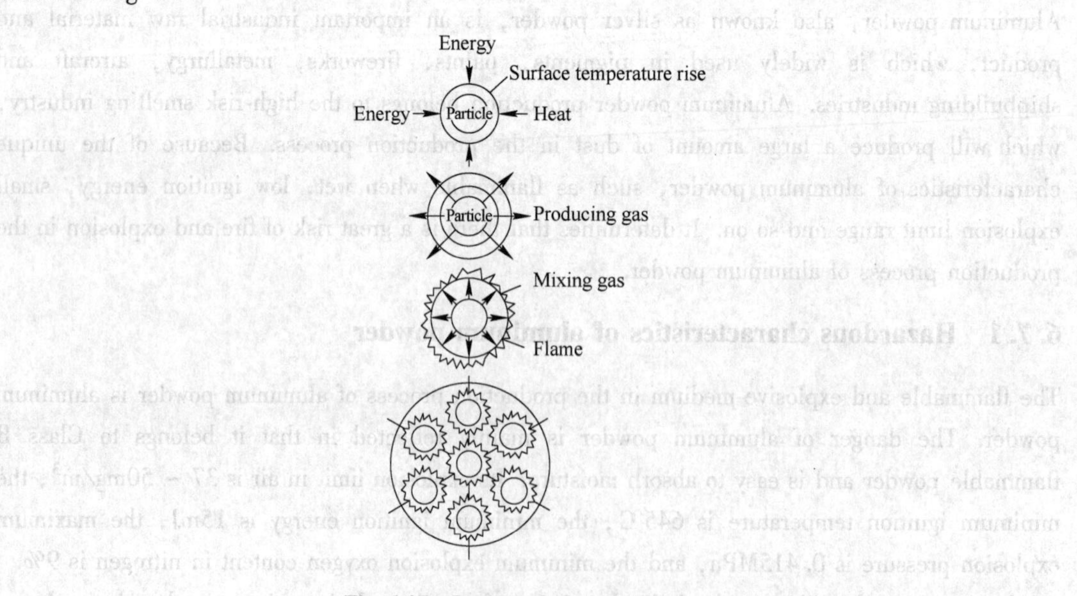

Fig. 6-27 Dust explosion mechanism

The surface of aluminum powder dust particles obtains energy from the fire source through heat conduction and heat radiation, so that the surface temperature rises sharply to reach the temperature of accelerated decomposition and evaporation of dust particles, forming dust vapor or decomposition gas, which is easy to cause ignition after mixing with air. In addition, the dust particles themselves are melted and gasified one after another, bursting out tiny sparks. It becomes the ignition source of the surrounding unburned dust, causing it to catch fire, thus expanding the explosion range. Electrostatic sparks can also trigger aluminum dust in the air, resulting in instantaneous deflagration and the release of large amounts of energy. Therefore, during the production of aluminum powder, measures must be taken to control the dust concentration in the working space and reduce the accumulation of static electricity and the generation of sparks. Cut off the chain reaction of dust explosion.

6.7.4 Hazards of aluminum dust explosion

Aluminum dust explosion belongs to explosive combustion, which is very harmful. When the dust in the air is premixed with a proper amount of air and reaches a certain concentration range, it will explode after ignition. Dust explosion is considered to be an instantaneous process in terms of mechanism, and the result of explosion may be extremely powerful and cause great damage. Because the temperature of the air produced by the explosion is as high as 2000~3000℃, or even higher. Usually, the heat generated by the explosion gas diffuses instantaneously, which will cause the combustion of the nearby combustible materials after high temperature, and then trigger the aluminum powder fire, aggravating the damage of the explosion.

For example, on August 2, 2014, an explosion occurred in the production process of the automobile hub polishing workshop of a metal products limited company, resulting in 75 deaths and 185 injuries. The cause of the accident was that the dust concentration exceeded the standard and the fire source exploded.

6.7.5 Effective extinguishing measures for aluminum powder fire

(1) Correct use of fire extinguishing agent. In order to effectively put out the aluminum powder fire, the fire extinguishing agent must be correctly selected and used.

1) It is prohibited to use water and foam to put out aluminum powder fire. The surface of aluminum powder leaked in the production process of aluminum powder is not oxidized, and the aluminum powder burning on the fire scene or under high temperature baking will react rapidly, releasing hydrogen with explosion and combustion risk and mixing with air to form explosive mixture.

2) Carbon tetrachloride and 1211 fire extinguishing agent (a kind of halon fire extinguishing agent) cannot be used to extinguish aluminum powder fire. Aluminum powder can combust with chlorine and bromine at room temperature, and can also react with Haloalkanes to produce a small amount of aluminum chloride, which often leads to explosion and combustion.

3) Aluminum powder fire shall not be put out with carbon dioxide and other gas fire

extinguishers. Aluminum powder is light in relative density and small in fineness. Once it is blown by wind or sprayed by air, it is very easy to fly and form explosive mixture in the air.

According to the above characteristics, chemical dry powder (such as alumina), dry sand, graphite powder and dry magnesium powder should be used to extinguish aluminum powder fires. Practice over the past few years has shown that the use of dry sand and aluminum silicate blankets is the most economical and effective, because these substances can cover the surface of burning aluminum powder and isolate it from the air. It can effectively prevent the aluminum powder from flying and mixing with the air, thus achieving the purpose of suffocation and fire extinguishing.

(2) Extinguishing measures for aluminum powder fire.

1) Dry sand and aluminum silicate blanket can be used to extinguish ground fire caused by aluminum powder. The method "f" one enclosure, two covers and three buri "ls" is adopted, that is, when besieging the fire, dry sand or powder must be carefully sprinkled with copper shovels or special fire extinguishing sand buckets, or the burning aluminum powder must be surrounded with dry sandbags, and then covered with aluminum silicate blankets or asbestos. Gently bury with dry sand (generally up to 30 ~ 50mm). Chemical dry powder can also be applied first, and then buried with dry sand when the fire weakens instantaneously. When discharging the dry powder, the powder shall be discharged to 1.5 ~ 2m around the burning aluminum powder, so that the dry powder can be involved in the burning area with the rising of the burning air wave and the flow of air to achieve the effect of suffocation and fire extinguishing. Remember not to directly discharge the burning aluminum powder. To prevent the dry powder from running away with the burning gas wave and losing the fire extinguishing effect.

2) If a dust explosion occurs in the working space and a large fire of dry aluminum powder is formed, it is difficult to extinguish the fire, so the initial fire must be controlled. At the beginning of the fire, the flame must be covered with a heat-insulating aluminum silicate blanket, and then covered with dry sand and dry inert powder for isolation. During operation, special attention must be paid to avoid flying of aluminum powder caused by airflow disturbance, so as to prevent secondary explosion accidents.

3) The method of first blocking and then extinguishing shall be adopted. First cool the workshop and other buildings on fire with water or foam to prevent the spread of the fire, and pay attention not to make water contact with aluminum powder; after preventing the spread of the fire and ensuring the safety of personnel and property concentrate on putting out the fire.

Exercises

1. Analyze the ignition conditions of combustible solid according to the heat balance on the surface of combustible solid.
2. What are the main physical and chemical parameters for assessing the fire hazards of combustible solid substances? Why are the flash points, ignition points and self-ignition points of thermally decomposed solid

substances related to the heating time?

3. Assuming that there is such a curtain with a thickness of 1.2mm, its density, specific heat capacity, thermal conductivity and its convective heat transfer coefficient to air are respectively 0.29g/cm^3, 1.1kJ/(kg · K), 0.15J/(m · K) and 15W/(m^2 · K). The initial temperature of the curtain is 20℃ and the ignition point is 270℃. Find the ignition time of the curtain in the following two cases:

   (1) The curtain is hung vertically in hot air at 310℃;

   (2) One side of the curtain is heated by radiation with a heat flux of 22kJ/(m^2 · s), and the other side is insulated (the absorptivity of the curtain is taken as 0.85).

4. How does the thermal inertia of a combustible solid affect its ignition? Which of the following cases of flame propagation over a solid surface is the fastest and the slowest, and why?

   (1) Horizontal propagation; (2) Vertical upward propagation; (3) Vertical downward propagation.

5. A plexiglass rod with the cross-sectional area of 0.8m^2 was ignited to produce stable combustion. Try to solve:

   (1) Ideal combustion speed and actual combustion speed;

   (2) The heat required by the solid phase reaction zone to maintain stable combustion (the thermal conductivity is 1.9×10^{4}kW/(m · K), assuming that the surface temperature of the glass breaking rod is 380℃ and the ambient temperature is 20℃ during combustion);

   (3) Heat release rate of combustion.

6. Give an example to illustrate under what circumstances can the smoldering of a combustible solid transition to flaming combustion?

7. Why is it necessary to break up the stack to further extinguish the internal flame after using water to extinguish the burning flame on the surface when putting out the fire of stacks of cotton, linen, grass, etc.? On the scene of a smoldering fire, why should we not suddenly open the doors and windows to extinguish the fire?

8. What are the factors that affect dust explosion? How do the particle size of dust, the mixing amount of combustible gas and the volume of explosion vessel affect the dust explosion respectively?

9. Why the finer dust has better dynamic stability and are easier to explode?

10. Calculate the oxygen balance rate for the following explosive:

    (1) Nitroglycerin; (2) TNT; (3) RDX.

References

[1] Du Wenfeng. Fire Combustion Science [M]. Beijing: Chinese People's Public Security University Press, 1997.

[2] Fu Weibiao. Combustion science[M]. Beijing: Higher Education Press, 1989.

[3] Zhou Lin. Fundamentals of Explosive Chemistry[M]. Beijing: Beijing Institute of Technology Press, 2005.

[4] Zhang Shouzhong. Explosive principle [M]. Beijing: National Defense Industry Press, 1988.

[5] Ji Heping, Cui Huifeng. Fire and explosion prevention technology [M]. Beijing: Chemical Industry Press, 2004.

[6] Xu Housheng, Zhao Shuangqi. Fire and explosion proof[M]. Beijing: Chemical Industry Press, 2004.